Lecture Notes in Mathematics

Edited by A. Dold and B. Eckmann

1212

Stochastic Spatial Processes

Mathematical Theories and Biological Applications
Proceedings of a Conference held in Heidelberg, September 10–14, 1984

Edited by P. Tautu

Springer-Verlag
Berlin Heidelberg New York London Paris Tokyo

Editor

Petre Tautu
Abtlg. Mathematische Modelle
Institut für Dokumentation, Information und Statistik
Deutsches Krebsforschungszentrum, Im Neuenheimer Feld 280
6900 Heidelberg 1, West Germany

Mathematics Subject Classification (1980): 60-06, 60G60, 60K35, 92-06, 92A15

ISBN 3-540-16803-6 Springer-Verlag Berlin Heidelberg New York
ISBN 0-387-16803-6 Springer-Verlag New York Berlin Heidelberg

Printing and binding: Druckhaus Beltz, Hemsbach/Bergstr.
2146/3140-543210

STOCHASTIC SPATIAL PROCESSES
MATHEMATICAL THEORIES AND BIOLOGICAL APPLICATIONS

TABLE OF CONTENTS

TABLE OF CONTENTS

FOREWORD

The present volume brings together 20 papers presented at the workshop "Stochastic Spatial Processes.Mathematical Models and Biological Applications", held in Heidelberg,September 10-14,1984, under the sponsorship of the German Research Society (Deutsche Forschungsgemeinschaft : DFG) and the assistance of the German Cancer Research Center (Deutsches Krebsforschungszentrum : DKFZ). The idea of holding this meeting grew out of the research activity in the Sonderforschungsbereich 123,"Stochastic Mathematical Models", and in the Department of Mathematical Models at DKFZ, where the central problem under mathematical investigation is the spatial growth and dispersal of interacting multitype cell systems.

In the workshop sessions the invited surveys and contributed papers were grouped under the following headings : 1)Spatial branching processes, 2)Infinite particle systems, 3)Measure-valued processes, 4)Percolation, 5)Statistical methods, 6)Related models, and 7)Biological applications. The choice of this large spectrum of mathematical theories may be justified on the one hand by the intention of acquainting the auditorium with the stage reached so far in the study of spatial processes and,on the other hand,by the intention of revealing the overlap and cross-fertilization within some present mathematical theories in the field of stochastic spatial processes.

Two round-table discussions on related subjects were additionally organized (moderated by Geoffrey Grimmett) : H.N.V.Temperley, J.van den Berg, H.Kersting and D.Mollison were the principal discussants.

Three contributed papers are not included in this volume, namely "Disordered Markov fields and percolation"(H.-O.Georgii), "On growth laws in cell biology"(L.Pilz) and "Renormalization of spatial processes" (H.Rost).

The participants at this workshop were mathematicians,physicists, and biologists. This might be the first step towards an extended participation (including,for example,chemists,mineralogists,geographers, sociologists,etc.) for a substantial interdisciplinary research. It must be admitted that,in the framework of such research,the pre-eminence of mathematicians and mathematical theories is unquestionable. One can easily demonstrate that,beyond the realm of mathematics,mathematical concepts and theories have supplied the essence of many scientific theoories in empirical sciences. As R.Thom recently stated, "the language of interdisciplinary research is...necessarily mathematical...Only when a concept of experimental origin has been made mathematical can it begin to play its role in interdisciplinary research" ("The Virtues and Dangers of Interdisciplinary Research", The Math.Intelligencer,1985). The organization of this meeting actually reflects this position.

During the editorial work,the necessity of writing an introductory article on stochastic spatial models in biology became evident. This is only a rough historical draft of the biomathematical work performed in the area from the beginning of this century, but its main purpose is to create the mental environment for all mathematical concepts emerging from this volume. One would say that it hints obliquely Nietzsche's idea that "the more abstract the truth is that you would teach, the more you have to seduce the senses to it".

This book appears with some (undesired) delay, but two volumes published in the interim period drew attention to it : "Infinite Particle Systems" by Thomas Liggett (Springer,1985) and "Particle Systems,Random Media and Large Deviations",the proceedings of the AMS-IMS-SIAM Joint Summer Research Conference on the Mathematics of Phase Transitions (R.Durrett ed.;Amer.Math.Soc.,1985). In both publications one can find the preoccupation with new,open problems in the field of interacting particle systems. For example,in the latter,ten open problems "for the 80's(and 90's)" in the domain of stochastic growth models (viewed as a class of interacting particle systems) are described by Richard Durrett. This is a valuable intention which again proves that mathematics is (or ought to be !) something to be actively done rather than passively learned or routinely applied. A repertoire of open problems suggested by the diverse investigations reported in this volume would have made it twice as large ; particularly,their placement in a proper biological context would have considerably increased the time needed to solve them. (Consider only cell systems in d-dimensions,with k-type biological particles !) Indeed,biology is a deep-reaching and complicated subject which is a rich source of mathematical problems. For this reason,S.Ulam counselled the biologists : "ask not what mathematics can do for you, ask what you can do for mathematics".

Fortunately,in the modern sciences something has changed and the bonds between abstract theories and their applications have become stronger. This may convert the biomathematical investigation in the line recently suggested :

"Mathematicians who study biology usually focus principally on properties which supply answers to specific biological questions. A bolder approach would be to study these problems with a view to finding theorems which answer questions which have not yet been asked, to look for and discover properties which have not been suggested. Ultimately this can mean not merely better answers to known biological questions,but new insights into biology itself,not to mention mathematics and other fields of application."(W.A.Beyer,P.H.Sellers,M.S.Waterman : Lett.Math.Phys.,1985)

I believe that this volume suggests this possibility.

P.Tautu

Acknowledgments. I should like to thank my young colleagues T.Götz,L.Pilz,W.Rittgen and G.Rosenkranz for their help in the organization of this workshop. Finally I would like to thank the staff of the Springer-Verlag for its support and patience.

LIST OF CONTRIBUTORS

BARBOUR,A.D. Universität Zürich,Institut für Angewandte Mathematik, CH-8001 Zürich,Schweiz

BOSE,A. Carleton University,Dept.of Mathematics and Statistics,Ottawa, Canada K1S 5B6

COX,J.T. Syracuse University,Dept.of Mathematics,Syracuse,N.Y.13210, USA

DAWSON,D.A. Carleton University,Dept.of Mathematics and Statistics, Ottawa,Canada K1S 5B6

DONNELLY,P. University College London,Dept.of Statistical Science, Gower Street,London WC1E 6BT,England

DURRETT,R. University of California-Los Angeles,Dept.of Mathematics, Los Angeles,CA 90024,USA

EAGLESON,G.K. CSIRO,Division of Mathematics and Statistics,Lindfield, NSW 2070,Australia

EDE,D.A. University of Glasgow,Dept.of Zoology,Developmental Biology Building,Glasgow G12 9LU,England

GÖTZ,T. Universität Heidelberg,Institut für Angewandte Mathematik,SFB 123, Im Neuenheimer Feld 294, D-6900 Heidelberg 1,BRD

GOROSTIZA,L.G. Centro de Investigacion y de Estudios Avanzados del IPN, Depart.de Matematicas,México 07000 D.F.,México

GREVEN,A. Universität Heidelberg,Institut für Angewandte Mathematik, Im Neuenheimer Feld 294, D-6900 Heidelberg 1,BRD

GRIFFEATH,D. University of Wisconsin,Dept.of Mathematics,Madison, Wisconsin 53706,USA

GRIMMETT,G. University of Bristol,School of Mathematics,Bristol BS8 1TW, England

HADELER,K.P. Universität Tübingen,Institut für Biologie,Lehrstuhl für Biomathematik, Auf der Morgenstelle 10, D-7400 Tübingen, BRD

IVANOFF,G. University of Ottawa,Faculty of Science and Engineering, Dept.of Mathematics,Ottawa,Canada K1N 9B4

KUULASMAA,K. National Public Health Institute,Dept.of Epidemiology, SF-00280 Helsinki,Finland

LIGGETT,T.M. University of California-Los Angeles,Dept.of Mathematics, Los Angeles,CA 90024,USA

NAGYLAKI,T. University of Chicago,Dept.of Molecular Genetics and Cell Biology,Chicago,Illinois 60637,USA

ROSENKRANZ,G. Universität Heidelberg,Institut für Angewandte Mathematik, SFB 123, Im Neuenheimer Feld 294, D-6900 Heidelberg 1, BRD

SULLIVAN,W.G. University College,Dept.of Mathematics,Belfield,Dublin 4, Ireland

TAUTU,P. Deutsches Krebsforschungszentrum,Abt.Mathematische Modelle, Im Neuenheimer Feld 280, D-6900 Heidelberg 1,BRD

WELSH,D.J.A. Merton College,Oxford OX1 4JD,England

INTRODUCTORY EDITORIAL

STOCHASTIC SPATIAL PROCESSES IN BIOLOGY : A CONCISE HISTORICAL SURVEY

P.Tautu

1.

Since the first report on the irregular movement of pollen grains and the accumulation of data about the migration of species, motion and spread in space have represented two phenomena of stimulating interest for biologists. However,their approach was methodologically and intelectually different from that accomplished by physicists. As J.G.Skellam (1951) pertinently remarked, biologists rarely formulated their problems in terms of abstract models at the end of the last century, while physicists proceeded from a mechanistic to a mathematical foundation. The great example is the work of J.C.Maxwell (often analyzed by philosophers of science) as the introduction of the differential equations of electromagnetic waves and the kinetic theory of gases urged on a new form of scientific realism, where mathematical thinking plays a formative role. The consequences of the structure of motion equations are randomness and irreversibility.

2.

In 1905,Albert Einstein developed the mathematical theory of Brownian motion, and Karl Pearson gave the first explicit description of a random walk(RW). One year later,in an extensive paper, written in collaboration with J.Blakeman, K.Pearson used a RW-model for describing the random migration of species. The walk was,obviously,a 2-dimensional one ; the straight lines of motion (the jumps) were equal in size,and the step directions were assumed to be isotropically distributed. In the same year,he studied the circular RW and gave a new proof to the solution found by J.C.Kluyver(1906) to this problem. The special case where the probability distribution of the terminal point of a 2-dimensional RW is bivariate normal was investigated by J.Brownlee(1911) who also considered the spread of an epidemic. Pearson's hypothesis of equal jumps was criticized by J.G.Skellam(1951) as being "somewhat artificial" : when the walkers are of a biological nature,the jumps vary in size. Indeed,it is plausible to assume that a "lone drunken Englishman who drops his house key on a long straight street and hunts for it"(Fisher,1984) will move in unequal jumps ! Thus,in 1982,R.Barakat introduced the hypothesis that the length of jumps is a random variable; the anisotropic distribution of step directions was considered by R.J. Nossal and G.H.Weiss(1974).

The continuous time RW, called "random flight", was reported by Lord Rayleigh(1919) ; he also gave the solution to the problem of uniform RW on a sphere and an approximation method for large samples. The reader is referred to Example(e) in W.Feller(1966,pp.32-33) and Problem 15 in F.Spitzer(1964,p.104). The biological example "Random dispersal of populations"(Example(c) in W.Feller,1966,p.255) is an empirical application of Example(b),"Random walks in d dimensions",(id.,p.255): it was actually borrowed from J.G.Skellam(1951,§2.3) who,in turn,used C.Reid's arguments(1899) about the spread of oak trees in prehistoric times.

The important step in the mathematical investigation was accomplished by G.Pólya(1921) : he investigated the RW on street networks (infinite 2-dimensional graphs) and in a "jungle gym" (a 3-dimensional infinite graph). His question can be formulated as follows : Is a walker who steps at random between nearest -neighbour sites (with all allowed steps being equally likely),certain to return to his starting point ? Pólya's answer emphasizes the role of the spatial dimension : a simple RW is recurrent (persistent) in d=1,2, and is transient for d>2. Applying his theorem to the problem of random vibrations posed by Lord Rayleigh, one can grasp that wind instruments are possible in our 3-dimensional world("Spaceland") but are not possible in Flatland (see also Doyle and Snell,1984). Walking on a map and starting from Rome, an ant will sooner or later return to Rome. The rough consequence in cell biology might be that a living cell which walks at random in a 3-dimensional tissue,is almost always "invasive" - if there is no restriction or attraction (Tautu,1978). The embryologists ought to make use of this theorem but they still ask,for example,"whether positional information in three dimensional structures is actually specified in detail in three dimensions,or whether two will suffice"(Bryant,1982). The "lamentable" gap in mathematical cell biology concerns cellular movements during morphogenesis : "the problem is obviously one of complexity ; morphogenetic movements define three-dimensional,asymmetrical trajectories that are not easily described by manageable expressions"(Campbell,1975).

A heuristic model for morphogenesis was suggested by S.Ullam as follows (see Finkel and Edelman,1985) :

> "Imagine that several different colored balls undergo a random walk on a lattice or tree structure,and that whenever two balls

land on the same (or adjacent) locations, they change colors according to a set of pre-specified rules. These rules could,for example,be quadratic transformations...It is also possible to consider changes in the number of balls as well as in their color. The fascinating part of the model is that it includes a "rule to change the rules". In other words,once a given situation is reached (e.g.,a certain density of red balls in a region, more than five generations of white balls produced,etc.) the actual set of transformation rules are modified...In this case,the modification of the rules correspond to the developmental process of induction."

The fact is that a large variety of natural phenomena involving movement,transport,change in structure(configuration),fluctuation in the state of a system,etc. can be mapped onto a random walk problem. As W.Feller(1968,p.356)remarked,"the simple random walk model does not appear realistic in any particular case,but fortunately the situation is similar to that in the central limit theorem. Under surprisingly mild conditions the nature of the individual changes is not important, because the observable effect depends only on their expectation and variance. In such circumstances it is natural to take the simple random walk model as universal prototype." The connection between random walk and Brownian motion (and related processes) was suggested by L.Bachelier in 1912 ; the reader is referred to the important paper by M.Kac(1947).

The investigation of cell movement is the first example. It was shown that fibroblasts in tissue culture are "pure" random walkers iff they are observed for long time intervals (Gail and Boone,1970). Initially, they persist in their direction of motion so that this movement can be accounted for in a correlated but not in a simple RW. Actually, a 2-dimensional "persistence" model with application to protozoa movements was still suggested fifty years ago (Fürth,1920) ; the development of correlated or biased RWs was succeeded in the 50's (Klein,1952;Patlak, 1953;Gillis,1955). The reader is referred to the book by M.N.Barber and B.W.Ninham(1970,p.76) or to the recent paper by E.A.Bender and L.B.Richmond(1984).

Experiments with bacteria (E.coli) in an isotropic medium showed that the direction of movements appears to be random and independent of the step length ; yet,during chemotaxis, the angles of the steps were still approximately random but the step lengths were longer if the bacterium moved (in a 3-dimensional space) towards the attractants. Then, the RW-models for cell movements must be restricted (correlated) or biased (Peterson,and Noble,1972;Nossal and Weiss,1974;Hall,1977;Alt, 1980). RW-models from the same family have been suggested for animal displacements (Skellam,1973) and for the rooting patterns of Sitka spruce (Henderson et al.,1984). Also,random walks of the "Brownian motion type" have been used to describe the movements of the larvae of a helminth,Trichostrongylus retortaeformis (Broadbent and Kendall,1953), of the larvae of codling moth (Williams,1961),and of humans (Yasuda, 1975).

The second example comes from genetics where a RW on Z and Z^d(d>1) has been used in the framework of neutral theory. J.F.C.Kingman (1976; 1980,p.28) constructed a "coherent" RW in order to describe the allelic frequencies in a finite but large population in which mutations are selectively neutral. The approach was suggested by a mutation model for estimating the number of electrophoretically detectable alleles (Ohta and Kimura,1973); it is called "charge-state model"(Evens,1979,§9.5) or "ladder-rung model". It is interesting to point out that a continuous-time Ohta-Kimura model was introduced by W.H.Fleming and M.Viot (1978) as a measure-valued stochastic process (see also Dawson and Hochberg,1982).

Restricted RWs (d=1,2,3) with absorbing barrier (resp.circular absorbing surface or spherical absorbing surface) have been introduced in neurophysiology as models of the fluctuation of membrane potential (Gerstein and Mandelbrot,1964). This potential is depicted as a Wiener process with drift. The reader is referred to the books by A.V.Holden (1976,Ch.6) and by G.Sampath and S.K.Srinivasan(1977).

In the case of dependent neuronal units activities, we deal with a neuronal network ; the stochastic problem is characteristic for all systems with local interactions. Under some specific assumptions,the model of random neural networks belongs to a class of Markov interaction processes (see,e.g.,Kryukov,1978). A continuous-time version of a homogeneous Markov networks is related to the contact processes (or contact interactions) introduced by T.Harris (1974). The reader should find in a paper by V.Cane(1967) interesting biological examples : (i)randomly connected neural nets, (ii)geographic spread of infections, (iii)stimulation of coral polyps, (iv)clutches of eggs of game birds.

It has already been stated that RW-problems came to be almost synonymous with lattice problems. There is a remarkable similarity between the generating function of an n-step RW on a periodic lattice and the correlation function of the spherical model of a ferromagnet and its variants. There is also an isomorphism between a particular random flight problem and a corresponding percolation problem (see Cummings and Stell,1983). The lattice -i.e.the medium into which one walks- can be defective,can have traps or difficult boundary conditions. The ran-

dom walk may involve many internal states or may have a coupled memory (e.g.,avoiding the sites already visited),etc. The reader is referred to the beautiful introduction written by D.B.Hughes and S.Prager(1983) or to the long review by G.H.Weiss and R.J.Rubin(1983).
Random walks on different structures were inventively investigated:on graphs(Göbel and Jagers,1974),on trees(Pearces,1980), on random trees (Moon,1973), on crystals(Hammersley,1953), on dodecahedrons(Letac and Takâcs,1980),etc.

Sometimes we find such a RW-problem behind a complex model,as for instance in the stochastic model for the spread of a malignant clone (Williams and Bjerknes,1972). This model was described by D.Richardson (1973,Ex.6) as follows:

> "The plane is divided into cells by the hexagonal or square tessellation. At time 0, the cell at the origin is black and all other cells are white. A certain event,'dividing',is occurring at intervals in the black cells. The hazard function for the occurrence of 'division' is constant. The waiting times between different 'divisions' of the same cell or of different cells are independent. When 'division' occures in a black cell,it picks a neighbour at random. The neighbour,if white,then becomes black. Once black,a cell remains black permanently."

T.Williams and R.Bjerknes assumed that the malignant cell divides faster than the normal cells by a factor κ. Clearly,if $\kappa=\infty$, only black cells divide (as in Richardson's example above). We can consider this cell as a black walker : the imbedded Markov chain obtained by considering the total number of black cells at the instant an event 'dividing' occurs is actually a self-avoiding RW with constant probabilities of moving to the nearest neighbour sites. The distribution of the time between steps explicitly depends on the length of the periphery at every step (Williams,1971). An analogous process was considered by M.Eden(1958, 1961;see Richardson,1973,Ex.9) : he conjectured that the configurations generated tended to be <u>round</u> and <u>solid</u>,and that the fraction of all possible configurations that are likely to be generated tends to zero as time goes to infinity. In their case,T.Williams and R.Bjerknes suspected that the generated pattern may have a fractal dimension. (For fractal RWs,see Shlesinger and Montroll,1983.) However,more recent work with this model suggests that the surface of the malignant focus is rough but not fractal-like(see,e.g.,Meakin,1986).

It is important to mention that the Williams-Bjerknes model can be considered as an example of a voter model,that is,a non-ergodic process with particular properties in the theory of interacting particle systems(Bramson and Griffeath,1980,1981). The "invasion model" introduced by P.Clifford and A.Sudbury in 1973 is a "basic" voter model (see Liggett,1985,Ch.5).By applying theorems about the range of a RW,one may have information about the size of the region occupied by the invaders (Sudbury,1976;see also Kelly,1977;Tautu,1978). The reader must be warned : the "swapping process" in the Clifford-Sudbury paper is not a voter model but an exclusion model,in terms of the theory of interacting particle systems (see Liggett,1985,Ch.8). The RW representation of such systems is already established (Brydges et al.,1982) but if one assumes that each particle undergoes an ordinary continuous-time RW (with exponential holding times of mean one,at each position),then the RWs are "coupled" (Liggett and Spitzer,1981). In the framework of the theory above mentioned, the systems of "coalescing" and "annihilating" RWs, arising as duals of the voter models are,in fact,systems of interferring walkers.

For particular genetic problems,a specific RW,namely a "strongly aperiodic finite-variance RW" in 1-and 2-dimensions was built up by S. Sawyer(1976).

3.

A new stage in modelling biological processes in space and time was reached during the period 1936-1937. We begin with the statistical analysis reported by W.G.Cochran(1936) about the propagation of a virus infection in a field,namely the spread of spotted wilt of tomatoes, carried by a species of thrips. The probles is stated as follows:

> "If every plant in the field has an equal and independent chance of becoming infected,the resulting distribution of infected plants over the area will be called a random one. The actual distribution may,however,deviate from a random one,owing to groups of diseased plants coming together more often that would occur by chance. The groups themselves may be scattered irregularly over the area,such as might happen if an insect carrying the disease had equal access to all plants in the field,but when feeding was able to infect several neighbouring plants at the same time. On the other hand,the deviation from randomness may be of a more regular type,infection being higher,for instance,near the borders than in the interior, or one side, owing to a source of infective insects near by."

As one can see in Fig.1,the field map is actually a 2-dimensional lattice with its non-marked sites occupied by non-infected plants and its marked sites occupied by the diseased plants. Three kinds of marks indicate the spread of infection over time. W.G.Cochran analyzed the existence of two mutually exclusive types of configurations ("doublets"), i.e.the links between two identically marked sites and the links between two differently marked sites.

Field Map of Diseased Plants.

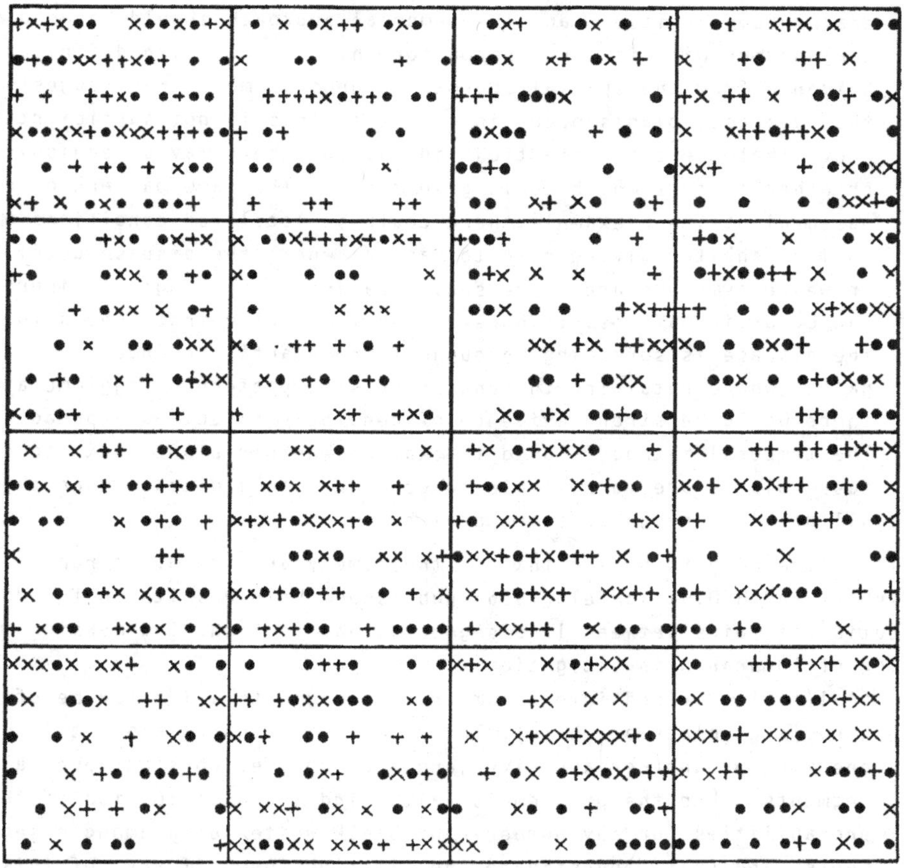

× = diseased at first count. + = diseased at second count.
● = diseased at a later count.

His aims were to find appropriate statistical tests for proving
(i)whether diseased plants tend to congregate in patches scattered over the area or in groups along or across the rows, and
(ii)whether the distribution of plants recently infected is related to that of plants previously infected.
As a matter of fact, Fig.1 might be viewed as the representation of a "lattice contagion model",that is,the basic model for the local spread of an infection(epidemic). Although the distinction between spatial and

temporal aspects is clearly exposed,both are necessary to understand the real process. Then,the future objective ought to be the creation of space-and-time stochastic models. The argument is lucidly stated by G.H.Freeman(1953):

> "Investigation of the spatial aspect of the matter is not in general enough by itself,and the temporal approach should be used in conjunction with it. The reason for this is that even definite evidence from the spatial approach cannot do more than suggests that diseased plants occur in clusters. This is not sufficient to prove that there is infectious spread,for there may be ecological or other factors which could give rise to the same pattern of diseased plants. For example,there could be localized conditions of soil or shelter giving rise to patches where the disease develops or where symptoms are expressed. Equally,a physiological disorder can occur in patches. If,however,the temporal method shows that the disease is spreading throughout the plantation under investigation,where retaining the characteristic pattern of agglomeration there would be strong evidence of genuinely infectious spread... The temporal method is used to examine whether plants next to diseased plants are more liable to go down with a disease than those all of whose neighbours are healthy."

It is interesting to notice that in the same year as W.G.Cochran, J.Wishart and H.O.Hirschfeld(1936) published their results on the distribution of joins between line segments,that is : the 1-dimensional analog of Cochran's investigation:

> "A line is divided into a number of n segments which may be of two kinds(e.g.,black or white). The probability that a segment is of one kind or another are given,and are supposed constant over all segments. Then the problem is (a)to find the distribution of the probabilities for any number r of black-white joins among n segments, and (b)to investigate the limiting form (if any) of this distribution for larger n."

The statistical approach to the analysis of the random associations of 2-coloured connected points (doublets,triplets,quadruplets) on 1- or 2-dimensional lattices was accomplished between 1947-1951 by the contributions of P.A.P.Moran(1947),D.J.Finney(1947),P.V.Krishna Iyer(1949, 1950) and B.V.Sukhatme(1951).

One of the results of these statistical investigations is the identification of the "neighbourhood" as an important space parameter. For example,by the inclusion of diagonal bonds one obtains a more efficient statistical test than by their exclusion,with the consequence of a smaller coefficient of variation. It was also noticed that the use of black--white doublets only gives a poor representation of a contagious process. Thus,in order to take into consideration the healthy carriers, G.H.Freeman(1953) assumed that they must be located on the vacant sites which will be "blackened" by the occurrence of a diseased plant.

Special mention must be made of P.A.P.Moran's 1948 paper on the geographical distribution of a biological (or pathological) attribute. The investigation was suggested by the work of D.B.Cruickshank(1947) on regional influences in the relative mortality from cancer. It was assumed that the putative neoplasia-generating factor(s) may be either (i)distributed statistically independently in different counties or (ii)neighbour-dependently. The results of significance tests show that they depend on the initial assumption : (I)the factor occurs independently in each county or (II)the factor exists in a given number of counties but these counties are randomly disposed. The thesis seconded by D.B.Cruickshank was that "there must be some general factor <u>other than regional</u> which operates to produce a definite incidence of cancer. Regional influences are not themselves the cause of cancer,but merely impress their effect,positively or negatively,upon this general factor".
The serious difficulty in investigations of this type is the correct definition of "neighbourhoods" ; as N.T.J.Bailey(1967)pointed out,they may be very complex and hard to define,particularly if there exists various forms of complicated social and geographical stratifications.

This inquiry reminds us of the old research about"cancer houses" (Pearson,1911) : the probability distribution of such houses appears to be not random - under the condition that the data are correct and strongly correspond to the parameters of a well-formulated hypothesis. By using old data,it seems that "cancer apparently avoids a house where it has paid one visit"(Pearson,1913) : this is the first image of a self-avoiding random walk before a formal definition.

The reader is also referred to the 1925 paper on the household infection by A.G.McKendrick : his 2-dimensional diagrams resemble to a lattice representation,with neighbourhoods and doublets, his house-to--house infection may be thought of as a kind of contact process. From four epidemics of bubonic plague in a certain village in India,he found that the probability of "internal" infection was 200 times as large as that of "external" infection. The plausible explanation was found in the behaviour of the disease vector : the disease is transmitted by fleas - a species which does not as a rule travel far from its own neighbourhood (see Irwin,1963).

4.

The celebrated problem of R.A.Fisher(1937) about the front wave of advantageous genes is stated as follows :

> "Consider a population distributed in a linear habitat,such as a shore line, which occupies with uniform density. If at any point of the habitat a mutation occurs,which happens to be in some degree,however slight,advantageous to survival,in the totality of its effects,we may expect the mutant gene to increase at the expense of the allelomorph or allelomorphs previously occupying the same locus. This process will be first completed in the neighbourhood of the occurrence of the mutation,and later,as the advantageous gene is diffused into the surrounding population,in the adjacent portions of its range. Supposing the range to be long compared with the distances separating the sites of offspring from those of their parents,there will be,advancing from the origin,a wave of increase in the gene frequency."

This represents the starting point in the development of the theory of geographically structured populations (which plays an important role in S.Wright's evolution theory : phase of interdemic selection and ecologic opportunity). J.G.Skellam(1973) analyzed Fisher's approach as an invasion process and gave an interesting description of what may happen near the edge of the geographical range of species(p.81).

In the same year,1937, A.Kolmogorov,I.Petrovsky and N.Piskounov (KPP) studied a similar problem :

> "We now consider a certain territory,or region,populated by some arbitrary species. First let us suppose that a dominant gene A is distributed over the territory,having constant concentration p ($0 \leq p \leq 1$). Let us suppose next that individuals with characteristic A (that is,belonging to genotypes AA and Aa) have an advantage in their struggle of existence against individuals who do not possess the characteristic A (belonging to genotype aa) ... Now let assume that the concentration p varies across the territory occupied by the species under consideration,that is,p depends on the x and y coordinates of the point in the plane ... Our first problem is to determine the speed of propagation of the gene A,that is to say the speed of movement of the boundary."

In his 1940 paper on the Volterra theory, W.Feller mentioned this model: "In the genetic work of Kolmogorov(1935) and of Kolmogorov,Petrovsky and Piskounov(1937) it is also the theory of continuous stochastic processes that plays a role,and to the same field belongs also the Theory of Diffusion. The solution of the diffusion equation satisfied by the space density of the diffusing substance can,by Einstein's procedure,also be regarded as the probability density that a moving particle subject to a certain chance mechanisms is located at the place in question."

R.A.Fisher discovered the remarkable fact that progressive waves are possible with any velocity greater than or equal to a certain minimum velocity determined by the diffusion coefficient and the selection intensity in favour of the mutant genes (see the computer simulations in Skellam,1973,and Gazdag and Canossa,1974). R.A.Fisher considered(1950) his paper as "an isolated paper which I have not followed up either practically ot theoretically",although he published,in the same year another paper where he postulated that the selection parameters are linearly dependent on the spatial position.

He neglected at least two papers which had appeared in the 40's, namely the study of T.Dobzhansky and S.Wright(1947) on the rate of diffusion of a mutant gene,and that by D.G.Kendall(1948) on wave propagation. The latter was suggested to the author by M.S.Bartlett in an epidemiologic context. The basic idea is stated as follows :

> "Since the equations governing the genetic and epidemic problem are non-linear, I considered instead a form of the equation of heat conduction which displays a similar phenomenon ; it is that appropriate to the description of a continuous linear population which is simultaneously undergoing a Brownian motion and growing geometrically."

In the epidemiologic situation,an initial spatial local concentration of infectives will generate two waves having asymptotically the minimal velocity. As D.G.Kendall(1965) conjectured,it is also likely that the high-speed waves are unstable. See also the previous paper by E.B.Wilson and J.Worcester(1945) and M.S.Bartlett(1956,§8). J.Canosa(1973) showed that this instability of waves disappears quickly : by instability is meant that under perturbation the waves "will break up into wavelets travelling with the minimum speed"(Kendall,1965). Or,such local perturbations disappear exponentially fast. The reader is referred to D.Mollison(1977) for a detailed presentation of this problem.

This approach could be regarded as a turning point in the analysis of complex stochastic population processes which involve - following the suggestive expression by J.F.C.Kingman(1973) - the twin elements of change and chance. It must be emphasized that this analysis did not progress before the principles and methods of convergence and approximation in the theory of stochastic processes had not been established and applied. The classical example of weak convergence (Erdös and Kac,1946; see Billingsley,1968) related random walks of smaller and smaller steps asymptotically to continuous Brownian motion. Further,W.Feller(1951) showed that a diffusion process is the limit of a sequence of Galton-Watson processes. After the theoretical results obtained by Yu.V.Prohorov(1956),the Ph.D.Thesis of C.J.Stone(1961) contained proofs of the convergence of a sequence of birth-and-death processes to a diffusion process (see Stone,1963). The reader will find in (Iizuka and Matsuda, 1982) the list of important papers devoted to the convergence of Markovian and non-Markovian models in genetics.

As J.Kingman remarked(1973),if we are faced with a complex situation, "any plausible model is likely to be very difficult to analyze,whether by mathematical and computational techniques,and a simple robust approximation may well be the most useful outcome of a stochastic study". Plentiful literature exists on this topic from, say P.Whittle(1957) to the latest survey by T.G.Kurtz(1981).

With regard to Fisher-KPP equations, I want to mention the linearized versions (also in R^2 and R^3) suggested by E.W.Montroll(1967;see Mollison, 1977,p.294) and particularly the result reported by T.G.Kurtz(1980) who obtained a solution of Fisher's equation as the deterministic limit of some Markov models for two competing species.

Additional information:for developments regarding Fisher-KPP equation, the reader is referred to H.P.McKean(1975),Y.Kametaka(1976),and D.A.Larson(1977); also consult the analyses performed by P.C.Fife(1979),H.F. Weinberger(1980),and A.Okubo(1980).

Let me revert to the population models with a spatial ("geographic") structure as they are built up in population genetics(see,e.g.,Nagylaki, 1977;Maruyama,1977). The main one is the "island model"(Wright,1951) useful for studying the interaction of genetic drift,population subdivision,and mutation. In his book,A.Okubo(1980,p.182) presents the corresponding model by E.Kuno(1968) for the spatial structure of populations of rice leafhoppers in paddy fields. See B.Latter(1973) and J.H.Gillespie(1976).

The "stepping-stone model" is a particular variant proposed independently by G.Malécot(1950) and M.Kimura(1953). G.Malécot(see 1966,p.207) pointed out the role of spatial dimension : the probability of two neighbouring individuals having a common ancestor essentially depends (for an identical density) on the number of dimensions of their spatial migration. The geometrical structure of stepping-stone models is the (infinite) d-dimensional lattice ($d\geq1$). A site on this lattice represents a "colony" of identical (or distinct) individuals. If the number of colonies is finite,the colonies at the boundary must be specially treated but this effect can be avoided by arranging the colonies in a circle or a torus. Results for the 3-dimensional lattice were obtained by G.Weiss and M.Kimura(1965), T.Nagylaki(1976),and S.Sawyer(1976,1977). If the geometrical structure is an infinite homogeneous tree,other dimensional differences canbe observed (Sawyer,1980) : the size,spread and survival of colonies depend on d (d=1,2 or d≥3).

Stepping-stone models can be set up with discrete generations and discrete time or as continuous birth-and-death processes. The latter case is closely related to the voter model(Liggett,1985,Ch.5) and then to the Williams-Bjerknes model(1973),the biased voter model.

The "diffusions with killing" are regular diffusion processes whose sample paths behave regularly until a possibly random,possibly infinite time when the process is killed (see Karlin and Taylor,1981,Sect.2,Ex.G and Sect.10). Such processes are governed not only be the usual infinitesimal drift and diffusion terms,but also by a state-dependent killing rate (which corresponds in genetics,for example,to the formation of a certain genotype). See also S.Karlin and S.Tavaré(1981,1983). Notice that the killing rate can depend on the position as well as on time.

5.

The dynamical equation for the spatial distribution of a biological population can generally be obtained by combining a growth term with a dispersal term. This idea clearly ensues from the paper on random dispersal in theoretical populations,published by J.G.Skellam in 1951. Following D.Aronson(1985), it represents "the first really systematic attempt at a critical examination of the role of diffusion in population biology". Indeed,J.G.Skellam shrewdly analysed the Malthusian and logistic laws of growth in 1- and 2-dimensional habitats (with different types of borders),including diffusion and asked about the stability of stationary states. Moreover,he introduced the competition process and examined the competition of two species of annual plants living in one or two habitats.

It is really worth while recalling his example concerning the geographical expansion of the muskrat (Ondatra zibethica L.) after its escape from captivity in Central Europe,following the data collected by J.Ulbrich between 1905 and 1927 (Fig.2). J.G.Skellam contended that the remarkable uniformity of spread (despite the obvious irregularities of the terrain) was to be theoretically expected "when diffusion and population growth occur simultaneously"(Skellam,1973,p.65). Undoubtedly,he was aware of the necessity of a complex,comprehensive model of spatial growth and spread.

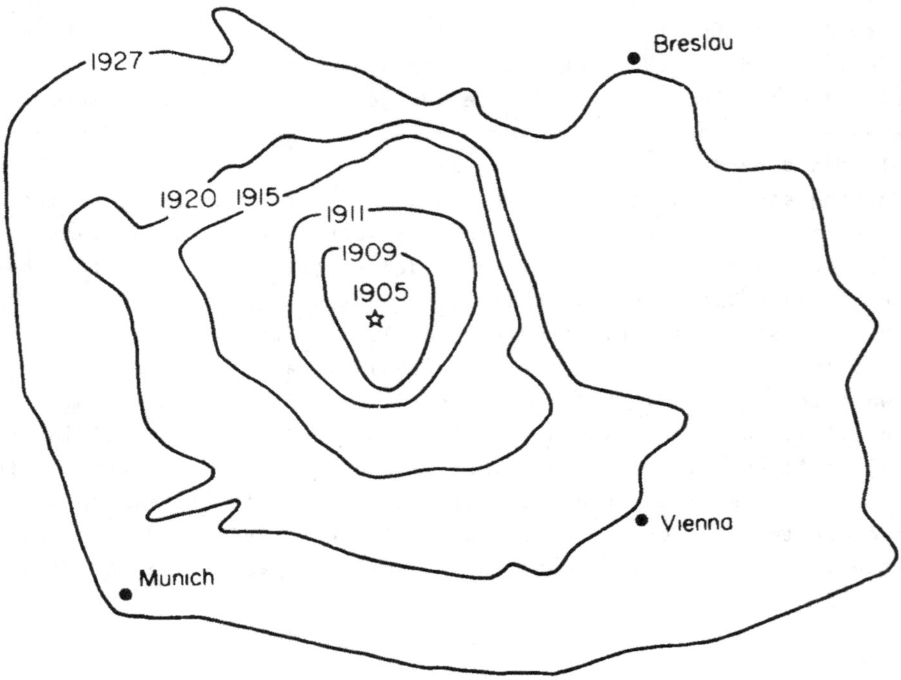

It is surprising that in the same year,1973,D.Richardson proved (the main theorem) that some biological populations grow and spread,as t increases,like a circle at a linear rate. The informed reader should compare the above picture with the graphical representation of the computer simulations of three different lattice growth models : Eden(1961), Williams- Bjerknes(1972) and Richardson(1973), realized by R.Kindermann and J.L.Snell(1981,pp.84-90).

Another point which must be mentioned is Skellam's 1973 statement that "the large-scale results of diffusionary processes are virtually independent of the fine texture of the medium". It is a risky remark which meets the universality problem, known by particular discrete dynamical systems,namely the deterministic cellular automata.

If we try to put through in a concise form the biological conception about appropriate stochastic spatial models, as imparted in the above paragraphs, we can frame the following six propositions :

(A) There are biological objects or particles (e.g.,viruses,genes, bacteria seeds,cells,small/big animals) moving in a discrete/continuous space and time.

(B) The space has/has not a regular/irregular/variable geometrical structure. (See Skellam,1951 : Rough approximations to circles are sometimes afforded by islands,hill tops,patches of woodland,etc.) From a dynamic point of view,one assumes a stable/variable environment. (Vide Skellam,1952 : Stable climax vegetation has a more uniform texture,but even here we find a mosaic of variability determined by the disposition of the more dominant organisms.) By "varying environment" one may understand a modification of spatial structure,changes in the neighbourhood relations or definite/indefinite factors influencing reproduction or motion. "Heterogeneity" may be of large- or small-scale.

(C) The biological objects reproduce identically or can change their characteristics (type or "colour") before or immediately after the movement. (Vide Skellam,1973 : It may seem unrealistic to a biologist to regard a population of diffusing animals as homogeneous.) The objects may die,too. Birth,death and "metamorphosis" are functions of demographic,temporal and spatial parameters (e.g.,population density,sex,age, position,etc.)

(D) The movement is either directed or at random. (See Skellam, 1951 : The random dispersal is at least approximately true for large numbers of terrestrial plants and animals.) "Dispersal" is a two-sided term:on the one hand it signifies "spread" and on the other han it means "aggregation". The directed motion ("taxis") occurs in different life cycle stages and is a response to some stimulus (demographic or spatial).

(E) There are biological particles (e.g.,infectious agents,genes, etc.) that can be carried locally (e.g.,by neighbourhood relationships or local contacts) or at distance. (See,e.g.,Hamilton,1971 : nearest-neighbour relationships join points into groups.) The general term "interaction" includes coexistence,cooperation or competition phenomena. It may be homogeneous/inhomogeneous.

(F) One of the results of dispersal is the achievement of a new spatial order (pattern) and maybe a new population structure (e.g., homogeneity by functional metamorphosis or speciation). (Vide Skellam, 1951 : In assessing the roles played by various processes in bringing about <u>speciation</u>,it might be borne in mind that fragmentation of the distribution area is by no means a necessary condition for the different accumulation of genetic <u>diversity</u> to arise in the same species in different parts of a comparatively uniform habitat.) Lattice structures are frequent in biology,especially if the objects are closely packed. In his 1941 book on retina,S.Polyak showed that the photoreceptors of macaque retina are closely packed with a triangle-like lattice pattern.

Other interesting points are:

(1)The application of Voronoi polygons to spatial problem of territories (Maynard Smith,1974;but the general problem goes back to W. D'Arcy Thompson). Observe the re-discovery of Voronoi polygons (or Dirichlet domains or tesselations,Theissen regions,S-mosaics,etc.)in the spatial structure of cell populations (see,e.g.,Ripley,1977,Sect.7 on Crick-Lawrence cell problem(1975)and the discussion thereon ; also Honda,1978;Diggle,1983)

(2)The comparison between the concepts of "territory" and of "functional unit"(animal vs cell populations). See R.Bachi(1973)for the definition and analysis of a territory.

(3)The process of aggregation. Recently,considerable interest in non-equilibrium growth and aggregation models was generated by the discovery by T.A.Witten and L.M.Sander(1981)that particle cluster-aggregation via Brownian(RW)particle trajectories leads to structures with a well-defined fractal dimensionality.

(4)The problem of reduction of dimensionality(Adam and Delbrück, 1968). This might occur in biochemical diffusion-limited association reactions*). Rather than increasing concentrations to ensure sufficient reaction rates in a 3-dimensional solution,the target could be associated with a 1- or 2-dimensional structure (e.g.,a linear polymer or a membrane surface). Is it a real phenomenon ? (McCloskey and Poo,1986).

(5)The distributions arising in large/small quadrants with a compact/non-compact plants arrangements. This is the problem in J.G.Skellam 1952 paper on "statistical ecology". In §3.8,J.G.Skellam discussed the problem in terms suggesting the existence of a simple point (counting) process : "Provided that the offspring develop independently of one another and that the expected number per family does not fluctuate widely among families,it is reasonable to assume that the actual number of offspring in a family has a Poisson distribution."

---

*)This process might also occur in morphogenesis,inducing important consequences:"At the later stage of morphogenetic development,it is more efficient from the standpoint of reaction-diffusion processes to grow in a fractal sense;i.e.,to keep the area (volume) constant while increasing the edge length (surface area) and to have the reactive sites distributed on the boundary of the system"(Hatlee and Kozak,1981)

(6)The problem of competition. See the discussion in P.Diggle(1976), also Gates,1980.

6.

One year after his study on multiple stationary time series,P.Whittle (1954) published an important paper on stationary processes in R^2. The introduction clearly shows the object of this investigation :

> "The disturbing effect of topographic correlation on the results on field experiments,forest and crop surveys,sampling surveys of populated area,etc.,is well known,and it is recognized that we have here examples of two-dimensional stochastic processes. The physicist also encounter higher dimensional processes (for instance, in the studies of turbulence and of systems of particles) and have indeed been the principal investigators of the subject...For many applications it is sufficient to consider only purely non-deterministic processes,and we shall restrict our attention to processes of this type,more particularly to linear autoregressions."

The fine point is that the author considers a 2-dimensional linear autoregression model,but the idea behind it is a certain correspondence between the statistical model and some spatial stationary processes. (In his classification,J.Besag(1974,§5.5) called this model "the stationary auto-normal process on an infinite lattice".) The reader will notice that the spatial autocorrelations were mentioned in Section 3 under "random associations". As regards the stationary time series,their study goes back to A.Khintchine(1934) and H.Wold(1938);in the same year as Whittle's paper,M.S.Bartlett(1954)published a survey on spectral analysis of stationary time series. However,the spatial series include an intrinsic difficulty,namely their "multilaterality" : the variate depends on events in all dimensions considered,while the variate of a time series is obviously influenced by past values.

It is interesting to note that,in the same year,P.J.Clark and F.C. Evens published their joint paper on the distance to nearest neighbour as a measure of spatial relationships in biological populations. Thus, in 1954 the two main problems in spatial pattern (see Bartlett,1974), namely (i)the detection of departures from randomness and (ii) their analysis in relation to some stochastic model,were propounded and investigated. See the useful references P.Holgate(1972) and W.G.Warren (1972).

If we are concerned with regular lattices -as it is usual in field studies- we can distinguish the following simple situations (Besag,1974):

(1)the lattice has point sites with (1a)discrete or (1b)continuous variables;

(2)the lattice contains regions with (2a)discrete or (2b)continuous variables.

The corresponding empirical situations were introduced by W.Cochran (1936) and G.H.Freeman(1953) for 1a (both mentioned in Section3),by L.D. Batchelor and H.S.Reed(1918) for 1b, by P.Greig-Smith(1952) for 2a, and by W.B.Mercer and A.D.Hall(1911) for 2b. In his 1954 paper,P.Whittle investigated both situations 1b and 2b with continuous variables,i.e. uniformity data on grain yields of wheat (Mercer-Hall) and uniformity trial on orange trees (Batchelor-Reed). See also Patankar(1954). The reader should note that Mercer-Hall data were re-examined bv J.Besag (1974,p.218) and B.Ripley(1981,p.88).

V.Heine(1955) continued Whittle's analytical approach ; he started by considering a stochastic partial differential equation,while P.Whittle considered a stochastic difference equation.

The data collected and analyzed by P.Greig-Smith(1952)[situation 2a above] stimulated H.R.Thompson(1955) to build up a 2-dimensional point process. He pointed out firstly that the simplest assumption which can be made about a community of plants,is that they are distributed at random in the Poisson distribution. Yet,a more realistic and useful model would take into consideration the random distribution of "parents" which become "centres of distribution" for the next generation of "offspring" whose positions depend on those of their respective parents. This description corresponds to that of clustering (aggregation) processes developed by J.Neyman and E.L.Scott,which is equivalent to a class of Markov linear self-exciting point processes. The reader should notice the contrast with a previous approach without the dispersal hypothesis by M.Thomas(1949). It was postulated a completely random distribution of parents,each producing independently a Poisson number of offspring. Yet, each offspring occupied that same site as its parents and it is also indistinguishable from them.

As far as the first assumption about Poisson distribution is concerned,it should be recalled that this is an old problem (see,e.g., Morant,1920) connected with the random division of an interval (e.g., Domb,1947;Moran,1947,1951). The Poisson process occurs in many limiting situations and represents the simplest possible stochastic mechanism for the generation of spatial point patterns. See J.L.Doob(1953,Ch.8, §5) on the application of the Poisson process to molecular and stellar distributions. The question was how particles (on R^1) can move to preserve the property of uniform distribution. One of Doob's examples was a system of finitely many particles : if they move independently, the system will be stationary, regardless of the number of particles.

Such a result may suggest the idea that Poisson processes should not be proper models for spatial growth and dispersal. Indded, this is the conclusion reached by J.F.C.Kingman(1977,§4). In fact, a Poisson equilibrium is usually inconsistent with the assumption of independent displacement. If that assumption is dropped, a Poisson equilibrium is possible.

The effect of displacements can be understood if we examine successively the problem of "controlled variability processes" in d($d\geq 2$), posed by D.R.Cox in 1971 (see Gâcs and Szâsz,1975) :

> "Consider a regular network of points in R^2, e.g., the set of lattice points, or the vertices of regular hexagons. Let each point be displaced by a random vector, displacements of different points being independent and identically distributed. Let A be a convex set, and let N(A) denote the number of displaced points contained in the set A. It is true that as the diameter of A tends to infinity,
>
> Var N(A) /(circumference of A)
>
> (i)is bounded above, (ii)is bounded away from 0, (iii)tends to a limit ?"

In 1-dimension,(i) holds for the ratio above, but not necessarily either (ii) or (iii); for d dimensions (ii) results under certain conditions, but it is not possible to give any affirmative answer to (iii) in general. (Compare with the results obtained by Kendall,1948, and Kendall and Rankin,1953.)

Under "Poisson ensembles of points", W.Feller(1968,p.159;1966,p.14) gathered together stars in space, raisins in cake, animal litters in fields, etc. His examples are the flying-bomb hits in London and the bacteria and blood counts (1968,Ex.7b and 7e) which correspond to situation 2a in (Besag,1974). The reader is also referred to two rather unknown papers by R.L.Dobrushin(1956) on Poisson laws for distributions of particles in space and by A.Prékopa(1960) on a Poisson spread model.

The first idea about a stochastic process of clustering occurred to J.Neyman in 1939 in connection with the distribution of the number of insects in equalized plots on the ground. The particles were larvae and the cluster centers were "egg masses" laid by moths more or less at the same time. The cluster size was represented by the number of those larvae, hatched from an egg mass located at a point x and later crawling around in search for food, that survived up to the time of counting (see Neyman and Scott,1972). The Neyman Type A contagious distribution corresponds to a non-orderly process of randomly distributed point clusters (Diggle,1983,p.54). Biological populations showing contagious distributions were some maritime and grassland communities as well as the European corn-borers,the Colorado potato beetles and the beet webworms (Evans,1953). For indices of contagion and measures of aggregation see J.E.Paloheimo and A.M.Vukov(1976). The spatial dispersal of infections follows frequently the clustering model with many subsequent "centres of dispersal" : e.g.,the spread of hepatitis (Brownlea,1972) or the simulation of virus plaque formation (foot and mouth virus). In the latter case it is assumed that the number of infective units (virus progenies) produced by a single infected cell is a Poisson distributed variable with mean λ. The radius of plaques increases linearly with time and the velocity of infection depends upon the values of λ (Schwöbel et al., 1966).

Two years after the publication of Thompson's paper, J.Neyman and E.L.Scott(1957) published their fundamental paper "On a mathematical theory of populations conceived as conglomerations of clusters". This is an enlightening stochastic theory of the spatial growth of a population : the authors viewed it as a specialization of the theory of general branching processes,but it remained (together with the Bartlett-Lewis clustering model) a particular class of point processes (see Lawrance,1972;Cox and Isham,1980). In 1974,A.G.Hawkes and D.Oakes found the Neyman-Scott model equivalent to a class of Markov linear self-exciting processes. In the presentation of their 1952 model of the spatial distribution of galaxies,at the Royal Society(1958), J.Neyman and E.L. Scott discussed the possiblity of introducing a kind of self-regulatory mechanism in the explanation of biological growth,namely density-dependent births and/or density-dependent deaths (actually suggested by J.G. Skellam in a discussion by this 1957 paper).

Two points must be mentioned :

(1) M.S.Bartlett remarked that density dependence "does not seem to me to provide any difficulty in principle with point-processes. It is replaced by dependence on near neighbours,but this is unfortunately very messy mathematically,as those who work on the molecular theory of fluids will testify". As a matter of fact,this suggests again the theory of interacting systems : Section 6 in the book on random fields by C.Preston(1976) is devoted to their treatment as point processes. In fact,any point process satisfying some definite condition is a "Gibbs process with local specifications". Moreover,the voter model on Z^d,$d\geq1$,converges to a linear combination of trivial point processes,if $d=1,2$, or to a nontrivial point process,if $d\geq3$,as t goes to infinity. Yet,the pioneering work on Gibbs distributions as equilibrium distributions of spatial point processes was done by A.N.Kolmogorov(1937).

In their paper on Markov point processes, B.D.Ripley and F.P.Kelly(1977) introduced a class of "pairwise interaction processes" which are linked in a natural way to the Markov random fields. Point processes with interactions were studied by X.X.Nguyen and H.Zessin(1976). This is a good opportunity to mention the work of E.H.Kerner on Gibbs ensembles in biology. For him, a large ecosystem is like an ideal gas ; the fundamental law of clustering of ecotypes is that cluster sizes have independent Poisson distributions(1978).

(2) For J.E.Moyal the problem of a population process with motion or diffusion of the individuals in space and with density-dependent birth and death rates "is solved in principle". Indeed, J.E.Moyal already published a paper on Markov jump processes(1957) and prepared his general theory of stochastic population processes(1962) which included cluster processes as an example. (The spatial variation in mortality is described by thinned point processes : Brown, 1979.)

Spatial point processes as models for epidemics were treated by M.S. Bartlett(1956). The reader is referred to the survey written by N.T.J. Bailey(1980). The problem of within-family contagion arises in those cases where, for a disease known to be communicable, the mechanism of its spread is not known : this is an example of a clustering process based on the data obtained by R.R.Puffer(1950) on the incidence of polyomielitis in a general population (Neyman and Scott, 1972).

For applications of point processes (particularly spatial) in neurophysiology, see A.V.Holden(1976, Ch.12). For a theory of "homogeneous cluster fields", the reader is referred to the book by K.Matthes, J.Kerstan and J.Mecke, "Infinitely Divisible Point Processes"(1978).

7.

Roughly speaking, the familiar representation of processes physicists are interested in is the model of particles walking on a lattice, while biomathematicians' usual device is the model of reproducing particles (in an indefinite habitat). Of course, the walk of a particle is subjected to many restrictions of a structural and positional nature (e.g., a deformed lattice, continuously occupied sites, traps, etc.), and also the reproduction of a particle may depend on physiological or demographic parameters (e.g., age, total size or density of population, etc.). As far as I know, the first spatial branching process considered in physics was the model for the transport and multiplication of neutrons. The reader should consult the classical book by T.Harris(1963, Ch.IV). See also S.Asmussen and H.Hering(1977, Sections 4 and 5). A theory of population transport in biology was later suggested by M.Rotenberg(1972) ; see also K.Gopalsamy(1976).

Yet,in the mid-fifties a new stage in the development of the theory of branching processes was attained. Besides the obvious theoretical source (analytic and measure-theoretical advancements) a distinct source contributed to this expansion. The Neyman-Scott models for galaxies and biological populations brought in a particular combination of two random processes,namely a multiplicative process and a motion (dispersion) process. In their first assumptions,J.Neyman and E.L.Scott suggested the birth-and-death process to be the multiplicative one and the diffusion process a model for dispersion. They already had in view a birth-and-death random walk in d-dimensions(1957:abstract). This line of investigation was continued by S.R.Adke and J.E.Moyal(1963) and further by J.E.Moyal(1964),S.R.Adke(1964) and A.W.Davis(1965).

In this construction,the branching part may be critical,super- or subcritical,with one or many types of biological particles,etc.,and the dispersal part may be a random walk,a Brownian motion,or a diffusion (with different properties). Immigration may be added (Ivanoff,1980) and the spatial process may develop in varying environments (Klebaner,1982; Ivanoff,1983).

Thus,in 1964,J.Neyman and E.L.Scott published their model for the spread of an epidemic. The space distribution of susceptibles and infectives is explicitly taken into account as well as their dispersal over a given plane habitat. The authors assumed that an epidemic may be generated either by a mutation in the population of infectious agents or by the occurrence of infected immigrants (with constant immigration rate). This is a position-dependent branching process in discrete time, called by N.T.J.Bailey(1980,p.250) a "general multiple-site model". The reader should compare it with another position-dependent branching process introduced by H.E.Conner(1961,1964,1967).

The excellent work by J.Neyman and E.L.Scott is limited by the assumption that the size of susceptibles is not reduced by each new infection. Thus,following N.T.J.Bailey(1980),the model could be "approximately valid" at the start of an epidemic but would become progressively less so as time passed. A sufficient condition allowing that the considered epidemic will be "heading for extinction" was given by W.Bühler(1966). A generalization was reported by L.P.Ammann and P.F.Thall(1977).

The spatial birth-and-death process with migration built up by N.T.J. Bailey(1968) should be considered as an exemplar of a spatial growth process. It can represent the growth of viral plaques as well as of bacterial colonies,the 3-dimensional tissue growth (e.g.,the local spread of malignant cell,the formation of metastases) as well as the spread of a species through consecutive niches,etc. A.W.Davis(1970) treated this model as a Markov branching diffusion process, and D.Dawson(1972,Appendix 1) treated a similar process as the solution of a stochastic evolution equation. Two subprocesses attracted special consideration : migration and interconnection. Although scrutinized for a long time (e.g., Bartlett,1949), migration was investigated in some variants of Bailey's model (Adke,1969;Renshaw,1972) and,as nonlinear open migration,by Peter Whittle(1967,1968). This promoted the construction of the theory of Markov population processes and of the flow models for biological populations (Kingman,1969). See F.P.Kelly(1979,Ch.2 and 6). Worthy of consideration is a general linear (open/closed) migration model (Radcliffe and Staff,1970) inspired by the model for attachment/detachment of antibodies to viruses,built up by J.Gani and R.C.Srivastava in 1968.

Interconnected birth-and-death processes were considered by P.S.Puri (1968) and E.Renshaw(1973) ; special interaction processes between two (Daley,1968) or many types of biological particles (Karlin and Kaplan, 1973),without spatial assumptions, are of particular interest. "Neutralization effects" were thought of in the latter case, and a self-annihilating branching process was introduced by K.B.Erickson(1973) to describe the antigenic behaviour of lymphoma cell populations. Interaction raises complex technical problems : W.Feller(1950) already draw attention to the difficulty of treating size and composition of a population with several interacting types (see also Kesten,1971). R.Durrett(1979) approached the problem by considering a spatially homogeneous infinite particle system with additive interactions as the superposition of supercritical branching random walks.

The finite spatial birth-and-death process was constructed by C.Preston(1975) as a jump process defined on a suitable space of particle configurations,while R.A.Holley and D.W.Strook(1978) treated an (infinite) nearest-neighbour birth-and-death process on R^1.

After the development of Neyman-Scott models,the domain of communicable diseases appears to be the privileged domain of spatial branching processes. The reader is referred to the papers by J.Radcliffe who in 1973 started his investigation on the geographical spread of different types of epidemics. As a matter of fact,his model for the spread of carrier-borne epidemics (1973) is the spatial generalization of the Pettigrew-Weiss model(1967). A model of branching diffusion with stopping was introduced by S.Asmussen and H.Hering(1977) - "stopping" means cure, isolation or death of an infected individual. The velocity of the infection wave was investigated by D.Mollison(1972) and H.E.Daniels(1977) ; see also M.J.Faddy and I.H.Slorach(1980).

The spatial dispersion of a neutral allele on an 1- or 2-dimensional lattice was examined by K.S.Crump and J.H.Gillespie(1976),using the theory of multitype branching processes ("types" represent the colonies between which individuals can migrate). On the basis of data obtained from Drosophila populations (Dobzhansky and Wright,1947),it appears that this spread is shortly ranged : if the neutral model is correct,the (accidentally) long-distance migrants do not significantly affect the genetic structure of the species. Under the name "branching random field", S.Sawyer(1975,1976) analyzed branching diffusion processes. The equation deduced in his Theorem 2.3 appeared in Bailey's and Dawson's models mentioned above. (See Ivanoff,1980,for the definition of branching random fields.) An interesting result concerning dimensionality is obtained by considering a Brownian branching random field : if the initial distribution of an allelic type is uniform Poisson, a bounded open set is visited at arbitrarily large times if $d \geq 2$ but not if $d=1$ (Sawyer and Fleischman,1979).

The model for carcinogenesis by N.Lenz attracts attention : the motion part is a random walk taking place on a mutation-and-deficiency space.

Addendum. The branching diffusion process introduced by B.A.Sevastyanov(1958) apparently had a pure mathematical motivation ; the author's basic tool was the nonlinear integral equation of Hammerstein type. The diffusion domain was restricted,with absorbing boundaries. A general Markov branching process with diffusion was reported by V.V.Skorohod in 1964 ; in the next year,P.Ney suggested a branching random walk for describing a cascade process (see the last chapter of the indispensable book by K.B.Athreya and P.Ney,1972).

Between 1965-1968,S.Watanabe and his colleagues,N.Ikeda and M.Nagasawa,developed a rigorous treatment of Markov branching processes. If one carefully reads the important series of three papers (1968), one will find in Part II the construction of a branching Brownian motion process and of a branching diffusion with and without absorbing boundaries (Ex.3.4 ; see also Davis,1967). Also,in Part III the authors introduced some transforming operations of these processes,e.g.,the killing of the non-branching part (Ex.5.2) and the transformation of branching laws (Ex.5.3).

Obviously,convergence and approximation methods are necessary. To study large population approximations for models involving spatial dependence,one needs to consider the theory of measure-valued processes (Jirina,1964;Dawson,1975). Frequently,the limiting process is a continuous-state process (Jirina,1958;Lamperti,1967;Watanabe,1968;Silverstein,

1969). See,for convergence,J.Radcliffe(1976), F.J.S.Wang(1980),and N. Lenz(1982) - to name but a few. The reader is referred to T.Watanabe (1969) and L.Gorostiza and R.Griego(1979) for the convergence of branching transport processes.

8.

A new mathematical theory,percolation,was originally suggested by S. R.Broadbent in his discussion of the paper "Poor Man's Monte Carlo" by J.M.Hammersley and K.W.Morton(1954) :

> "A square (in two dimensions) or a cubic (in three) lattice consists of "cells" at the interstices joined by "paths" which are either open or closed,the probability that a randomly-chosen path is open being p. A "liquid" which cannot flow upwards or a "gas" which flows in all directions penetrates the open paths and fills a proportion $\lambda_r(p)$ of the cells at the r-th level. The problem is to determine $\lambda_r(p)$ for a large lattice."

The authors replied that "Mr.Broadbent's problem is very fascinating and difficult" and gave a preliminary answer. Indeed,the problem is still captivating,"it is a source of fascinating problems of the best kind a mathematician can wish for : problems which are easy to state with a minimum of preparation,but whose solutions are (apparently) difficult and require new methods"(Kesten,1982).

In their joint paper,S.R.Broadbent and J.M.Hammersley(1957) formally defined percolation ; the name was suggested by the fact that Broadbent's liquid flowing downwards through his maze was like coffee in a percolator (Hammersley,1983). One of their examples (Ex.5) was the spread of an infection (blight) in a plantation ; it was re-considered in the fundamental paper by J.M.Hammersley and D.J.A.Welsh (1965) on first-passage percolation (Ex.1.1.1 and 1.1.2). In the same year,R.W.Morgan and D.J.A. Welsh(1965) set up a model of the spread of an infection through an orchard (a 2-dimensional lattice) "in the presence of a very strong south-west wind". They assumed that the first-passage time has an exponential distribution and conjectured that the limit of the ration between the mean number of infected("black")sites and the time tends towards a finite constant as t goes to infinity. The conjecture was verified by J. Hammersley(1966) - see also R.T.Smythe and J.C.Wierman(1978,p.163).

Some spatial biological growth processes are percolation processes "in a wider sense" (Welsh,1977),as,e.g.,Richardson's simple growth model,an analog growth model with deaths,infection models as above,or some cell growth models like the Eden or Williams-Bjerknes model. An important point is that the Morgan-Welsh model is a Markovian process on the oriented square lattice,while the Williams-Bjerknes model is such a process on an unoriented lattice. Oriented and regular percolation belong to different universality classes : the nature of the infinite cluster and the behaviour of the system for p near critical probability differ in the two models (Durrett,1984).

Before describing the Eden model,I should like to briefly discuss the connections between percolation and other random processes. Originally, it was emphasized that the multidimensional randomness in the medium (i.e. in the system of bonds) is characteristic for percolation,while in a diffusion process the (1-dimensional) randomness is in the motion of the fluid itself. This difference led J.Hammersley(1983) to distinguish Markov processes and Markov random fields following the dimensionality of randomness. The connection with branching processes is based on the graph structure : if we consider the classical problem of surnames in the theory of Galton-Watson processes,we observe that it may be represented as a bond percolation on a family tree. The surname is the "fluid" which flows down all open bonds from the single ancestral source (Frisch and Hammersley,1963;Hammersley and Walters,1963). For the relation between "maximal" branching processes and "long-range" percolation, the reader is referred to J.Lamperti(1970)[and the appendix by H.Kesten].

The correspondence between statistical mechanical models and percolation was proved by P.W.Kasteleyn and C.M.Fortuin(1969) ; in particular, the bond percolation can be formulated in terms of the Potts model and the site percolation as the limit of a Potts model with multi-site interactions. Moreover,certain "percolation substructures" (i.e. particular random graphs) serve to define particle systems and to deduce "associate" (or dual) processes (see,e.g.,Donnelly and Welsh,1983).

The Eden model may,in my opinion,be thought of as a stochastic version of a Turing model ; it was pictured as follows (Eden,1961) :

> "Given a single cell or a homogeneous population of cells,how can such a cluster develop into a structure of lower order of symmetry or no symmetry at all,that is,into a structure of recognizable and characteristic shape,without involving special structural properties of the cells themselves ? ...Starting from a single cell which may divide,its daughter cells divide again and again,what are the structural properties of the resulting colony of cells and how do various possible constraints effect the architecture ?"

In his paper on a probabilistic model for morphogenesis,M.Eden(1958) pointed out the combinatorial aspects of his model and the resemblance with combinatorial problems in statistical mechanics (e.g.,Kac and Ward, 1954),as order-disorder problem. In the simplest symmetrical Eden model,

an arrangement of cells that contains exactly k connected cells will be called a k-configuration, c^k. The biological examples given by M.Eden for this model refer to the growth of the common sea lettuce, Ulva lactuca, which apparently grows only at its periphery, and that of the gamete-forming thalli of Prasiola stipitata which exhibit a pattern almost identical to that obtained for k-configurations with small values of k.
See also the description of the Eden model by D.Richardson(1973), D.J.A. Welsh(1977) and H.Kesten(1985). Two generalized Eden processes were introduced by K.Schürger(1981).

Three results must be mentioned particularly (Eden,1958) :

(i) configurations with many short branches are more probable than those with a few long branches;

(ii) large configurations will be essentially circular in outline;

(iii) they will have a high density, i.e. they will contain very few "holes" and short "tentacles".

The essential Richardson's theorem that can be applied for Eden's model is valid only under certain conditions : percolation processes on Z^d ($d \geq 2$) obey the law of large numbers (specifically "the strong law of large configurations" ; see Schürger,1979) provided their time-coordinate distribution satisfies certain moment condition (Schürger,1980; Cox and Durrett,1981 - also Vahidi,1979; Branvall,1980).

The relation between percolation and random graphs can easily be conceived (Welsh,1977; McDiarmid,1981). Bond and site percolation processes are locally dependent random graphs, i.e. directed graphs with random coloring of the bonds (see for definition Kuulasmaa,1982). Thus, infectious processes (epidemics) can be appropriately represented as random graphs (Tautu 1974/1977, Gertsbakh,1977; Kuulasmaa,1982; Kuulasmaa and Zachary, 1984). Vide B.Bollobas(1985,p.370). As far as I know, epidemics are the only spatial growth processes studied as random graphs, so that K.Schürger 1976 paper on the evolution of random graphs over expanding square lattices may be the starting point in the study of spatial cellular growth.

Supplement. The doublet problem evoked in Paragraph 3 is connected with a problem posed in the famous paper by E.Ising(1925). He asked for the number of ways of obtaining a total number of runs (without regard to length) from arrangements of two kinds of points. This is a combinatorial problem which arose in chemistry as the "dimer" problem : find the number of ways in which diatomic molecules (dimers) can cover a doubly periodic lattice (of lattice spacing equal to a dimer length), so that each dimer covers two adjacent lattice points and no lattice point remains uncovered (Fowler and Rushbrooke,1937). As H.N.V.Temperley stated(1979),equilibrium statistical mechanics is primarily concerned with problems of enumerating graphs of a given specification. The reader is referred to the book by J.K.Percus(1971,Ch.2) for the counting and enumeration problems on the regular lattice.

Enumerative combinatorial analysis (based on the fundamental counting theorem stated by G.Pōlya in 1937) was applied to trees and graphs and led to the "cell growth problem" (or creation of "lattice animals"). Following F.Harary et al.(1975),the name stems from an analogy with an "animal" which,starting from a single "cell" of some specified basic polygonal shape,grows step by step in the plane by adding at each step a cell of the same shape to its periphery. The fundamental combinatorial problem concerning these animals is that of determining exactly how many distinct animals there are with a given number of cells,where two animals are regarded as distinct if one can be brought into coincidence with the other by rotations,translations or reflection. This problem was posed to Frank Harary(1964,p.213) by G.E.Uhlenbeck who wanted to determine the number of different shapes of paving blocks and by a(n) (unnamed) biologist who was curious about the number of different shapes of animals having a given number of cells.

If the basic shape is a square,the animals are "polyominoes" (generalized dominoes) studied by S.W.Golomb(1954,1965). This represents,in addition,the combinatorial problem of the Eden model : if C^k is the set of all k-configurations (unique under translation,rotation or reflection),find lower and upper bounds of the number of elements in C^k (Eden, 1961;see the improvements by D.A.Klarner,1965,1967)*).

During the last years it has been observed that the animal problem is related to the study of diluted branching polymers in a good solvent. The connection of isotropic animals with random field models has recently been revealed (Parisi and Sourlas,1981);moreover,anisotropic directed lattice animals may model dilute branched polymers in a suitably flowing solvent,and are also closely related to Markov branching processes with a single ancestor (Redner and Coniglio,1982). The directed animal problem is soluble (Dhar et al.,1982)

Polyominoes played an important role in the creation of games of life

*)As M.Eden affirmed in a letter to D.A.Klarner(1970),"the cell growth problem was original with me and I began to work on it in 1953 at Princeton. It is related to the polyomino problem that Golomb has worked on. I cannot say whether he derived the problem from my work or initiated it independently."

created by J.H.Conway around 1970.

9.

A new chapter was added to the history of biological growth processes by the development of the theory of cell automata (CA). In the last section of his paper given at the International Congress of Mathematicians in 1952,S.M.Ulam reported the creation of a general model for systems of an infinite number of interacting elements,"considered by von Neumann and the author". The closed finite subsystems formed by introducing simple (deterministic) neighbourhood rules were called "automata" or "organisms". "One aim of the theory -wrote S.Ulam- is to establish the existence of subsystems which are able to multiply,i.e.create in time other systems identical ("congruent") to themselves." The reader should consult other papers by S.Ulam about recursive growth processes ,e.g., "On some mathematical problems connected with patterns of growth of figures"(1962),"Some ideas and prospects in biomathematics"(1972),and his book,"A Collection of Mathematical Problems"(1960,p.85).

Within the framework of automata theory,self-reproduction was considered to be a special case of universal construction. J.von Neumann asked how complex a cell (tessellation*)) automaton must be in order to exhibit universal computational power (theory of the Turing machine) and to mimic reproduction,and how this complexity is to be assessed in terms of tessellation structure. Consequently,he sought a system with not too many states : it was E.F.Codd(1968,Ch.4) who showed that eight states are sufficient for the cells to endow a CA which is computation-construction-universal. A necessary condition for computation universality is that,by choice of initial configuration,it is possible to obtain "boundable propagation" as far out from this configuration as desired (Codd, 1968,Proposition 3,p.24). (See also the Propositions 6 and 8 on rotation symmetry and computation universality.) [Remark:Universality is a sufficient condition for self-reproduction,but not a necessary one (Langton, 1984.] The CA invented by J.H.Conway ("game of Life") has the Moore template and 2-state cells. The computation universality was informally proved by J.H.Conway and independently by R.W.Gosper. (See Smith III, 1976,p.409 : "Their formal proof would be very long and tedious ; hence they will probably never write it down." Vide Baer and Martinez,1974.)

---

*)A plethora of names is to find in this field : "tessellation structures"(Moore,1962),"tessellation automata"(Yamada and Amoroso,1969),"iterative realizations"(Arnold et al.,1970),"homogeneous structures" (Aladyev,1974),etc.

It is stated (Wolfram,1984) that no general finite algorithm can predict whether a particular initial configuration in a computationally universal CA will evolve to the null configuration after a finite time, or will generate persistent structures,so that occupied sites will exist at arbitrarily large times. A unique configuration generates an infinite sequence of propagating structures : this is the "glider". It seems that in this case the required initial configuration is quite large and is very difficult to find (Wolfram,1984,p.31;see Berlekamp, Conway and Guy,1982,pp.831-847 for more details about the "glider" and its behaviour). L.S.Schulman and P.E.Seiden(1978) studied a discrete stochastic system of particles whose evolution is governed by rules similar to those in Game of Life. I mention two results : (i)the system shows phase transitions and (ii)it develops structures and correlations ("order") from an initially random state.

J.von Neumann considered five different kinds of models of self-reproduction,namely kinematic,cellular,excitation-threshold-fatigue,continuous,and probabilistic. The first two models differ in the treatment of motion : in kinematic systems self-reproduction results from organized motion,whereas in CAs motion is a special case of self-reproduction (Burks,1970,p.49;Myhill,1964). Although J.von Neumann contemplated the problem of stochastic automata,their study was initiated in other (related) fields of research : the information theory (see the 1948 book by C.E.Shannon and W.Weaver),the theory of Turing machines (De Leuw et al.,1956),cybernetics (e.g.,the 1958 book by R.W.Ashby),and the theory of Markov processes (Davis,1961:Markov chains as random input automata). The specific development of the stochastic automata theory begun in the sixties ; for example,in a single year,1963,the papers by M.O.Rabin, J.W.Carlyle,L.Lofgren,M.Tsetlin,V.I.Varshavskii and I.P.Vorontsova,and G.N.Tsertsvadze appeared,followed in 1965 by O.Onicescu and S.Guiasu, and P.H.Starke.

The book by A.Paz(1971) does not contain biological examples of SAs, so that I will refer to two authors,namely L.Lofgren and M.Tsetlin. The former posed the growth problem in an unusual way : he studied the <u>extinction</u> of configurations and concluded that an "automaton wave" can be immortal if it is at least 2-dimensional and if the configuration has a certain complexity. L.Lofgren pointed out that the elementary construction-destruction processes considered in stochastic automata(SAs) show some similarities with the creation-annihilation processes of quantum mechanics. The work of M.L.Tsetlin(1961,1963) seems to be more influential. Starting from considerations about the behaviour of neural centers, he assumed that CAs should work in a random environment. For instance,

this environment is stationary if the functioning of a SA is described by a Markov chain and determines the probabilities of "winning" and "losing". The behaviour of an SA with "evolving structure" in random media can be described by a nonhomogeneous Markov chain (see also Tsetlin,1973). The collective behaviour of SAs is then interpreted as a game: automaton games are defined by specifying not only the systems of payoff functions,but also the structures of the participating automata. (The reader should also have knowledge of the game introduced by T.Kitagawa, 1974.) Because the automata taking part in games do not possess any a priori information about the game,the actions of each SA are determined only by its gains and losses in the course of the game (Tsetlin,1973,p. 44). This approach led to the creation of "ballot" models with random errors (Schmookler,1975),also translated as "voting" models (Vasilyev et al.,1969) and as "majority vote systems" (Dawson,1977;see Liggett, 1985,Example 4.3(e),p.33 and Example 2.12,p.140). They must not be mixed up with the voter models of the theory of interacting particle systems; they actually are spin models. There exists a long series of papers dealing with the analogies between infinite systems of 2-state CAs,homogeneous Markov networks and infinite particle systems (Vasershtein, 1969;Toom,1976,1978;Galperin,1976,1977,1978;Vallander,1980,etc.).
I quote N.B.Vasilyev(1969) for an example :

> "By a "random medium" we mean a system of automata,each of which can be in one of two states,0 or 1;at time t+1,the state of an automaton is determined probabilistically by the states of certain "neighboring" automata at time t (the choice made by each automaton is independent of that others). In our example of a random medium, the automata are arranged in a chain."

The evolution of such a network is described in terms of the evolution of the probability distributions on the space of the network states. The graphical representation (the "planar scheme") constructed by A.L.Toom (1968) is the natural approach to describing a network of formal neurons. It must be pointed out that the biological prototypes are,in general, excitable cell systems (see in Tsetlin,1973 : motoneuron groups,motor units,cardiac tissue) - but A.M.Leontovich et al.(197o) suggested the possibility of applications in embryology. (The reader is referred to M.Arbib(1972) for the general problem.) We have to take notice of the fact that the theory deals with <u>homogeneous</u> locally interacting components (e.g.,atoms in points of crystal lattices,interacting finite-state automata,logic-informational elements,queue systems,etc.) : the models of morphogenesis or carcinogenesis with evolving many cell types are still intractable (see Tautu,1975).

In the literature of late years d-dimensional SAs were studied under the name of crystal growth models (Welberry and Galbraith,1973);their temporal development was found equivalent to the equilibrium statistical mechanics of (d+1)-dimensional Ising models (Verhagen,1976;Domany and Kinzel,1984). Recent work (Kinzel,1985) relates that SAs with one absorbing state show phase transitions : if the system is random either in space or in time,then randomness changes the critical properties in any dimension ; if the system is random both in space and time,then the critical properties are the same as those of the pure CA (with averaged reaction probabilities). It is interesting to note that strong randomness with competing reaction rates leads to a behaviour characteristic for a system studied in equilibrium statistical mechanics,the spin glass,a disordered magnetic system. Spin glasses are also of biological interest. As P.W.Anderson(1983) pointed out,stability and diversity are properties of both life models and spin glass systems. His model of prebiotic evolution follows the construction of such a system. The idea goes back to J.H.Holland(1976) who defined an "artificial universe" (with kinetic and biological operators) which begins in "chaos" and,after a lapse of time,contains "life" in the sense of self-replicating systems undergoing heritable adaptations.

Additional information. Deterministic CAs have frequently been used as models for biological processes (e.g.,Stahl,1966;Arbib,1966,1969, 1972;Wassermann,1973;Laing,1975,etc.). Aristid Lindenmayer(1968) considered the development of filamentous organisms as growing cellular arrays of finite automata,with or without inputs coming from neighbouring cells. It was shown that these CAs have a universal computing ability (Herman,1971;Herman and Rozenberg,1975,§17.2). This approach became rapidly a very actively investigated topic,especially after the reformulation of 1-dimensional systems in terms of formal language theory (Lindenmayer,1971;Herman,1974;Rozenberg,1974).

The set of automaton states is then interpreted as the alphabet of a language generating system,the state transition function becomes the set of productions,and the starting array the axiom. The production rules must be applied simultaneously to all symbols of an array ("string"), making them "parallel rewriting systems". [Note:A similar construction, the τ_n-grammars,was reported by V.Aladyev in 1974,with the interpretation that all τ_n-languages constitute a subset of the set of all Lindenmayer languages.]

The Lindenmayer(L)systems were frequently viewed as deterministic CAs and designated as DOL systems (0 for zero-sided interactions). If interaction is assumed,it can only be 1- or 2-sided (i.e.D1L or D2L systems). Long-range interactions were also postulated : in the k,l systems,there are k left neighbouring cells and l right neighbours (see Herman and Rozenberg,1975,Ch.6). In the case of L systems with interactions (IL systems) we deal with context-sensitive grammars with parallel rewriting (e.g.,Culik II and Opatrný,1974). It is to be expected that the modern investigation of IL systems may bring new and interesting results. The equivalence of CAs to Ising models and directed percolation (Domany and Kinzel,1984) is proposed for consideration.

DOL systems were used to understand the role that cell lineages and equal/unequal cell divisions play in morphogenesis. See the simulation of the embryonic development of the snail Limnea stagnalis (Bezem and Raven,1975;Raven and Bezem,1976). DOL systems may be appropriate models in the first developmental period,but in a second period the course of cell divisions is modulated by the "induction field",while cell differentiation might result from the interaction of the axial gradient field and the induction field. This should lead to the change of the considered DOL system with a TOL("table" OL)system which describes the environmental effects : TOL systems are OL systems with a kind of control imposed on the use of productions (see Herman and Rozenberg,1975,Ch.5). This operation has an equivalent in the "transitions of growth rules" taken into account by T.Kitagawa(1974) as the change of growth rules according to the different stages of growth. This goes back to S.M.Ulam (1972) who suggested the existence of "rules to produce the change of rules".

The growth of strings generated by DOL systems was intensively studied (Salomaa,1973,1976;Vitányi,1973,1974;see for a survey Herman and Rozenberg,1975,Ch.15). DOL growth functions are always exponential,polynomial or a combination of the two (Paz and Salomaa,1973),while the growth function of an IL is at most exponential and at least logarithmic (Rozenberg,1974,Th.6.1). More precisely,logarithm functions and fractional powers (e.g.,growth type 2 1/2) are D2L growth functions (Karhumäki,1974). The reader should consult the paper by S.J.Willson (1981) on growth patterns of ordered CAs.

Attempts to generalize L systems to trees and graphs were made by A.K.Joshi and L.S.Levy(1974),K.Culik II and A.Lindenmayer(1974),M.Nagl (1975,1976) - to name but a few. Multidimensional DOL systems were imagined by B.H.Mayoh(1974,1976).

In his 1975 survey,A.Lindenmayer admitted that the mathematics of developmental systems with interactions is not as far advanced yet ; as much one can say about the mathematics of stochastic L systems. As far as I know,there exists only one published paper on stochastic OL systems (Jürgensen,1976). [At the same meeting,my paper "On stochastic growth processes in space",was read by title : stochastic L systems with local interactions were suggested as examples within the framework of the theory of interacting particle systems.]

R E F E R E N C E S

(Note to the reader : This is only a selected list of references since the great number of papers mentioned exceeds the space reserved for this article.)

ADAM,G.,DELBRÜCK,M.(1968) Reduction of dimensionality in biological diffusion processes. In:Structural Chemistry and Molecular Biology (A. Rich,N.Davidson eds.),pp.198-215. New York:W.H.Freeman&Co.

ADKE,S.R.,MOYAL,J.E.(1963) A birth,death and diffusion process. J.Math. Anal.Appl.7,209-224

ALADYEV,V.(1974) Survey of research in the theory of homogeneous structures and their applications. Math.Biosci.,22,121-154

ANDERSON,P.W.(1983) Suggested model for prebiotic evolution:The use of chaos. Proc.Natl.Acad.Sci.USA,80,3386-3390

ARBIB,M.A.(1972) Automata theory in the context of theoretical embryology. In:Foundations of Mathematical Biology (R.Rosen ed.),Vol.II, p.141-215. New York:Academic Press

ASMUSSEN,S.,HERING,H.(1977) Some modified branching diffusion models. Math.Biosci.,35,281-299

BAILEY,N.T.J.(1968) Stochastic birth,death and migration processes for spatially distributed populations. Biometrika,55,189-198

BAILEY,N.T.J.(1980) Spatial models in the epidemiology of infectious diseases. In:Biological Growth and Spread (W.Jäger,H.Rost,P.Tautu eds.),p.233-261 (LN in Biomath.,Vol.38). Berlin-Heidelberg-New York: Springer

BARAKAT,R.(1959) A note on the transient stage of the random dispersal of logistic populations. Bull.Math.Biophys.,21,141-151

BARAKAT,R.(1973) Isotropic random flights. J.Phys.A,6,796-804

BARNES,R.B.,SILVERMAN,S.(1934) Brownian motion as a natural limit to all measuring processes. Rev.Modern Phys.,6,162-192

BARTLETT,M.S.(1949) Some evolutionary stochastic processes. J.Roy.Statist.Soc.Ser.B.11,211-229

BARTLETT,M.S.(1956) Deterministic and stochastic models for recurrent epidemics. Proc.3rd Berkeley Symp.Math.Statist.Probability,Vol.IV, p.81-109. Berkeley-Los Angeles:Univ.California Press

BARTLETT,M.S.(1974) The statistical analysis of spatial pattern. Adv. Appl.Probability 6,336-358

BEASG,J.(1974) Spatial interaction and the statistical analysis of lattice systems. J.Roy.Statist.Soc.Ser.B,36,192-236

BROADBENT,S.R.(1954) Discussion on the paper "Poor man's Monte Carlo" by J.M.Hammersley and K.W.Morton. J.Roy.Statist.Soc.Ser.B,16,68

BROADBENT,S.R.,KENDALL,D.G.(1953) The random walk of Trychostrongylus retortaeformis. Biometrics,9,460-466

BROWNLEE,J.(1911) The mathematical theory of random migration and epidemic distribution. Proc.Roy.Soc.Edinburgh,31,262-289

BURKS,A.W.(1970) Von Neumann's self-reproducing automata. In:Essays on Cellular Automata (A.W.Burks ed.),p.3-64. Urbana:Univ.Illinois Press

BÜHLER,W.(1966) A theorem concerning the extinction of epidemics. Biom. Z.,8,10-14

CANOSA,J.(1973) On a nonlinear equation of evolution. IBM J.Res.Develop. 17,307-313

COCHRAN,W.G.(1936) The statistical analysis of fields counts of diseased plants. J.Roy.Statist.Soc.,Suppl.3,49-67

CONNER,H.E.(1964) Extinction probabilities for age- and position-dependent branching processes. J.Soc.Indust.Appl.Math.,12,899-909

COX,J.T.,DURRETT,R.(1981) Some limit theorems for percolation processes with necessary and sufficient conditions. Ann.Probability,9,538-603

CRAMÉR,H.(1940) On the theory of stationary random processes. Ann.Math., 41,215-230

CRUICKSHANK,D.B.(1947) Regional influences in cancer. Brit.J.Cancer,1, 109-128

CRUMP,K.S.,GILLESPIE,J.H.(1976) The dispersion of a neutral allele considered as a branching process. J.Appl.Probability,13,208-218

DANIELS,H.E.(1977) The advancing wave in a spatial birth process. J.Appl. Probability,14,689-701

DAVIS,A.S.(1961) Markov chains as random input automata. Amer.Math. Monthly,68,264-267

DAVIS,A.W.(1965) On the theory of birth,death and diffusion processes. J.Appl.Probability,2,293-322

DAVIS,A.W.(1967) Branching-diffusion processes with no absorbing boundaries.I,II. J.Math.Anal.Appl.,18,276-296;19,1-25

DAVIS,A.W.(1970) Some generalizations of Bailey's birth,death and migration model. Adv.Appl.Probability,2,83-109

DAWSON,D.A.(1972) Stochastic evolution equations. Math.Biosci.15,287-316

DAWSON,D.A.,IVANOFF,G.(1978) Branching diffusions and random measures. In:Branching Processes(A.Joffe,P.Ney eds.),p.61-103. New York:M.Dekker

DHAR,D.,PHANI,M.K.,BARMA,M.(1982) Enumeration of directed site animals on two-dimensional lattices. J.Phys.A,15,L279-L284

DOBZHANSKY,T.,WRIGHT,S.(1947) Genetics of natural populations.XV.Rate of diffusion of a mutant gene through a population of Drosophila pseudoobscura. Genetics,32,303-339

DOMANY,E.,KINZEL,W.(1984) Equivalence of cellular automata to Ising models and directed percolation. Phys.Rev.Lett.,53,311-314

DONNELLY,P.,WELSH,D.(1983) Finite particle systems and infection models. Math.Proc.Camb.Phil.Soc.,94,167-182

DURRETT,R.(1979) An infinite particle system with additive interactions. Adv.Appl.Probability,11,355-383

DVORETZKY,A.,ERDÖS,P.(1951) Some problems on random walk in space. Proc. 2nd Berkeley Symp.Math.Statist.Probability,p.353-367. Berkeley-Los Angeles:Univ.California Press

EDEN,M.(1958) A probabilistic model for morphogenesis. In:Symp.on Information Theory in Biology (H.P.Yockey ed.),p.359-370. New York:Pergamon Press

EDEN,M.(1961) A two-dimensional growth process. Proc.4th Berkeley Symp. Math.Statist.Probability,Vol.IV,p.223-239. Berkeley-Los Angeles: Univ.California Press

ERDÖS,P.,KAC,M.(1946) On certain limit theorems in the theory of probability. Bull.Amer.Math.Soc.,52,292-302

FELLER,W.(1939) Die Grundlagen der Volterrasche Theorie des Kampfes um Dasein in wahrscheinlichkeitstheoretischer Behandlung. Acta Biotheor. 5,11-40. (Partially reprinted in:Applicable Mathematics of Non-physical Phenomena (F.Oliveira-Pinto,B.W.Connolly eds.),p.140-165. Chicester:E.Horwood&Wiley,1982)

FELLER,W.(1951) Diffusion processes in genetics. Proc.2nd Berkeley Symp. Math.Statist.Probability,p.227-246. Berkeley-Los Angeles:Univ.California Press

FELLER,W.(1959) The birth and death processes as diffusion processes. J.Math.Pures Appl.,38,301-345

FISHER,R.A.(1937) The wave of advance of advantageous genes. Ann.Eugenics 7,355-369 (Re-published in:R.A.Fisher:Contributions to Mathematical Statistics,London-New York:Wiley&Chapman-Hall,1950)

FISHER,R.A.(1950) Gene frequencies in a cline determined by selection and diffusion. Biometrics,6,353-361

FREEMAN,G.H.(1953) Spread of diseases in a rectangular plantation with vacancies. Biometrika,40,287-296

FÜRTH,R.von(1920) Die Brownsche Bewegung bei Berücksichtigung einer Persistenz der Bewegungsrichtung.Mit Anwendungen auf die Bewegung lebender Infusorien. Z.Phys.,2,244-256

GANI,J.,SRIVASTAVA,R.C.(1968) A stochastic model for the attachment and detachment of antibodies to virus. Math.Biosci.,3,307-321

GERSTEIN,G.L.,MANDELBROT,J.(1964) Random walk models for the spike activity of a single neuron. Biophys.J.,4,41-68

GERTSBAKH,I.B.(1977) Epidemic process on a random graph:Some preliminary results. J.Appl.Probability,14,427-438

GILLESPIE,J.H.(1976) The role of migration in the genetic structure of populations in temporally and spatially varying environments. II. Island models. Theor.Popul.Biol.,10,227-238

GILLIS,J.(1955) Correlated random walk. Proc.Camb.Phil.Soc.,51,639-651

GOLOMB,S.W.(1965) Polyominoes. New York:Scribner

GOROSTIZA,L.G.,GRIEGO,R.J.(1979) Convergence of branching transport processes to branching Brownian motion. Stoch.Proc.Appl.,8,269-276

GRANT,P.R.(1968) Polyhedral territories of animals. Amer.Nat.,102,75-80

HAMMERSLEY,J.M.(1953) Markovian walks on crystals. Compositio Math.,11, 171-186

HAMMERSLEY,J.M.(1983) Origins of percolation theory. In:Percolation Structures and Processes (G.Deutscher,R.Zallen,J.Adler eds.),p.47--57. Bristol:A.Hilger

HAMMERSLEY,J.M.,WALTERS,R.S.(1963) Percolation and fractional branching processes. J.Soc.Indust.Appl.Math.,11,831-839

HARARY,F.(1964) Combinatorial problems in graphical enumeration. In: Applied Combinatorial Mathematics (E.F.Beckenbach ed.),p.185-217. New York:Wiley

HAWKES,A.G.,OAKES,D.(1974) A cluster process representation of a self-exciting process. J.Appl.Probability,11,493-503

HEINE,V.(1955) Models for two-dimensional stationary stochastic processes. Biometrika,42,170-178

HERMAN,G.T.,ROZENBERG,G.(1975) Developmental Systems and Languages. Amsterdam:North-Holland

IKEDA,N.,NAGASAWA,M.,WATANABE,S.(1968-1969) Branching Markov processes. I-III. J.Math.Kyoto Univ.,8,233-278;8,365-410;9,95-160

IVANOFF,G.(1980) The branching random field. Adv.Appl.Probability,12, 825-847

IVANOFF,G.(1980) The branching diffusion with immigration. J.Appl.Probability,17,1-15

JIRINA,M.(1958) Stochastic branching processes with continuous state space. Czech.Math.J.,8,292-313

JIRINA,M.(1964) Branching processes with measure-valued states. Trans. 3rd Prague Conf.Information Theory,Statist.,Decision Fcts.,Random Proc.,p.333-357. Prague: Publ.House Czech.Acad.Sci.

JÖRGENSEN,H.(1976) Probabilistic L systems. In:Automata,Languages,Development (A.Lindenmayer,G.Rozenberg eds.),p.211-225. Amsterdam:North-Holland

KAC,M.(1947) Random walk and the theory of Brownian motion. Amer.Math.

Monthly,54,369-391

KAMETAKA,Y.(1976) On the nonlinear diffusion equations of Kolmogorov--Petrovsky-Piskunov-type. Osaka J.Math.,13,11-66

KARLIN,S.,KAPLAN,N.(1973) Criteria for extinction of certain population growth processes with interacting types. Adv.Appl.Probability,5,183--199

KARLIN,S.,TAVARE,S.(1982) Linear birth and death processes with killing. J.Appl.Probability,19,477-487

KENDALL,D.G.(1948) A form of wave propagation associated with the equation of heat conduction. Proc.Camb.Phil.Soc.,44,591-594

KENDALL,D.G.(1948) On the number of lattice points inside a random oval. Quart.J.Math.(Oxford Ser.),19,1-26

KENDALL,D.G.(1965) Mathematical models of the spread of infection. In: Mathematics and Computer Science in Biology and Medicine,p.213-225. London:Her Majesty's Stationery Office

KERNER,E.H.(1978) Multiple speciation,competitive exclusion,evolutionary pattern,and the grand ensemble in Volterra eco-dynamics. Bull.Math. Biophys.,40,387-410

KESTEN,H.(1971) Some nonlinear stochastic growth models. Bull.Amer.Math. Soc.,77,492-511

KHINTCHINE,A.(1934) Korrelationstheorie der stationäre stochastische Prozesse. Math.Ann.,109,604-615

KIMURA,M.,WEISS,G.(1964) The stopping stone model of population structure and the decrease of genetic correlation with distance. Genetics, 49,561-576

KINGMAN,J.F.C.(1969) Markov population processes. J.Appl.Probability, 6,1-18

KINGMAN,J.F.C.(1973) Discussion on the paper "Central limit analogues for Markov population processes" by D.R.McNeil and S.Schach. J.Roy. Saatist.Soc.Ser.B,35,15

KINGMAN,J.F.C.(1977) Remarks on the spatial distribution of a reproducing population. J.Appl.Probability,14,577-583

KINZEL,W.(1985) Phase transitions of cellular automata. Z.Phys.B--Condensed Matter,58,229-244

KLEIN,G.(1952) A generalization of the classical random walk problem and a simple model of Brownian motion based thereon. Proc.Roy.Soc. Edinburgh,63,268-279

KLUYVER,J.C.(1906) A local probability theorem. Ned.Akad.Wet.Proc.A,8, 341-350

KOLMOGOROFF,A.N.,PETROVSKY,I.G.,PISCOUNOFF,N.S.(1937) Etude de l'équation de la diffusion avec croissance de la quantité de matière et son application à un problème biologique. Bull.Univ.d'Etat Moscou(Sér. Intern.),Sect.A,1(6),1-25.(Partially re-published in "Applicable Mathematics of Non-physical Phenomena"(F.Oliveira-Pinto,B.W.Conolly eds.), p.171-184.Chicester:E.Horwood&Wiley,1982)

KRISHNA IYER,P.V.(1950) The theory of probability distributions of points on a lattice. Ann.Math.Statist.,21,198-217.(Corr.:id.,1961,32,619)

KURTZ,T.G.(1980) Relationships between stochastic and deterministic population models. In:Biological Growth and Spread(W.Jäger,H.Rost,P.Tautu eds.),p.449-467 (LN in Biomath.,Vol.38) Berlin-Heidelberg-New York: Springer

KUULASMAA,K.(1982) The spatial general epidemic and locally dependent random graphs. J.Appl.Probability,19,745-758

LAMPERTI,J.(1970) Maximal branching processes and 'long-range percolation'. J.Appl.Probability,7,89-98

LENZ,N.(1982) Poisson convergence on continuous time branching random walks and multistage carcinogenesis. J.Math.Biol.,14,301-307

LEONTOVICH,A.M.,PYATETSKII-SHAPIRO,I.I.,STAVSKAYA,O.N.(1970) Certain mathematical problems related to morphogenesis. Avt.i Telemech.,4, 94-107 (Re-published in:Mathematical Models for Cell Rearrangement (G.D.Mostow ed.),p.7-24. New Haven:Yale Univ.Press,1975)

LIGGETT,T.M.(1985) Interacting Particle Systems. New York-Berlin-Heidelberg-Tokyo:Springer

LIGGETT,T.M.,SPITZER,F.(1981) Ergodic theorems for coupled random walks and other systems with locally interacting components. Z.Wahrscheinlichkeitstheorie verw.Gebiete,56,443-468

LINDENMAYER,A.(1968) Mathematical models for cellular interactions in development.I,II. J.Theor.Biol.,18,280-299;300-315

MALÉCOT,G.(1966) Probabilités et Hérédité. Paris:Presses Univ.de France

McCLOSKEY,M.A.,POO,M-m(1986) Rates of membrane-associated reactions: Reduction of dimensionality revisited. J.Cell Biol.,102,88-96

McDIARMID,C.(1981) General percolation and random graphs. Adv.Appl. Probability,13,40-60

McKean,H.P.(1975) Application of Brownian motion to the equation of Kolmogorov-Petrovskii-Piscunov. Comm.Pure Appl.Math.,28,323-331

MOLLISON,D.(1977) Spatial contact models for ecological and epidemic spread. J.Roy.Statist.Soc.Ser.B,39,283-326

MONTROLL,E.W.(1967) On nonlinear processes involving population growth and diffusion. J.Appl.Probability,4,281-290

MONTROLL,E.W.,WEST,B.J.(1973) Models of population growth,diffusion, competition and rearrangement. In:Cooperative Phenomena in Multi--Component Systems(H.Haken ed.),p.143-156. Berlin-Heidelberg-New York:Springer

MORANT,G.(1921) On random occurrences in space and time,when followed by a closed interval. Biometrika,13,309-337

MORAN,P.A.P.(1947) Random associations on a lattice. Proc.Camb.Phil.Soc., 43,321-328

MORAN,P.A.P.(1947-1953) The random division of an interval.I-III. J.Roy. Statist.Soc.Ser.B,9,92-98;13,147-150;15,77-80

MORAN,P.A.P.(1948) The interpretation of statistical maps. J.Roy.Statist. Soc.Ser.B,10,243-251

MORGAN,R.W.,WELSH,D.J.A.(1965) A two-dimensional Poisson growth process. J.Roy.Statist.Soc.Ser.B,27,497-504

MOYAL,J.E.(1962) The general theory of stochastic population processes. Acta Math.,108,1-31

MOYAL,J.E.(1964) Multiplicative population processes. J.Appl.Probability, 1,267-283

NAGYLAKI,T.(1974) The decay of genetic variability in geographically structured populations. Proc.Natl.Acad.Sci.USA,71,2932-2936

NAGYLAKI,T.(1978) The geographical structure of populations. In:Studies in Mathematical Biology(S.A.Levin ed.),Vol.2,p.588-624. (MAA Studies in Math.,Vol.16) Mathematical Association of America

NEYMAN,J.,SCOTT,E.L.(1957) On a mathematical theory of populations conceived as conglomerations of clusters. Proc.Cold Spring Harbor Symp. Quant.Biol.,22,109-120

NEYMAN,J.,SCOTT,E.L.(1957) Birth and death random walk process in s dimensions (Abstr.). Ann.Math.Statist.,28,1071

NEYMAN,J.,SCOTT,E.L.(1964) A stochastic model of epidemics. In:Stochastic Models in Medicine and Biology(J.Gurland ed.),p.45-83. Madison:Univ. Wisconsin Press

NEYMAN,J.,SCOTT,E.L.(1972) Processes of clustering and applications. In:Stochastic Point Processes:Statistical Analysis,Theory and Applications(P.A.W.Lewis ed.),p.646-681. New York:Wiley

NOSSAL,R.J.,WEISS,G.(1974) A generalized Pearson random walk allowing for bias. J.Statist.Phys.,10,245-253

OHTA,T.,KIMURA,M.(1973) A model of mutation appropriate to estimate the number of electrophoretically detectable alleles in a finite population. Genet.Res.,22,201-204

PARISI,G.,SOURLAS,N.(1981) Critical behavior of branched polymers and the Lee-Yang edge singularity. Phys.Rev.Lett.,46,871-874

PATLAK,C.S.(1953) Random walk with persistence and external bias. Bull. Math.Biophys.,15,311-338

PEARSON,K.(1905) The problem of the random walk. Nature,72,294

PEARSON,K.,BLACKEMAN,J.(1906) Mathematical contributions to the theory of evolution.XV.Mathematical theory of random migration. Draper's Co. Res.Mem.Biom.Ser.III,No.15,p.1-54

PEARSON,K.(1913) Multiple cases of disease in the same house. Biometrika, 9,28-33

PETERSON,S.C.,NOBLE,P.B.(1972) A two-dimensional random-walk analysis of human granulocyte movement. Biophys.J.,12,1048-1055

PETTIGREW,H.M.,WEISS,G.(1967) Epidemics with carriers:the large population approximation. J.Appl.Probability,4,257-263

POLYA,G.(1921) Über eine Aufgabe der Wahrscheinlichkeitsrechnung betreffend die Irrfahrt im Strassennetz. Math.Ann.,83,149-160

PREKOPA,A.(1960) On the spreading process. Trans.2nd Prague Conf.Information Theory,Statist.,Decision Fcts.,Random Processes,p.521-529. Prague:Academia

PRESTON,C(1975) Spatial birth-and-death processes. Bull.Intern.Statist, Inst.,46,Book 2,371-391

PURI,P.S.(1968) Interconnected birth and death processes. J.Appl.Probability,5,334-349

RACZ,Z.,PLISCHKE,M.(1985) Active zone of growing clusters:Diffusion-limited aggregation and the Eden model in two and three dimensions. Phys.Rev.A,31,985-994

RADCLIFFE,J.(1976) The convergence of a position-dependent branching process used as an approximation to a model describing the spread of an epidemic. J.Appl.Probability,13,338-344

RAYLEIGH,Lord(1905) The problem of random walk. Nature,72,318

RAYLEIGH,Lord(1919) On the problem of random vibrations and of random flights in one,two,or three dimensions. Phil.Mag.,37,321-347

RENSHAW,E.(1972) Birth,death and migration processes. Biometrika,59, 49-60

RENSHAW,E.(1974) Stepping-stone models for population growth. J.Appl. Probability,11,16-31

RICHARDSON,D.(1973) Random growth in a tessellation. Proc.Camb.Phil. Soc.,74,515-528

RIPLEY,B.D.(1981) Spatial Statistics. Chichester:Wiley

ROTENBERG,M.(1972) Theory of population transport. J.Theor.Biol.,37, 291-305

SAWYER,S.(1976) Results for the stepping stone model for migration in population genetics. Ann.Probability,4,699-728

SAWYER,S.(1976) Branching diffusion processes in population genetics. Adv.Appl.Probability,8,659-689

SAWYER,S.(1980) Random walks and probabilities of genetic identity in a tree. In:Biological Growth and Spread(W.Jäger,H.Rost,P.Tautu eds.), p.290-295. Berlin-Heidelberg-New York:Springer

SCHÜRGER,K.(1976) On the evolution of random graphs over expanding square lattice. Acta Math.Acad.Sci.Hung.,27,281-292

SCHÜRGER,K.(1980) On the asymptotic geometrical behaviour of percolation processes. J.Appl.Probability,17,385-402

SCHWÖBEL,W.,GEIDEL,H.,LORENZ,R.J.(1966) Ein Modell der Plaquebildung. Z.Naturforsch.,21b,953-959

SHIGA,T.(1980) An interacting system in population genetics.I,II. J.Math. Kyoto Univ.,20,213-242;20,723-733

SKELLAM,J.G.(1951) Random dispersal in theoretical populations. Biometrika,38,196-218

SKELLAM,J.G.(1952) Studies in statistical ecology.I.Spatial pattern. Biometrika,39,346-362

SKELLAM,J.G.(1973) The formulation and interpretation of mathematical models of diffusionary processes in population biology. In:The Mathematical Theory of the Dynamics of Biological Populations (M.S.Bartlett,R.W.Hiorns eds.),p.63-85. London:Academic Press

STONE,C.(1963) Limit theorems for random walks,birth and death processes, and diffusion processes. Illinois J.Math.,7,638-660

SUDBURY,A.(1976) The size of the region occupied by one type in an invasion process. J.Appl.Probability,13,355-356

TAUTU,P.(1977) The evolution of epidemics as random graphs. Proc.5th Conf.Probability Theory,Brasov 1974,p.287-291. Bucuresti:Ed.Academiei

TEMPERLEY,H.N.V.(1979) Graph theory and continuum statistical mechanics. In:Applications of Graph Theory (R.J.Wilson,L.W.Beineke eds.),p.121--148. London:Academic Press

THOMAS,M.(1946) A generalization of Poisson's binomial limit for use in ecology. Biometrika,36,18-25

THOMPSON,H.R.(1955) Spatial point processes,with applications to ecology. Biometrika,42,102-115

TSETLIN,M.L.(1973) Automata Theory and Modeling of Biological Systems. New York:Academic Press

ULAM,S.M.(1952) Random processes and transformations. Proc.Intern.Congr. Mathematicians,1950,p.264-275. Providence:AMS

ULAM,S.M.(1962) On some mathematical problems connected with patterns of growth of figures. Proc.Symp.Appl.Math.,14,215-224 (Re-published in: Essays on Cellular Automata(A.W.Burks ed.),p.219-231.Urbana:Univ.of Illinois Press,1970)

VASILYEV,N.B.(1969) On the limiting behavior of a random medium. Probl. Transmission Information,5,64-72

WANG,F.J.S.(1980) The convergence of a branching Brownian motion used as a model describing the spread of an epidemic. J.Appl.Probability, 17,301-312)

WATANABE,T.(1969) Convergence of transport processes to diffusion. Proc. Japan Acad.,45,470-472

WEISS,G.H.(1965) On the spread of epidemics by carriers. Biometrics,21, 481-490

WEISS,G.H.,KIMURA,M.(1965) A mathematical analysis of the stepping stone model of genetic correlation. J.Appl.Probability,2,129-149

WEISS,G.H.,RUBIN,R.J.(1983) Random walks:theory and selected applications. Adv.Chem.Phys.,52,363-505

WELSH,D.J.A.(1977) Percolation and related topics. Sci.Prog.(Oxford), 64,65-83

WHITTLE,P.(1954) On stationary processes in the plane. Biometrika,41, 434-449

WHITTLE,P.(1957) On the use of the normal approximation in the treatment of stochastic processes. J.Roy.Statist.Soc.Ser.B,19,268-281

WHITTLE,P.(1962) Topographic correlation,power-law covariance functions, and diffusions. Biometrika,49,305-314

WILLIAMS,E.J.(1961) The distribution of larvae of randomly moving insects. Austral.J.Biol.Sci.,12,598-604

WILLIAMS,T.,BJERKNES,R.(1972) Stochastic model for abnormal clone spread through epithelial basal layer. Nature,236,19-21

WILLSON,S.J.(1981) Growth patterns of ordered cellular automata. J.Comp. System Sci.,22,29-41

WILSON,E.B.,WORCESTER,J.(1945) Damping of epidemic waves. Proc.Natl.Acad. Sci.USA,31,294-298

WISHART,J.,HIRSCHFELD,H.O.(1936) A theorem concerning the distribution of joins between line segments. J.London Math.Soc.,11,227-235

WOLD,H.(1948) Sur less processus stationnaires ponctuels. Coll.Intern. CNRS,13,75-86

YASUDA,N.(1975) The random walk model of human migration. Theor.Popul. Biol.,7,156-167

TESTS FOR SPACE-TIME CLUSTERING

A.D.Barbour
Institut für Angewandte Mathematik
Universität Zürich
CH-8001 ZUERICH

G.K.Eagleson
Avalon, N.S.W.
AUSTRALIA

Introduction. Knox (1964) observed that cases of childhood leukaemia tend to occur in clusters, both spatially and in time. In order to distinguish whether the clustering arises because of infection rather than from environmental and seasonal variability in incidence rates, he proposed the following simple statistic. For data consisting of N cases, for each of which the times $\{T_i\}_{i=1}^N$ of presentation and places $\{X_i\}_{i=1}^N$ of residence are available, Knox suggested the statistic

$$K := \sum_{i<j} I[|T_i - T_j| < \tau] I[|X_i - X_j| < \delta] \qquad (1)$$

as a measure of contagion, where the time threshold τ and the space threshold δ are suitably chosen. K is large only if many cases occur close together in space and time, and frequent coincidences of this sort are evidence for a possible infection process. Mantel (1967) suggested using smooth functions to describe closeness, rather than the indicator functions used by Knox, and in particular an analogous statistic

$$M_1 := \sum_{i<j} \{|T_i - T_j| + \tau\}^{-1} \{|X_i - X_j| + \delta\}^{-1} \ ;$$

a numerical comparison of these and other statistics, including

$$M_2 := \sum_{i<j} \{|T_i - T_j| + \tau\}^{-2} \{|X_i - X_j| + \delta\}^{-2},$$

was carried out by Siemiatycki (1978).

Similar ideas have been used to test for geographical association, based on the papers of Moran (1948) and Geary (1954), and developed in the monograph of Cliff and Ord (1973). In this framework, there are N states in a federation, the geographical association between states i and j being described by a non-negative quantity w_{ij} and x_i denoting a measure of some factor of interest in state i. Thus w_{ij} might take the value 1 if states i and j had a common border and zero otherwise, and x_i could represent the prevalence of hypertension in state i. The statistics used to analyse association are typically of the forms

$$I := \sum_{i<j} w_{ij}(x_i-\bar{x})(x_j-\bar{x}) \qquad (2)$$

and

$$C := \sum_{i<j} w_{ij}(x_i-x_j)^2 \ . \qquad (3)$$

In order to determine when values of these statistics are strikingly large, it is necessary to be able to approximate their distribution under reasonable null hypotheses. Siemiatycki (1978) observed that, in his simulations, the distributions of the statistics were not typically like the normal, and recommended fitting a Pearson curve using their first four moments. Cliff and Ord (1973), pp36-37, state conditions under which the geographical statistics are to be asymptotically normally distributed, to which Barton (unpublished manuscript) has exhibited counter-examples. In this paper, the extent to which normal behaviour should or should not be expected is explored.

General setting.

The statistics under discussion can all be expressed in the form

$$T := \sum_{i,j} A_{ij} B_{ij} \ ,$$

where A and B are symmetric, and the notation $\sum_{i,j}$ denotes the sum over $1 \leq i \neq j \leq N$. In particular, if A is a zero-one matrix,

$$T = 2 \sum_{(i,j) \in G} B_{ij} \ ,$$

for the graph G defined by the incidence matrix A: if B is also the incidence matrix of a graph H, $T=2|G \cap H|$. The simplest null hypothesis to consider would be

H_0: $A_{ij}=a_0+a(X_i,X_j)$ and $B_{ij}=b_0+b(Y_i,Y_j)$, where the arrays of independent and identically distributed elements $(X_i)_{i\geq 1}$ and $(Y_i)_{i\geq 1}$ are mutually independent; the functions a and b are symmetric,

$$\mathbb{E}\, a(X_1,X_2) = \mathbb{E}\, b(Y_1,Y_2) = 0 \text{ and } \mathbb{E}\, a^2(X_1,X_2) < \infty,\ \mathbb{E} b^2(X_1,X_2) < \infty.$$

However, in the geographical context, it is more natural to consider the weights w_{ij} as fixed, in which case the null hypothesis

H_1: A is generated from a sequence of independent and identically distributed elements $(X_i)_{i\geq 1}$ as under H_0, and B is a fixed symmetric matrix,

is more appropiate. Finally, for many purposes, and in particular for the determination of thresholds, it is convenient to work with the very weak null hypothesis

H_2: B is a fixed symmetric matrix and $A_{ij} = C_{\pi(i)\pi(j)}$, where C is a fixed symmetric matrix and π is uniformly distributed on the permutation group S_N.

This, for instance, is the hypothesis used by Knox (1964), when testing for infectivity: the effect is to condition on the observed places and times of presentation of cases, and to test only their association.

The problem is to approximate the distribution of T under the various null hypotheses. This is frequently accomplished by obtaining an estimate of the distance ρ between the distribution of a normalized version of T and the standard normal distribution, from which limit theorems as $N \to \infty$ can be deduced. The distance ρ used here is defined by

$$\rho(F,G) := \sup_{h \in C_1} \left| \int h \, dF - \int h \, dG \right| / \|h\|_1 ,$$

where $\|h\|_1 = \sup_x |h(x)| + \sup_x |h'(x)|$.

The distribution of T under H_0.

Suppose that H_0 holds. Then we can write

$$T - \mathbb{E}\, T = \sum_{i,j} t(Z_i, Z_j),$$

where $Z_i := (X_i, Y_i)$ and t is a symmetric function: thus T has the structure of a two-dimensional U-statistic. In particular, using Hájek's projection technique (Lehmann (1975), section 5 of Appendix),

$$T - \mathbb{E}\, T = U + V,$$

where U and V are uncorrelated, and

$$V := \sum_i \mathbb{E}\,\{T - \mathbb{E}\, T \mid Z_i\} = 2(N-1) \sum_i \mathbb{E}\,\{t(Z, Z_i) \mid Z_i\}$$

is a sum of independent and identically distributed zero mean random variables. The corresponding decomposition of the variance of T is

$$s^2 := \text{var}\, T = \text{var}\, U + \text{var}\, V = 2N(N-1)\sigma^2(1-2\theta) + 4N(N-1)^2\theta\sigma^2,$$

where $\sigma^2 = \text{var}\{t(Z_1, Z_2)\}$ and $\theta = \text{corr}\{t(Z_1,Z_2),\ t(Z_1,Z_3)\}$. Thus, for $\theta \neq 0$ and large N, $\text{var}\, T = \text{var}\, V\{1 + O(N^{-1})\} \sim 4N^3\theta\sigma^2$, and the distribution of T is dominated by that of V, which by the central limit theorem is approximately normal: more precisely, it is easy to show, using Stein's (1970) method, that

$$\rho\{\mathcal{L}(s^{-1}(T - \mathbb{E}\, T)), \mathcal{N}\} \le K_0\{\delta_0 + s^{-1}(\text{var}\, U)^{\frac{1}{2}}\} , \qquad (4)$$

where $\mathcal{N}$ denotes the standard normal distribution, K_0 is a universal constant and $\delta_0 := s^{-3}N^4\,\mathbb{E}\,|t(Z_1,Z_2)|^3$. Thus, if $\mathbb{E}\,|t(Z_1,Z_2)|^3 < \infty$, it follows that

$$\rho\{\mathcal{L}(s^{-1}(T - \mathbb{E}\, T)), \mathcal{N}\} = O(N^{-\frac{1}{2}}),$$

and that, for large N, T is asymptotically normally distributed.

If, on the other hand, $\theta = 0$, $s^{-1}(\mathrm{var}U)^{\frac{1}{2}} = 1$, and estimate (4) is of no use. In this, the degenerate case, it can be shown that, as $N\to\infty$, $\mathcal{L}(s^{-1}(T-\mathbb{E}T))$ converges to the distribution of a weighted sum of independent centred χ_1^2 random variables: see Gregory (1977) and Serfling (1980), section 5.5.2. In particular, the limit is not normal.

The distribution of T under H_1.

Suppose that H_1 holds. Then T takes the form

$$T - \mathbb{E}T = \sum_{i,j} B_{ij} a(X_i, X_j) \;,$$

and can be considered as a weighted U-statistic or as a B^2 statistic. Projection yields the decomposition

$$T - \mathbb{E}T = U + V \;,$$

where U and V are uncorrelated, and

$$V := \sum_i \mathbb{E}(T - \mathbb{E}T | X_i) = 2\sum_i (\sum_{j\neq i} B_{ij}) \mathbb{E}\{a(X, X_i) | X_i\}$$

is a weighted sum of independent and identically distributed zero mean random variables. The corresponding variance decomposition is

$$s^2 := \mathrm{var}\, T = \mathrm{var}\, U + \mathrm{var}\, V$$
$$= 2(\sum_{i,j} B_{ij}^2)\sigma^2(1 - 2\theta) + 4\sum_i (\sum_{j\neq i} B_{ij})^2 \theta\sigma^2 \;,$$

where $\sigma^2 := \mathrm{var}\{a(X_1, X_2)\}$ and $\theta = \mathrm{corr}\,\{a(X_1, X_2),\; a(X_1, X_3)\}$.

As before, if $\mathrm{var}\, U$ is small compared to $\mathrm{var}\, V$, the distribution of T can be expected to be approximately normal; here,

$$\rho\{\mathcal{L}(s^{-1}(T-\mathbb{E}T)), \mathcal{N}\} \leq K_1\{\delta_1 + s^{-1}(\mathrm{var}\; U)^{\frac{1}{2}}\}, \qquad (5)$$

where K_1 is a universal constant and

$$\delta_1 := s^{-3} \sum_i |\sum_{j\neq i} B_{ij}|^3 \,\mathbb{E}|a(X_1, X_2)|^3 .$$

Thus, for a sequence of such statistics T, $\mathcal{L}(s^{-1}(T-\mathbb{E}T))$ converges to the standard normal if $\delta_1\to 0$ and $s^{-1}(\mathrm{var}\; U)^{\frac{1}{2}}\to 0$; however, for a particular fixed T, when it may not be at all obvious into which sequence of matrices B should be imbedded, (5) is still a useful guide as to whether approximate normality is to be expected or not.

In contrast to the previous case, degeneracy - in the sense that $s^{-1}(\mathrm{var}\; U)^{\frac{1}{2}}$ is not small - does not occur only when $\theta = 0$: it may also happen because the matrix B has small enough column sums, for instance because of balancing positive and negative terms, that $\sum_i (\sum_{j\neq i} B_{ij})^2 = O(\sum_{i,j} B_{ij}^2)$. In these circumstances, (5) is of no help: yet it turns out, in some cases, that T is still approximately normally distributed. Using the B^2 structure, it follows from Barbour and Eagleson (1984), Theorem 2.1, that

$$\rho\{\mathcal{L}(s^{-1}(T-\mathbb{E}\,T)),\mathcal{N}\} \le K_1'\varepsilon_1, \qquad (6)$$

where K_1' is a universal constant, and

$$\varepsilon_1 := s^{-3}\sum_{i,j}|B_{ij}|\{\sum_k(|B_{ik}|+|B_{jk}|)\}^2\,\mathbb{E}|a(X_1,X_2)|^3.$$

Thus, for sequences of statistics T, $\varepsilon_1\to 0$ is sufficient to establish asymptotic normality. In general,

$$\delta_1 \le \varepsilon_1 \le Ks^{-3}\{N^4(B_3+B_4)+N^3B_5+N^2B_6\}\,\mathbb{E}|a(X_1,X_2)|^3,$$

where the third order absolute array moments B_3, B_4, B_5 and B_6 are defined by

$$B_3 := \frac{1}{(N)_4}\sum_{i,j,k,m}|B_{ij}B_{ik}B_{im}|,$$

$$B_4 := \frac{1}{(N)_4}\sum_{i,j,k,m}|B_{ij}B_{jk}B_{km}|,$$

$$B_5 := \frac{1}{(N)_3}\sum_{i,j,k}|B_{ij}^2B_{ik}|,$$

$$B_6 := \frac{1}{(N)_2}\sum_{i,j}|B_{ij}|^3,$$

all sums being taken over distinct suffices: and δ_1 can only be of strictly smaller order than ε_1 if $\sum_i|\sum_{j\ne i}B_{ij}| = o\{\sum_i(\sum_{j\ne i}|B_{ij}|)\}$. On the other hand, there is no general inequality possible between ε_1 and $s^{-1}(\mathrm{var}\,U)^{\frac{1}{2}}$: indeed, it is possible to construct a sequence of statistics T for which $\varepsilon_1\to 0$ and $s^{-1}(\mathrm{var}\,U)^{\frac{1}{2}}\not\to 0$, and another for which $\delta_1+s^{-1}(\mathrm{var}\,U)^{\frac{1}{2}}\to 0$ and $\varepsilon_1\not\to 0$.

Limiting distributions other than the normal can also easily occur in the degenerate case, as under H_0: the range of possibilities is rather greater, and is similar to that obtainable under H_2.

The distribution of T under H_2

If H_2 holds, define

$$B_0 := \frac{1}{(N)_2}\sum_{i,j}B_{ij};$$

$$b_{ii} := 0;\quad b_{ij} := B_{ij}-B_0,\quad i\ne j;$$

$$b_{i*} := \frac{1}{N-2}\sum_{j\ne i}b_{ij}.$$

Define C_0, c_{ij} and c_{i*} similarly. Then $\mathbb{E}\,T = N(N-1)B_0C_0$, and $T := \sum_{i,j}B_{ij}C_{\pi(i)\pi(j)}$ can be projected in the form

$$T - \mathbb{E}T = U + V = \sum_{i,j} \tilde{b}_{ij}\tilde{c}_{\pi(i)\pi(j)} + 2(N-2)\sum_i b_{i*}c_{\pi(i)*} ,$$

where $\tilde{b}_{ij} = b_{ij} - b_{i*} - b_{j*}$ and $\tilde{c}_{ij}$ is defined similarly: note that $\sum_i b_{i*} = 0$. The corresponding variance decomposition is

$$s^2 := \text{var } T = \text{var } U + \text{var } V$$
$$= \{2N(N-1)^2/(N-3)\}B_2C_2 + \{4N^2(N-2)^2/(N-1)\}B_1C_1 ,$$

where $B_1 := N^{-1}\sum_i b_{i*}^2$ and $B_2 := \frac{1}{(N)_2}\sum_{i,j} \tilde{b}_{ij}^2$, and C_1 and C_2 are defined similarly. Now, if $s^{-1}(\text{var } U)^{\frac{1}{2}}$ is small, the distribution of T is dominated by that of V, to which the Wald-Wolfowitz theorem is applicable. Applying Stein's method in the Wald-Wolfowitz context, it can be shown that

$$\rho\{L(s^{-1}(T - \mathbb{E}T)), N\} \leq K_2\{\delta_2 + s^{-1}(\text{var } U)^{\frac{1}{2}}\} , \qquad (7)$$

where K_2 is a universal constant and

$$\delta_2 := s^{-3}N^2\{\sum_i |b_{i*}|^3\}\{\sum_i |c_{i*}|^3\} ,$$

and hence a sequence of such statistics T converges to the normal if $\delta_2 \to 0$ and $s^{-1}(\text{var } U)^{\frac{1}{2}} \to 0$.

As under H_1, it is also possible to have approximate normality when $s^{-1}(\text{var } U)^{\frac{1}{2}}$ is not small. It is shown in Barbour and Eagleson (1985), by using Stein's method, that

$$\rho\{L(s^{-1}(T - \mathbb{E}T)), N\} \leq K_2'\varepsilon_2 , \qquad (8)$$

where K_2' is a universal constant, and

$$\varepsilon_2 := s^{-3}\{N^4(b_1^3 + b_1b_2 + b_3 + b_4)(c_1^3 + c_1c_2 + c_3 + c_4) +$$
$$+ N^3(b_1b_8 + b_5)(c_1c_8 + c_5) + N^2b_6c_6\} :$$

the absolute array moments b_3, b_4, b_5 and b_6 are as previously defined, but with b in place of B,

$$b_1 := \frac{1}{(N)_2}\sum_{i,j} |b_{ij}|$$

$$b_2 := \frac{1}{(N)_3}\sum_{i,j,k} |b_{ij}b_{ik}|$$

$$b_8 := \frac{1}{(N)_2}\sum_{i,j} b_{ij}^2$$

are the lower order absolute array moments, and $c_1 - c_8$ are defined similarly. Thus $\varepsilon_2 \to 0$ is also sufficient for asymptotic normality. Once again, $\delta_2 \leq K\varepsilon_2$, but no comparison is in general possible between ε_2 and $s^{-1}(\text{var } U)^{\frac{1}{2}}$.

Convergence to non-normal limits under H_2 is also investigated in Barbour and Eagleson (1985), under some assumptions on the structure of the sequences of matrices b(N) and c(N). Both are assumed to be close to being generated by L_2 kernels: that is,

$$b_{ij}(N) \simeq N^2 \iint_{I_{Ni}\times I_{Nj}} b(u,v)\, dudv ,$$

for some square integrable symmetric function b, where $I_{Ni}:=\{u: \frac{i-1}{N} < u \le \frac{i}{N}\}$, and similarly for c(N). In order to obtain non-normal limits, at least one of the kernels, say b, must be degenerate, in the sense that $\int_0^1 b(u,v)\, dv = 0$ for all u, in which case U dominates the distribution of T. The limiting distributions obtained are in general rather complicated, though if both kernels are degenerate they have a representation like that of the limits under H_0.

Practical applications.

Consider the Knox statistic K defined in (1). The distribution of the space variables $(X_i)_{i=1}^N$ tends in practice to be spread over the study area in a non-uniform way, due to the aggregation of population in cities and towns. Letting $B_{ij}:=I[|X_i - X_j| < \delta]$, and letting θ denote the proportion of pairs of cases that are spatially close, the spatial data may be described by saying that, for each i, there are $(N-1)f_i\theta$ elements B_{ij} with the value 1 and the remainder have the value zero, where $N^{-1}\sum_{i=1}^N f_i = 1$. Suppose that the quantities $(f_i)_{i=1}^N$ are representative of an N-sample from a well behaved distribution F. Then it is easy to estimate that

$$\begin{aligned}
&B_0 = \theta ;\\
&b_{i*} \sim (f_i-1)\theta , \quad 1 \le i \le N ;\\
&B_1 \sim \theta^2 \{\frac{1}{N}\sum_{i=1}^N (f_i-1)^2\} = O(\theta^2) ; \qquad (9)\\
&b_{ij} = B_{ij} - \theta\{1 + (\frac{N-1}{N-2})(f_i + f_j - 2)\} ;\\
&B_2 = \theta\{1 + O(\theta)\} .
\end{aligned}$$

Hence typical orders of magnitude of B_1 and B_2 are θ^2 and θ respectively. A similar argument, if the distribution of the time variables $(T_i)_{i=1}^N$ is also non-uniform, as may be the case if seasonal effects are significant, leads to estimates ϕ^2 and ϕ for the orders of magnitude of C_1 and C_2 respectively, where $C_{ij}:=I[|T_i - T_j| < \tau]$ and ϕ denotes the proportion of pairs of cases that are temporally close. Thus the orders of magnitude of var U and var V are $N^2\phi\theta$ and $N^3\phi^2\theta^2$ respectively, and so

$s^{-1}(\text{var } U)^{\frac{1}{2}}$ is only small if $(N\phi\theta)^{\frac{1}{2}}$ is large. For practical values of the parameters, say $N = 500$ and $\phi = \theta = 1/20$, this is clearly not the case. Increasing the size of the study area would typically increase N, leave ϕ constant and decrease θ in proportion to the increase in N, so that $N\phi\theta$ would remain unchanged. Increasing the length of the study would increase N, reduce ϕ in proportion to the increase in N, and leave θ unchanged, again leaving $N\phi\theta$ unchanged. Hence degeneracy is highly plausible in such applications. If, as may well be the case for some diseases, the underlying distribution of the time variables is uniform, it can be seen by analogy with (9) that C_1 can be of smaller order than $\phi^2 - \phi^3$ is typical - further increasing the likelihood of degeneracy. Turning to the alternative condition (8) for normal approximation, the quantity ε_2 is found to be small if $N(\phi\theta)^{3/2}$ is small and $N(\phi\theta)^{\frac{1}{2}}$ is large. For the parameter values given above, $N(\phi\theta)^{3/2} = 1/16$ and $N(\phi\theta)^{1/2} = 25$, suggesting that a normal approximation may not be too bad: however, increasing θ and ϕ to 1/10 each would increase $N(\phi\theta)^{3/2}$ to 1/2. Increasing the size of the study area and increasing the length of the study both have beneficial effects, reducing $N(\phi\theta)^{3/2}$ in proportion to the square root of the increase in N and increasing $N(\phi\theta)^{\frac{1}{2}}$ by a similar factor. Thus, the larger the study, the better the normal approximation should be expected to fit.

The requirement that $N(\phi\theta)^{\frac{1}{2}}$ should be large for a normal approximation to be feasible is quite natural, since $N(\phi\theta)^{\frac{1}{2}}$ is essentially the square root of the expected number, under H_2, of ones in the sum defining K. In Knox's (1964) study of childhood leukaemia, the expected number of ones was not large, and he proposed instead a Poisson approximation to the distribution of K. This approximation is discussed in Barbour and Eagleson (1983).

To illustrate the geographical applications of statistics such as in (2) and (3), take B_{ij} to be one if two states abut and zero otherwise, and let $C_{ij}(I) = (x_i-\bar{x})(x_j-\bar{x})$ and $C_{ij}(C) = (x_i-x_j)^2$. Because of geographical constraints, $\sum_{j\neq i} B_{ij}$ tends to remain of the order of six or less for each i, whatever the size of N: hence, letting $m = NB_0$, the orders of B_1 and B_2 are typically $(m/N)^2$ and (m/N) respectively. Assuming, as is also often the case, that the $(x_i)_{i=1}^N$ are typical of a sample from a distribution with unit variance and without long tails, $C_1(C)$ and $C_2(C)$ are both of order 1, and so, for the statistic C under H_2, the variance decomposition gives $s^{-1}(\text{var } U)^{\frac{1}{2}} \asymp m^{-\frac{1}{2}}$, which does not become small, even for large N. In the case of I, c_{i*} is of order N^{-1} for each i, and so, whereas $C_2(I)$ is of order 1, $C_1(I) = O(N^{-2})$: in this case, $s^{-1}(\text{var } U)^{\frac{1}{2}} \sim 1$, and the projection estimate (7) can never be expected to be useful in justifying an assumption of normality.

Considering the alternative condition (8) for approximate normality, ε_2 can be shown to be small, provided that $m^{3/2}N^{-\frac{1}{2}}$ is small and $(Nm)^{\frac{1}{2}}$ is large: so, clearly, increasing the size of the study area improves the accuracy of normal approximation. However, taking the départements of France (excluding Corsica) as an example, $N = 94$ and $m = 5$, so that $m^{3/2}N^{-\frac{1}{2}}$ is larger than one, giving little support for a normal approximation. It is possible that, for graphs as regular as in such an example, the right measure of approximate normality would be $(m/N)^{\frac{1}{2}}$: even then, $(m/N)^{\frac{1}{2}} \approx 1/4$ is still not strikingly small, and it would be in no way surprising if a normal approximation turned out to be poor.

Barbour, A.D. and Eagleson, G.K. (1983) Poisson approximation for some statistics based on exchangeable trials. Adv. Appl. Prob. 15, 585-600

Barbour, A.D. and Eagleson, G.K. (1985) Multiple comparisons and sums of dissociated random variables. Adv. Appl. Prob. 17, 147-162.

Barbour, A.D. and Eagleson, G.K. (1985) The asymptotics of the permutation distribution of some measures of spatial correlation. Ann. Statist. (submitted).

Barton, D.E. (unpublished manuscript).

Cliff, A.D. and Ord, J.K. (1973) Spatial autocorrelation. Pion, London.

Geary, R.C. (1954) The contiguity ratio and statistical mapping. The Incorporated Statistician 5, 115-145.

Gregory, G.G. (1977) Large sample theory for U-statistics and tests of fit. Ann. Statist. 5, 110-123.

Knox, G. (1964) Epidemiology of childhood leukaemia in Northumberland and Durham. Brit. J. Prev. Soc. Med. 18, 17-24.

Lehmann, E.L. (1975) Nonparametrics: statistical methods based on ranks. Holden Day, San Francisco.

Mantel, N. (1967) The detection of disease clustering and a generalized regression approach. Cancer Research 27, 209-220.

Moran, P.A.P. (1948) The interpretation of statistical maps. J. Roy. Statist. Soc. B, 10, 243-251.

Serfling, R.J. (1980) Approximation theorems of mathematical statistics.

Wiley, New York.

Siemiatycki, J. (1978) Mantel's space-time clustering statistic: computing higher moments and a comparison of various data transforms. J. Statist. Comp. Simul. 7, 13-31.

Stein, C. (1970) A bound for the error in the normal approximation to the distribution of a sum of dependent random variables. Proc. VIth Berkley Symp. Math. Statist. Prob. 2, 583-602.

AGE DISTRIBUTIONS IN BIRTH AND DEATH PROCESSES

A. Bose*
Department of Mathematics and Statistics
Carleton University
Ottawa, CANADA K1S 5B6

1. Introduction

We define the age distribution, X_t, of a birth and death process to be the frequency distribution of ages of particles alive at time t while the total number of particles, N_t, evolves according to a birth and death process starting initially from one or more particles. Note that the age distribution, in our terminology, is not a normalized quantity.

From the various possible representations for the age distribution, we have chosen to express X_t as a sum of discrete measures. That is, we represent the age distribution as a sum of the form $\sum_{j=1}^{n} k_j \delta_{x_j}$ with k_j denoting the frequency of particles of age $x_j \geq 0$. All the variables n, k_j and x_j $(j=1,\ldots,n)$ are functions of the time variable t and the chance parameter ω. However, this dependence will not be made explicit. One consequence of this representation is that the total number of particles alive at time t, N_t, can now be expressed as an integral of X_t, namely $\int_0^\infty X_t(da)$. This simple observation will be useful later on. For a different approach to age distribution in terms of point processes, the reader may consult Brillinger [1981].

At birth, N_t increases by one. But there are two possibilities for X_t. One may replace the mother cell by two daughter cells of age zero. Or alternatively, following Kendall [1949], one may add one daughter cell of age zero. In either case, the equality $N_t = \int_0^\infty X_t(da)$ is preserved. Note however that the Kendall's model takes into consideration the natural phenomena that the birth of an offspring does not automatically lead to the death of the mother. In fact, both may coexist and moreover the mother may still reproduce in future. Here, we propose to adopt the Kendall's model. A nice comparison of these

* Research supported in part by NSERC grant No. A3453.

two models could be found in Harris [1963, p.159].

At death, we have to decrease N_t by one. We assume that the incidence of death is independent of actual age. Thus, at death we remove one particle at "random" i.e. with uniform probability from the population. This is a simplification as realistically deaths should be age specific.

2. Axioms

Before introducing the assumptions underlying the age distribution, it is convenient to recall first the axioms for the critical birth and death process N_t.

Denote the σ-field $\sigma\{N_u: 0 \leq u \leq t\}$ by σ_t then for small positive h one has,

i) $P[N_{t+h} = N_t+1 | \sigma_t] = \lambda N_t h + o(h)$,

ii) $P[N_{t+h} = N_t-1 | \sigma_t] = \lambda N_t h + o(h)$,

iii) $P[N_{t+h} = N_t | \sigma_t] = 1 - 2\lambda N_t h + o(h)$.

Notation: Let $\langle\mu,f\rangle$ represent the integral $\int_0^\infty f d\mu$. Thus, $\langle X_t, \underline{\underline{1}}\rangle$ equals N_t where $\underline{\underline{1}}$ denotes the constant function with value one defined on the positive half of the real axis.

Let us introduce the σ-fields F_t to be $\sigma\{X_u: 0 \leq u \leq t\}$. Then for small positive h one has,

1) $P[X_{t+h} = X_t + \delta_o | F_t] = \lambda\langle X_t, \underline{\underline{1}}\rangle h + o(h)$.

Here, δ_o is the unit mass at the origin indicating the instantaneous age at birth.

2) $P[X_{t+h} = X_t - \delta_y | F_t] = \lambda \underline{\underline{1}}_y(X_t)h + o(h)$.

Here, $\underline{\underline{1}}_y(X_t)$ equals one if $y \in \text{supp}(X_t)$ and equals zero otherwise.

3) $P[X_{t+h} = S_h X_t | F_t] = 1 - 2\lambda\langle X_t, \underline{\underline{1}}\rangle h + o(h)$

where S_h is the shift operator mapping $\sum_{j=1}^{n} k_j \delta_{x_j}$ to $\sum_{j=1}^{n} k_j \delta_{(x_j+h)}$.

To realize X_t as a measure-valued process, it is probably easier to start from the following set of postulates due to Kendall [1949]:

a) The sub-population generated by two co-existing individuals develop in complete independence of one another.

b) An individual existing at epoch t has a chance

$$\lambda h + o(h)$$

of producing a new individual of age zero during the subsequent time interval of length h .

c) An individual existing at epoch t has a chance

$$\lambda h + o(h)$$

of dying during the subsequent time interval of length h .

d) An individual ages linearly in time.

The motion of X_t can be described in terms of two independent exponential clocks with parameter $\lambda\langle X_t, \underline{\underline{1}}\rangle$. The process ages until a clock rings. If it is a birth time then we add δ_o to X_t. Otherwise, we remove δ_y for $y \in \mathrm{Supp}(X_t)$ with uniform probability. Thus by sampling from exponential and uniform distributions appropriately, we construct typical sample paths of the process. It is possible then to assign to this set of paths a probability measure so that we obtain a measure-valued Markov process. As the critical birth and death process, N_t , is non-explosive, we find that for every $t \geq 0$, X_t is a bounded measure on $[0,\infty]$.

3. Moments

We first consider the first moment in detail and relate the associated deterministic differential equation to the balance equation due to Von Foerster. A thorough treatment of the deterministic theory may be found in Oster [1977].

From the axioms a) - d) it is clear that the lifetime distribution has density $\lambda e^{-\lambda t}$. Then for $\phi \in C([0,\infty])$, the set of continuous real-valued functions on $[0,\infty]$,

$$\begin{aligned} &E[\langle X_t,\phi\rangle | X_0 = \delta_x] \\ &\quad = \phi(x+t)\ P[\text{initial particle is alive at time } t] \\ &\quad\quad + \int_0^t \phi(y)\ P[\text{particle of age } y \text{ alive at time } t]dy \\ &\quad = \phi(x+t)e^{-\lambda t} + \int_0^t \phi(y)\ P[\text{age } 0 \text{ at time } t-y \text{ and lives } y \text{ unit}]dy \\ &\quad = \phi(x+t)e^{-\lambda t} + \int_0^t \phi(y)\lambda e^{-\lambda y}dy\ . \end{aligned} \tag{1}$$

Another way to derive the formula for the conditional expectation is to consider $E[\langle X_{t+h},\phi\rangle | F_t] - \langle X_t,\phi\rangle$ for positive h. Using the axioms 1) - 3) obtain

$$E[\langle X_{t+h},\phi\rangle \mid F_t] = \langle X_t + \delta_0,\phi\rangle\lambda\langle X_t,\underline{\underline{1}}\rangle h$$

$$+ [\langle X_t,\phi\rangle\langle X_t,\underline{\underline{1}}\rangle - \langle X_t,\phi\rangle]\lambda h$$

$$+ \langle S_h X_t,\phi\rangle[1 - 2\lambda\langle X_t,\underline{\underline{1}}\rangle]h + o(h).$$

Hence the difference $E[\langle X_{t+h},\phi\rangle \mid F_t] - \langle X_t,\phi\rangle$ may be expressed as $h\langle X_t,\phi' - \lambda\phi + \lambda\phi(0)\underline{\underline{1}}\rangle + o(h)$ and we obtain

$$\frac{\partial}{\partial t} E\langle X_t,\phi\rangle = E\langle X_t,L\phi\rangle \tag{2}$$

where $L\phi = \phi' - \lambda\phi + \lambda\phi(0)$ with prime denoting the derivative.

We now provide a heuristic connection between the above relation and the Von Foerster equation. Write $E\langle X_t,\phi\rangle$ as $\int_0^\infty \phi(y)\alpha(t,y)dy$ with $\alpha(t,y)$ representing the expected mass at age y at time t. Then $\alpha(t,0)$, the expected mass at zero at time t, equals $\lambda E(N_t)$ which equals $\lambda\int_0^\infty \alpha(t,y)dy$. Then from equation (2), after integration by parts and a formal interchange of integral and derivative, obtain the Von Foerster equation

$$\frac{\partial}{\partial t}\alpha(t,y) + \frac{\partial}{\partial y}\alpha(t,y) = -\lambda\alpha(t,y) .$$

One may treat the second moment in a similar fashion. However, it is possible to obtain a "closed-form" expression for the characteristic functional itself. By conditioning on the first ring, one can, as was done by Bartlett and Kendall [1951], write down the integral equation for the characteristic functional. Let

$$\pi(t,\phi,\delta_x) := E[e^{i\langle X_t,\phi\rangle} \mid X_0 = \delta_x] .$$

Then one has,

$$\pi(t,\phi,\delta_x) = e^{i\phi(x+t)}e^{-2\lambda t} + \int_0^t [1+\pi(t-u,\phi,\delta_{x+u})\pi(t-u,\phi,\delta_0)]\lambda e^{-2\lambda u}du$$

along with the initial condition $\pi(0,\phi,\delta_x) = e^{i\phi(x)}$. The solution according to Kendall [1950] is

$$\pi(t,\phi,\delta_x) = 1 + \frac{(e^{i\phi(x+t)}-1)e^{-\lambda t} + \int_0^t (e^{i\phi(y)}-1)g(y)dy}{1 - \int_0^t (e^{i\phi(y)}-1)h(y,t)dy}$$

where $g(y) = \lambda e^{-\lambda y}$ and $h(y,t) = [1 + \lambda(t-y)]g(y)$.

It is easy to obtain the formulas for the moments. In fact

$$E[\langle X_t,\phi\rangle | X_0 = \mu] = \langle \mu, T_t\phi\rangle \tag{3}$$

where $\mu = \sum_{j=1}^{n} k_j \delta_{x_j}$ and $(T_t\phi)(\cdot) := \phi(\cdot + t)e^{-\lambda t} + \int_0^t \phi(y)g(y)dy.$ (4)

Also, $E[\langle X_t,\phi\rangle\langle X_t,\psi\rangle | X_o = \mu]$

$$= \langle\mu,T_t\phi\rangle\langle\mu,T_t\psi\rangle - \langle\mu,(T_t\phi)(\cdot)(T_t\psi)(\cdot)\rangle + \langle\mu,(T_t\phi\psi)(\cdot)\rangle$$

$$+ \langle\mu,T_t\phi\rangle \int_0^t \psi(y)h(y,t)dy + \langle\mu,T_t\psi\rangle \int_0^t \phi(y)h(y,t)dy. \tag{5}$$

It can be shown that the family $\{T_t : t \geq 0\}$ form a semigroup on $V := \{\phi | \phi : [0,\infty] \to R, \text{ continuous}\}$ with the sup-norm topology and its generator L is given by $L\phi = \phi' - \lambda\phi + \lambda\phi(0)$.

The renewal argument used to obtain the characteristic functional may be employed to compute the generator of the age distribution on the convergence determining class of the form $F(\langle\mu,\phi\rangle)$ with $F : R \to R$ smooth. We have

$$GF(\langle\mu,\phi\rangle)$$

$$= [F(\langle\mu,\phi\rangle - \phi(0)) - F(\langle\mu,\phi\rangle)] \lambda\langle\mu, \underline{\underline{1}}\rangle$$

$$+ \int_0^\infty [F(\langle\mu,\phi\rangle - \phi(y)) - F(\langle\mu,\phi\rangle)]\mu(dy)$$

$$+ F'(\langle\mu,\phi\rangle)\langle\mu,\phi'\rangle. \tag{6}$$

4. Scaling limits

It is easy to see that the expectation semigroup has an invariant measure with density $\lambda e^{-\lambda x}dx$ on $[0,\infty[$. We will denote the invariant measure as $\lambda e^{-\lambda x}dx$. The main aim is to study the fluctuations in the age distribution around this deterministic measure.

Suppose ν_n is a sequence of measures of the form $\frac{1}{n}\sum_{j=1}^{n} k_j^{(n)} \delta_{x_j^{(n)}}$ converging weakly to $\lambda e^{-\lambda x}dx$ on $[0,\infty]$, $\nu_n \overset{W}{\to} \lambda e^{-\lambda x}dx$. In what follows, we will drop the superscripts. Corresponding to each ν_n, define μ_n to be $\sum_{j=1}^{n} k_j\delta_{x_j}/n$ so that $\langle\nu_n,\phi\rangle = \langle\mu_n,\phi_n\rangle$ with $\phi_n(x) := \frac{\phi(nx)}{n}$. Now define a sequence of processes $\{Y_n(t,\lambda)\}$ by

$\langle Y_n(t,\lambda),\phi\rangle := \langle X(\frac{t}{n},n\lambda),\phi_n\rangle$ so that

$$E[<Y_n(t,\lambda),\phi>|Y_n(0)=\nu_n] = E[<X(\tfrac{t}{n},n\lambda),\phi_n>|X_n(0)=\mu_n] = <\nu_n,T_t\phi>.$$

Put $\tilde{\phi}(x) = \phi(x) - \phi_o$ where $\phi_o = <\lambda e^{-\lambda x}dx,\phi>$. Then as $n \to \infty$ $E[<Y_n(t,\lambda),\tilde{\phi}>|Y_n(0)=\nu_n] = <\nu_n,T_t\phi> - <\nu_n,\phi_o> \to 0$. Also from (5),

$$E[<Y_n(t,\lambda),\tilde{\phi}>^2|Y_n(0)=\nu_n]$$

$$= <\nu_n,T_t\tilde{\phi}>^2 - \frac{1}{n}<\nu_n,(T_t\tilde{\phi})^2> + \frac{1}{n}<\nu_n,(T_t\tilde{\phi}^2)>$$

$$+ \frac{2}{n}<\nu_n,T_t\tilde{\phi}> \int_0^t \tilde{\phi}(z)h(z,t)dz \to 0 \quad \text{as} \quad n \to \infty.$$

This suggests that on a suitable path space with Skorohod topology, one would expect that $Y_n(\cdot,\lambda) \xrightarrow{\mathcal{D}} \lambda e^{-\lambda x}dx$ provided $Y_n(0) \xrightarrow{w} \lambda e^{-\lambda x}dx$. We also wish to consider the fluctuations limit, $Z(\cdot) := \mathcal{D}\text{-}\lim_{n\to\infty} \sqrt{n}[Y_n(\cdot,\lambda)-\lambda e^{-\lambda x}dx]$.

A formal calculation suggests that the limit process is such that

$$F(<Z(t),\tilde{\phi}>) - \int_0^t F'(<Z(s),\tilde{\phi}>)<Z(s),L\tilde{\phi}>ds$$

$$- \frac{\lambda}{2}<\lambda e^{-\lambda x}dx,\ \tilde{\phi}^2 + \tilde{\phi}^2(0)\underline{\underline{1}}> \int_0^t F''(<Z(s),\tilde{\phi}>)ds$$

is a P-martingale. For $F(x) = x$, we have

$$[<Z(t),\tilde{\phi}> - \int_0^t <Z(s),\tilde{\phi}>ds]^2 - \lambda t<\lambda e^{-\lambda x}dx,\tilde{\phi}^2 + \tilde{\phi}^2(0)\underline{\underline{1}}>$$

is a P-martingale. That is,

$$<Z(t),\tilde{\phi}> - \int_0^t <Z(s),L\tilde{\phi}>ds = <B_t,\phi>$$

where B_t is a centered Brownian motion such that

$$E[<B_t,\tilde{\phi}>^2] = \lambda t<\lambda e^{-\lambda x}dx,\ \tilde{\phi}^2 + \tilde{\phi}^2(0)\underline{\underline{1}}>.$$

This formula dictates that for our path space we consider $D([0,\infty[,(S \oplus R)')$ with $(S \oplus R)'$ denoting the topological dual of the topological direct sum of Schwartz's S space and the Euclidean space R. For $\phi \in S$, note also that $L\phi = \phi' - \lambda\phi + \lambda\phi(0) \notin S$.

The conditions for a sequence of probability measures on $D([0,\infty[,(S \oplus R)')$ to be tight are analogous to those given by Holley and Stroock [1981].

5. References

1. M.S. Bartlett and D.G. Kendall, On the use of the characteristic functional in the analysis of some stochastic processes occurring in physics and biology, Proc. Cambridge. Philos. Soc., 47(1951), 65-76.

2. D.R. Brillinger, Modern population mathematics, Can. J. Stat., 9 (1981), 173-194.

3. D.G. Kendall, Stochastic processes and population growth, J. Roy. Stat. Soc. Ser B, 11(1949), 230-264.

4. ___________ , Random fluctuations in the age distribution of a population whose development is controlled by the simple "Birth-and-Death" process, J. Roy. Stat. Soc. Ser B, 12(1950), 278-285.

5. T.E. Harris, The theory of branching processes, Springer-Verlag, New York, 1963.

6. R.A. Holley and D.W. Stroock, Generalized Ornstein-Uhlenbeck processes and infinite particle branching Brownian Motions, Publ. R.I.M.S. Kyoto Univ., 14(1978), 741-788.

7. G. Oster, "Lectures in population dynamics" in: Modern Modeling of continuum phenomena, ed. by R.C. Diprima, Lectures in applied mathematics, Vol. 16, Amer. Math. Soc., 1977.

CRITICAL CLUSTERING IN THE TWO DIMENSIONAL VOTER MODEL

by

J. Theodore Cox

Syracuse University

and

David Griffeath

University of Wisconsin

1. Introduction.

The voter model of Clifford and Sudbury [4] and Holley and Liggett [5] is one of the simplest stochastic processes which exhibits *spatial clustering*. For the basic model this phenomenon occurs in dimensions one and two. Arratia's work [1] provides a very good description of the one dimensional behavior. Our purpose here is to announce some new results which give information about "critical clustering" in dimension two.

Let us begin by describing the process. The basic voter model $\{\eta_t\}_{t\geq 0}$ is the strong Markov process on $\{0,1\}^{\mathbb{Z}^d}$ with rates specified by

$$\eta_t(x) \rightarrow 1 - \eta_t(x) \quad \text{at rate} \quad (2d)^{-1}\#\{y: |x-y|=1,\ \eta_t(y)\neq\eta_t(x)\} .$$

Liggett [8] is an excellent source for a complete technical description of η_t, and for general background on infinite particle systems. To keep matters simple we will always assume that η_t has initial distribution μ_θ for some $0<\theta<1$, where μ_θ is product measure with density θ, i.e. $\mu_\theta\{\eta(x)=1\}=\theta$ for all $x\in\mathbb{Z}^d$. The intuitive description of the model is as follows. At each "site" x there is a "voter" holding opinion $\eta_t(x) = 0$ or 1. Opinions at time $t=0$ are chosen independently, θ the probability for opionion 1. Each voter then changes opinion randomly as time goes on, the rate of change being proportional to the number of nearest neighbors with the opposite opinion at any given time.

The fundamental behavior of the voter model, discovered independently by Clifford and Sudbury [4] and Holley and Liggett [5], is described by

Theorem 0. $\eta_t \Rightarrow \eta_\infty$ as $t \to \infty$, where

$$P(\eta_\infty \in \cdot) = (1-\theta)\mu_0 + \theta\mu_1 \qquad d = 1,2 ,$$

$$= \nu_\theta \qquad d \geq 3 .$$

Here $\Rightarrow$ denotes weak convergence, μ_0 and μ_1 concentrate on "all 0's" and "all 1's" respectively, and the ν_θ are nontrivial. This last means, for example, that $\{\eta_\infty(x)\}_{x\in\mathbb{Z}^d}$ are neither independent nor totally correlated with respect to ν_θ . See e.g. [2] and [5] for more information about the ν_θ . Theorem 0 asserts that η_t approaches a nontrivial equilibrium if $d\geq 3$, whereas η_t approaches "complete consensus" for $d=1,2$. Consensus means that for any $x,y \in \mathbb{Z}^d$,

$$\lim_{t\to\infty} P(\eta_t(x) \neq \eta_t(y)) = 0 \qquad d = 1,2 ,$$

indicating the formation of larger and larger *clusters* over time. One way to try to describe this clustering is to scale space appropriately with time.

Such scalings have been studied by Arratia [1] and Bramson and Griffeath [3] for the one dimensional voter model, and by Holley and Stroock [6,7] for the voter model and many other particle systems. The next theorem is a sampling of the results in [1] and [3].

Theorem 1. Let d=1. As $t \to \infty$,

(i) $$P(\eta_t(x\sqrt{t}) \neq \eta_t(y\sqrt{t})) \to 2\theta(1-\theta)[1 - \frac{1}{\pi}\int_0^1 ds\ e^{-|x-y|^2/4s}/\sqrt{s(1-s)}]$$
$(x \neq y)$,

(ii) $$\{\eta_t(x\sqrt{t})\}_{x\in\mathbb{Z}} \Rightarrow \{\eta_\infty(x)\}_{x\in\mathbb{Z}},$$

(iii) $$t^{-1/2} \sum_{|x|\le\sqrt{t}} \eta_t(x) \Rightarrow Y .$$

The limits η_∞ and Y are nontrivial, but are too complicated to allow much in the way of explicit computation. The factor $\sqrt{t}$ is the "natural scale" in the following sense. If at time t we replace $\sqrt{t}$ by $f(t)$, and if $f(t)/\sqrt{t} \to 0$ or $+\infty$, then the limits in Theorem 1 will be "trivial". For instance, $\{\eta_\infty(x)\}$ will be totally correlated or independent.

Arratia's work [1] shows that at large times t the one dimensional η_t consists of large blocks of 0's and 1's, each block of length $\approx \sqrt{t}$. See [1] and [3] for precise results and proofs. However the techniques of those papers do not apply in two dimensions, so that new methods are needed.

In fact the behavior of the voter model is strikingly different when d=2. Our purpose here is to state a counterpart of Theorem 1, and to give a sketch of its derivation. Much more detailed results and their rather lengthy proofs will appear elsewhere. Our mystery guest in the scenario of critical clustering is the Fisher-Wright diffusion $\{Y_t\}_{t\ge0}$ on $[0,1]$ with generator

$$Gf(r) = \frac{1}{2} r(1-r) f''(r) \qquad 0 < r < 1 ,$$

and probability transition semigroup

$$q_t(r,\cdot) = P_r(Y_t \in \cdot) .$$

We are able to obtain the following characterization of clustering in the two dimensional voter model.

Theorem 2. Let d=2. For each $0<\alpha<1$, as $t\to\infty$,

(i) $P(\eta_t(xt^{\alpha/2})\neq\eta_t(yt^{\alpha/2})) \to 2\theta(1-\theta)\alpha \qquad (x\neq y)$,

(ii) $\{\eta_t(xt^{\alpha/2})\}_{x\in\mathbb{Z}^2} \Rightarrow \{\eta_\infty(x)\}_{x\in\mathbb{Z}^2}$, where η_∞ is an exchangeable family of {0,1}-valued random variables with de Finetti mixture $q_{\log(1/\alpha)}(\theta,\cdot)$,

(iii) $\frac{1}{4t^\alpha}\sum_{\|x\|\le t^{\alpha/2}} \eta_t(x) \Rightarrow Y \qquad (\|\ \| = \text{box norm})$,

where Y has distribution $q_{\log(1/\alpha)}(\theta,\cdot)$.

The contrast with Theorem 1 is striking. There is no longer a unique "natural scale" for η_t, but rather a continuuum of scales which produce nontrivial limit fields η_∞. The "honest" spatial dependence of Theorem 1(i) is not present in Theorem 2(i), however. It is satisfying that in the two-dimensional setting the limits

η_∞ and Y can be discribed so explicitly. Our work suggests that at time t the two dimensional voter model consists of regions of 0's and 1's of size $t^{\alpha/2}$ for *all* $0<\alpha<1$. We have run extensive computer simulations which are also suggestive of this phenomenon (see the figure).

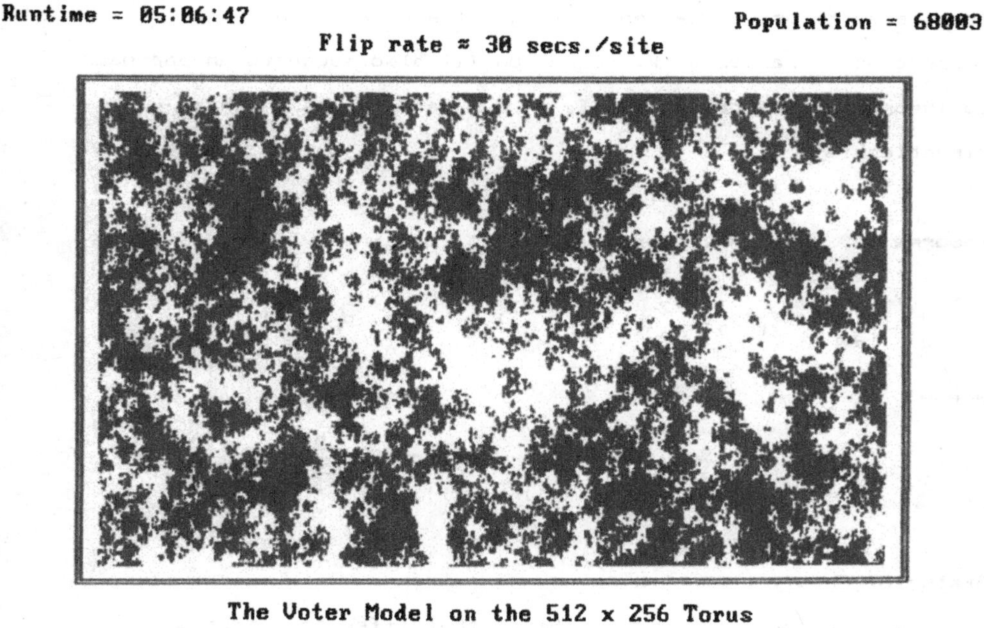

Figure. Computer Simulation of the Voter Model

2. Methods.

We briefly describe the proofs for parts (i) and (ii) of Theorem 2, and explain how the Fisher-Wright diffusion enters the picture. Our chief tool is *duality*. The voter model is dual to a coalescing random walk system ξ_t defined as follows. Starting from each $x \in \mathbb{Z}^2$ particles do simple rate 1 random walks. The particles act independently, except that when any two meet they coalesce into one particle which then continues the random walk. For $A \subset \mathbb{Z}^d$, let $\xi_t(A)$ denote the set of positions at time t of particles which started from A. The duality equation ([4,5]) is

$$(1) \qquad P(\eta_t(x) \equiv 1 \text{ on } A) = E\,\theta^{\#\xi_t(A)} \qquad (A \text{ finite}) .$$

It is easy to prove Theorem 0 using (1) and the recurrence properties of random walk. Equation (1) also suggests an approach to Theorem 2 via scaling of ξ_t . Our main result in this direction is

Theorem 3. Let $d=2$. For $n\geq 2$, $1\leq k\leq n$, distinct $x_i \in \mathbb{Z}^2$, $0\leq\alpha\leq 1$,

$$\lim_{t\to\infty} P(\#\xi_t(x_1 t^{\alpha/2},\ldots,x_n t^{\alpha/2}) = k) \;=\; p_{n,k}(\alpha) \quad ,$$

where

$$p_{n,k}(\alpha) \;=\; \sum_{j=k}^{n} \frac{(-1)^{j+k}(2j-1)(j+k-2)!}{k!(k-1)!(j-k)!} \; \frac{\binom{n}{j}}{\binom{n+j-1}{j}} \; \alpha^{\binom{j}{2}} \quad .$$

<u>Sketch of Proof:</u>

(a) The case $n=2$, $k=1$. Since the difference process $\xi_t(x_1 t^{\alpha/2}) - \xi_t(x_2 t^{\alpha/2})$ is a rate 2 random walk up until the particles meet,

$$P(\#\xi_t(x_1 t^{\alpha/2}, x_2 t^{\alpha/2}) = 1) \;=\; P_{xt^{\alpha/2}}(\tau_0 \leq 2t) \quad ,$$

where $x = x_1 - x_2$ and τ_0 is the first hitting time of 0 for a rate 1 random walk. Using a "last time at 0" decomposition, and writing $p_u(x,y) = P(\xi_u(x)=y)$ we see that

$$P_{xt^{\alpha/2}}(\tau_0 \leq 2t) \;=\; p_{2t}(0,xt^{\alpha/2}) + \int_0^{2t} p_u(0,xt^{\alpha/2}) P_{(1,0)}(\tau_0 > 2t-u)\, du \;.$$

Standard random walk estimates (cf. [9]) yield

$$P(\#\xi_t(x_1 t^{\alpha/2}, x_2 t^{\alpha/2}) = 1) \;\sim\; \int_1^t \frac{e^{-|x|^2 t^{\alpha}/u}}{\pi u} \, \frac{\pi}{\log(2t-u)}\, du$$

$$\sim\; \int_0^1 e^{-|x|^2 t^{\alpha-s}} \, \frac{\log t}{\log(2t-t^{s})}\, ds$$

$$\to\; 1-\alpha \;\stackrel{\text{def}}{=}\; p_{2,1}(\alpha) \quad .$$

Note that there is no dependence on x_1-x_2 in the limit.

(b) The next step is to show that for $n\geq 3$, $0<\alpha\leq\gamma<\infty$, as $t\to\infty$,

(2) $$P(\#\xi_{t^\gamma}(x_1t^{\alpha/2},\ldots,x_nt^{\alpha/2}) = n) \to (\alpha/\gamma)^{\binom{n}{2}} \overset{\text{def}}{=} p_{n,n}(\alpha/\gamma) .$$

We will not indicate our proof of (2) here, but merely point out that this is the "right" answer if one believes that the events $\{\#\xi_{t^\gamma}(x_it^{\alpha/2},x_jt^{\alpha/2}) = 2\}$ are asymptotically independent, since each of these $\binom{n}{2}$ events has probability $\sim \alpha/\gamma$ by (a). It is helpful to recast (2) as follows. Let

$$\tau = \inf\{u\colon \#\xi_u(x_1t^{\alpha/2},\ldots,x_nt^{\alpha/2}) = n-1\}$$

be the time of the first collision. Then (2) implies that as $t\to\infty$,

(3) $$P(\tau\in d(t^\gamma)) \to \binom{n}{2}\alpha^{\binom{n}{2}}\gamma^{-\binom{n}{2}-1}\,d\gamma \qquad \gamma\geq\alpha .$$

(c) We now proceed by induction. Assume that the limit in Theorem 3 exists when $n=N-1$. For $k<N$,

(4) $$P(\#\xi_t(x_1t^{\alpha/2},\ldots,x_Nt^{\alpha/2}) = k)$$
$$= \int_0^t P(\tau\in ds,\text{the remaining } N-1 \text{ particles coalesce to } k \text{ particles by time } t) .$$

If we change variables (let $s=t^\gamma$) and assume that the remaining $N-1$ particles at time t^γ are spaced about $t^{\gamma/2}$ apart, then (3) and the induction hypothesis suggest that (4) is approximately

(5) $$\int_\alpha^1 \binom{n}{2}\alpha^{\binom{n}{2}}\gamma^{-\binom{n}{2}-1} p_{N-1,k}(\gamma)\,d\gamma \overset{\text{def}}{=} p_{N,k}(\alpha) .$$

(d) One can check by induction that (5) gives rise to our explicit formula for the $p_{n,k}$, so our sketch is finished.

It should be clear that the duality equation (1) and Theorem 3 imply convergence in (i) and (ii) of Theorem 2. What is not yet clear is why the Fisher-Wright diffusion represents the de Finetti mixture. Perhaps the easiest way to understand this is to introduce one more process.

Let $\{D_t\}_{t\geq 0}$ be the pure death process on $\{1,2,...\}$ with jump rates

$$n \to n-1 \qquad \text{at rate} \quad \binom{n}{2} \qquad\qquad (n\geq 2) ,$$

and 1 absorbing. It is well-known in the mathematical genetics literature (cf. Tavaré [9]) that D_t and the Fisher-Wright diffusion Y_t satisfy the duality equation

$$(6) \qquad E_\theta[Y_t^n] = E_n[\theta^{D_t}] \qquad\qquad n\geq 1,\ 0\leq\theta\leq 1 .$$

Note the similarity to (1). Thus η_t and ξ_t are dual, Y_t and D_t are dual, and it remains only to connect ξ_t with D_t .

The intriguing connection is

$$(7) \qquad P_n(D_t = k) = p_{n,k}(e^{-t}) \qquad\qquad 1\leq k\leq n .$$

(Recall Theorem 3.) To get (7) differentiate (5) and note that, except for the time change $t = \log(1/\alpha)$, the result is precisely the system of backward equations for the probability transition semigroup of D_t . The proof that η_∞ has de Finetti mixture $q_{\log(1/\alpha)}(\theta,\cdot)$ is now just three lines:

$$P(\eta_\infty(x_i)=1\ ,\ 1\le i\le n) = \sum_{k=1}^{n} \theta^k p_{n,k}(\alpha) \qquad \text{by (1) and Theorem 3}$$

$$= E_n\ \theta^{D_{\log(1/\alpha)}} \qquad \text{by (7)}$$

$$= \int_0^1 s^n\ q_{\log(1/\alpha)}(\theta,ds) \qquad \text{by (6)}\ .$$

REFERENCES

[1] Arratia, R. (1982) Coalescing Brownian motions and the voter model on $\mathbb{Z}$. Unpublished manuscript.

[2] Bramson, M. and Griffeath, D. (1979) Renormalizing the 3-dimensional voter model. *Annals of Probability* **7**, 418-432.

[3] Bramson, M. and Griffeath, D. (1980) Clustering and dispersion rates for some interacting particle systems on $\mathbb{Z}$. *Annals of Probability* **8**, 183-213.

[4] Clifford, P. and Sudbury, A. (1973) A model for spatial conflict. *Biometrika* **60**, 581-588.

[5] Holley, R. and Liggett, T.M. (1975) Ergodic theorems for weakly interacting infinite systems and the voter model. *Annals of Probability* **3**, 643-663.

[6] Holley, R. and Stroock, D. (1978) Generalized Ornstein-Uhlenbeck Processes and Infinite Particle Branching Brownian Motions. *Publ. RIMS, Kyoto Univ.* **14**, 741-788.

[7] Holley, R. and Stroock, D. (1979) Central limit phenomena of various interacting systems. *Ann. Math.* **110**, 333-393.

[8] Liggett, T.M. (1985) *Interacting Particle Systems.* To appear.

[9] Spitzer. F. (1976) *Principles of Random Walk.* 2nd ed. Springer-Verlag, New York.

[10] Tavaré, S. (1984) Line-of-descent and genealogical processes, and their applications in population genetics models. To appear.

MEASURE-VALUED PROCESSES
CONSTRUCTION, QUALITATIVE BEHAVIOR AND STOCHASTIC GEOMETRY

Donald Dawson
Department of Mathematics and Statistics,
Carleton University,
Ottawa, Canada.

1. INTRODUCTION.

The objective of this paper is to review recent developments and applications of measure-valued stochastic processes. We begin by reviewing the construction and characterization of measure-valued processes and also illustrate the method of duality which is a useful tool in the study of some special classes of measure-valued processes. We then consider the qualitative behavior of probability-measure-valued processes, that is, processes in which the total measure is conserved. A classification of the qualitative behavior of these processes is introduced, namely, recurrent, ergodic, transient-diffusive and transient-wandering. In the next part we consider possibly infinite measure-valued processes including spatially homogeneous measure-valued processes. In this part we emphasize stochastic geometric aspects of the behavior of the processes. For example, we discuss the set-valued support process and questions involving clustering at both microscopic and macroscopic levels. Finally these questions are discussed in detail for measure-valued branching processes.

2. ORIGIN AND EXAMPLES OF MEASURE-VALUED DIFFUSION PROCESSES.

Beginning with the work of Feller (1951) the diffusion approximation has played an important role in stochastic population modelling. In particular in the diffusion limit certain questions concerning extinction and fixation probabilities can be answered in terms of Feller's boundary classification. In this section we consider the extension of the diffusion approximation to the space-time context. In this case the limiting diffusion is a measure-valued Markov process which is characterized as the solution of a measure-valued martingale problem. In order to introduce the notion of diffusion approximation for spatially distributed systems we present three simple but natural models which lead to these considerations.

MODEL I: POPULATION OF BRANCHING RANDOM WALKERS.

Consider a finite population of individuals who inhabit the one dimensional lattice $\varepsilon^{1/2}Z = \{0, \pm\varepsilon^{1/2}, \pm 2\varepsilon^{1/2}, \pm 3\varepsilon^{1/2}, \ldots\}$ where $\varepsilon > 0$ is a small parameter. We assume that there are initially $x(0)/\varepsilon$ individuals at the origin and that all individuals have mass ε. The dynamics are assumed to be continuous time binary branching and nearest neighbour symmetric random walk. The expected time between random walk jumps and also between binary fissions for each individual are assumed to be equal to ε. The branching offspring distribution is assumed to be given by:

Probability of death (0 offspring) = $(\frac{1}{2} - 2a\varepsilon)$

Probability of two offspring = $(\frac{1}{2} + 2a\varepsilon)$.

For $t \geq 0$ and $A \in \beta(R^1)$, the Borel subsets of R^1,

$X^\varepsilon(t,A)$:= Total mass in the set A at time t.

Then X^ε is a measure-valued Markov process. The total mass at time t is

$$x^\varepsilon(t) := X(t,R^1).$$

Feller (1951) proved that as $\varepsilon \to 0$,

$x^\varepsilon(.) \to x(.)$ (weak convergence of processes)

where $x(.)$ is a one dimensional diffusion process with generator of the form:

$$(2.1) \quad A = ax\, d/dx + \tfrac{1}{2}\gamma^2 x d^2/dx^2 .$$

This process can also be characterized as the unique strong solution of the stochastic differential equation:

$$(2.2) \quad dx(t) = a\, x(t)dt + \gamma x^{1/2}(t)dw(t), \qquad \gamma \geq 0,$$

where w is a standard Brownian motion. Subsequently in population biology this equation has been generalized by the addition of "environmental stochasticity" and possibly non-linear growth leading up to the stochastic differential equation

$$(2.3) \quad dx(t) = f(x(t))dt + \gamma_1 x^{1/2}(t)dw_1(t) + \gamma_2 x(t)dw_2(t)$$

where $\gamma_1 \geq 0$, $\gamma_2 \geq 0$, and w_1 and w_2 are independent Brownian motions.

In order to describe at the heuristic level the diffusion limit as $\varepsilon \to 0$ of the spatially distributed model, let

$X(t,x)$:= limiting density of the population at x at time t.

The space-time analogue of (2.3) is the stochastic partial differential equation (SPDE):

$$(2.4) \qquad dX(t,x) = (\Delta + a)X(t,x)dt + \gamma[X(t,x)]^{\frac{1}{2}}\, dW(t,x)$$

where Δ denotes the Laplacian operator and W is a S'-valued Brownian motion with covariance

$$(2.5) \qquad E(\langle W(t),\phi\rangle\langle W(t),\psi\rangle) = t\int \phi(x)\psi(x)dx$$

for $\phi,\psi \in S(R)$, the Schwartz space of C^{∞}-rapidly decreasing functions and where $\langle W,\phi\rangle$ denotes the canonical bilinear form on $S'(R^1)\times S(R^1)$. In other words, $dW(t)/dt$ is a space-time white noise. There remains the question as to whether or not Equation (2.4) really makes sense, that is, does it have a solution. The question can be answered in the affirmative under the assumption that the Brownian motion W is somewhat more regular. This is contained in the following result of Viot.

THEOREM (Viot (1976)). The SPDE (2.4) has a unique strong solution provided that the Brownian motion W has a bounded measureable covariance density function $q(.,.)$, that is,

$$E(\langle W(t),\phi\rangle\langle W(t),\psi\rangle) = t\iint\phi(x)\psi(y)q(x,y)dxdy .$$

Note that in the case of (2.5), $q(x,y) = \delta(x-y)$ so that it does not fall under the purview of this theorem. In addition if W satisfies the conditions of the theorem then the multiplicative property of the underlying branching mechanism is lost. Furthermore, in Dawson (1975) it is demonstrated that (2.4) cannot have a strong solution in dimensions $d > 1$ if W satisfies (2.5).

In spite of these fundamental difficulties, the formalism of stochastic partial differential equations remains useful and can be reinterpreted in a meaningful way. Equation (2.4) can be generalized to the form:

$$(2.6) \qquad dX(t) = A^*X(t)dt + B(X(t))dt + Q^{\frac{1}{2}}(X(t))dW(t) ,$$

where

(i) A^* is the generator of a semigroup of operators on $S'(R^d)$ or some appropriate subspace of $S'(R^d)$,

(ii) $B: M(R^d) \to M(R^d)$ where $M(R^d)$ denotes the family of measures on R^d and B satisfies a Lipschitz or monotonicity condition, and

(iii) Q denotes the "diffusion coefficient" and will be discussed in detail below.

Under appropriate hypotheses on the coefficients A^*, B and Ω it is possible to establish the existence and uniqueness of strong solutions in an appropriate Hilbert space of functions (or generalized functions) on R^d. The reader is referred to Krylov-Rozovskii (1981) for a survey of results in this direction.

MODEL II: THE STEPWISE MUTATION MODEL OF OHTA AND KIMURA.

This model arises in population genetics and was originally motivated by the study of electrophoretically detectable alleles. For our purposes we regard it as a model for the distribution of a quantitative characteristic such as height in a population.

In this model the total number of individuals is conserved and the dynamics involves two mechanisms, finite population sampling and mutation. Mutation is modelled by a random walk on Z^d when the population involves d quantitative characteristics. To model random sampling of the gene pool individuals who die are replaced by a new individual whose quantitative type is chosen at random by sampling from the current empirical distribution of types.

The type of an individual is denoted by $k \in Z^d$. We set

$$p(t,k) := N^{-1}.(\text{Number of individuals of type } k \text{ at time } t)$$

where N denotes the total population. Then $p(t,.)$ forms a countable state space Markov jump process with generator:

$$\text{(2.7)}\quad L_N f(p) = \sum_{i\neq j}[\gamma^2 p(i)p(j) + Dp(i)\theta_{ij}](f(p^{ij}) - f(p)), \quad \text{(a) } \underline{\text{or}},$$

$$= \sum_{i\neq j}[\gamma^2 p(i)p(j) + Dp(i)\sum_k p(k)\theta_{kj}](f(p^{ij}) - f(p)) \quad \text{(b)},$$

where

$$\begin{aligned} p^{ij}(k) &= p(k) - N^{-1} \quad \text{if } k = i,\\ &= p(k) + N^{-1} \quad \text{if } k = j,\\ &= p(k) \quad \text{otherwise, and} \end{aligned}$$

$$\begin{aligned} \theta_{ij} &= 1 \text{ if } |i - j| = 1,\\ &= -2d \text{ if } i = j, \text{ and}\\ &= 0 \text{ otherwise.} \end{aligned}$$

Form (a) of (2.7) corresponds to the situation in which a change of type is allowed during the lifetime of an individual whereas in form (b) change of type (mutation) occurs only at the time of birth of an individual.

We now consider a rescaling of space and time based on the notion that a single mutation will result in a small change of order $N^{-\frac{1}{2}}$ in the quantitative characteristic. This is accomplished by the rescaling

(2.8) $$X_N(t,A) := \sum_{k/N^{1/2} \in A} p_N(N^2 t,\{k\}) .$$

As $N \to \infty$, the processes X_N converge to a limiting measure-valued diffusion process. The resulting process known as the Fleming-Viot is characterized in Section 3 .

MODEL III: THE MULTILEVEL ENSEMBLE OF INTERACTING PARTICLES.

Consider a system consisting of M interacting ensembles each consisting of N interacting individuals. The dynamics are governed by the systems of MN stochastic differential equations:

(2.9)
$$\begin{aligned} dx_{ij}(t) &= b(x_{ij}(t))dt + \sigma_0 dw_{ij}(t) + \sigma_1 dw_i(t) \\ &+ \theta_1(\int xX_i(t,dx) - x_{ij}(t))dt \\ &+ \theta_2(\int\int x\mu(dx)X_{MN}(t,d\mu) - \int xX_i(t,dx))dt \end{aligned}$$

where

$$X_i(t) := N^{-1} \sum_{j=1}^{N} \delta_{x_{ij}(t)} \in M_1(R^1) \quad \text{for } i = 1,2,\ldots,M, \text{ and}$$

$$X_{MN}(t) := M^{-1} \sum_{i=1}^{M} \delta_{X_i(t)} \in M_2 := M_1(M_1(R^1))$$

and $M_1(S)$ denotes the space of probability measures on S.
The term with coefficient θ_1 models the interaction within each of the M ensembles and the term with coefficient θ_2 denotes the interaction between the M ensembles.

In the multilevel law of large numbers (Dawson (1984)),

(2.10) $$X_{MN}(.) \Longrightarrow X(.) \qquad \text{as } M,N \to \infty, \text{ (weak convergence of processes)}$$

where $X(t)$ denotes the probability law of the $M_1(R^1)$-valued Markov process which is associated with the SPDE

(2.11) $$\langle Z(t),\phi\rangle - \langle Z(0),\phi\rangle - \int_0^t \langle Z(s),L_{Z(s)}\phi\rangle\, ds = \sigma_1\int_0^t (\nabla Z(s),dw(s))$$

where

$$L_\mu\phi = b(x)\partial\phi/\partial x + \theta_1(\int y\mu(dy) - x)\partial\phi/\partial x + \tfrac{1}{2}(\sigma_0^2 + \sigma_1^2)\partial^2\phi/\partial x^2.$$

In (2.11) w denotes a standard Brownian motion and ∇ denotes the gradient. The term on the right hand side of (2.11) is a special case of a stochastic flow of mass. Other models of this type have arisen in modelling turbulence and have been studied by Chow (1978) and also in the study of wave propagation in random media (Dawson and Papanicolaou (1984).

3. SOME GENERAL THEORY FOR MEASURE-VALUED PROCESSES.

3.1. BASIC DEFINITIONS.

Let (S,ρ) be a locally compact separable metric space (LCSM) and let $\bar{S}$ denote its one point compactification. Let $M(\bar{S})$ $(M_1(\bar{S}))$ denote the family of Borel (probability) measures on $\bar{S}$ furnished with the topology of weak convergence. Then $M_1(\bar{S})$ is compact and $M(\bar{S})$ is locally compact. Let $M_r(S)$ denote the space of Radon measures on S furnished with the topology of vague convergence. In the special case $S = R^d$, let $\bar{R}^d := R^d \cup \{\tau\}$ where τ is an isolated point. Iscoe (1983) introduced the space of p-tempered measures as follows. Let

$$M_p(\bar{R}^d) := \{\mu \in M_r(\bar{R}^d): \langle\mu,\phi_p\rangle < \infty\},$$

where

$$\phi_p(x) := 1/(1+|x|^p) \quad \text{for } x \in R^d$$
$$\phi_p(\tau) := 1.$$

$M_p(\bar{R}^d)$ when furnished with the weakest topology making the maps $\mu \to \langle\psi,\mu\rangle$ continuous for $\psi \in C_c(R^d) \cup \{\phi_p\}$ is locally compact.

As a canonical probability space we adopt $\Omega = D([0,\infty),M(S))$ with the usual Skorohod topology and set

$$X(\omega,t) := \omega(t) \quad \text{for } \omega \in \Omega,\ t \geq 0.$$

Further Ω is equipped with the standard right continuous filtration, F_t, completed where appropriate.

A measure-valued Markov processes is prescribed by a family of probability measures $\{P_\mu : \mu \in M(S)\}$ on Ω such that

(3.1) (i) $P_\mu(X(0) = \mu) = 1$, and

(ii) (Strong Markov property) for every F_t-stopping time τ with $P_\mu(\tau < \infty) = 1$,

$$P_\mu(X(\tau+t) \in B|F_\tau) = P_{X(\tau)}(X(t) \in B).$$

Let $C(M(S))$ denote the space of bounded continuous functions on $M(S)$ with the supremum norm. To a Markov process X, $t \geq 0$ and $F \in C(M(S))$, let

$$T_tF(\mu) := E_\mu(F(X(t))). \tag{3.2}$$

The process is said to be a Feller process if $T_t: C(M(S)) \to C(M(S))$ in which case it yields a semigroup of contractions on $C(M(S))$. If it is uniformly stochastically continuous, then it forms a strongly continuous semigroup. The <u>infinitesimal generator</u> with domain $D(G)$

is defined by:

$$D(G) := \{F: \lim_{t\downarrow 0} (T_t F - F)/t \quad \text{exists in} \quad C(M(S))\}$$

and for $F \in D(G)$,

(3.3) $$G\,F(\mu) := \lim_{t\downarrow 0} (T_t F(\mu) - F(\mu))/t \quad .$$

The family $\{P_\mu : \mu \in M(S)\}$ is then uniquely characterized by the pair $(D(G), G)$.

3.2. EXISTENCE AND CHARACTERIZATION OF MEASURE-VALUED PROCESSES.

Let E be a closed subset of $M(S)$. A pregenerator is defined by $D_0(G) \subset C(E)$ and a mapping $G: D_0(G) \to C(E)$.

A probability measure P_μ on Ω is said to be a solution to the martingale problem for G if for every pair (f,g) with $f \in D_0(G)$ and $g = Gf$,

(3.4) $$f(X(t)) - \int_0^t g(X(s))ds$$

is a P_μ-martingale. P_μ is said to have initial condition μ if (3.1) is satisfied. The importance of this notion is a consequence of the following theorem.

THEOREM (STROOCK AND VARADHAN)

Let D be dense in $C(E)$. Assume that for $t \geq 0$, $F \in D$, there exists $\Lambda_{F,t} \in L^\infty(E)$ such that for any $\mu \in E$ and any solution P_μ of the martingale problem for G,

(3.5) $$E_\mu(F(X(t))) = \Lambda_{F,t}(\mu).$$

Then there is at most one solution family $\{P_\mu : \mu \in M(E)\}$ of the martingale problem for G. If a solution exists, then it is measurable and satisfies the strong Markov property.

3.3. HEURISTICS ON PREGENERATORS FOR MEASURE-VALUED DIFFUSIONS.

Consider the formal SPDE

(3.6) $$dX(t) = A^*X(t)dt + B(X(t))dt + Q^{1/2}(X(t))dW(t) \quad .$$

If this equation had a solution, $\phi \in C(S) \cap D(A)$, where $D(A)$ denotes the domain of A, then

(3.7) $$M_t^\phi := (X(t) - \int_0^t A^*X(s)ds - \int_0^t B(X(s))ds \,,\phi\,)$$

would be a local martingale with increasing process (quadratic variation),

(3.8) $$\langle\langle M^{\phi}\rangle\rangle_t = \int_0^t \iint \phi(x)\phi(y)\ \Omega(X(s):dx\times dy)\ ds ,$$

where $\Omega:M(S) \to Q$ where Q denotes the space of bilinear forms Ω on $D(\Omega)\times D(\Omega)$, $D(\Omega) \subset C(M(S))$ such that $\Omega(\mu:\phi\times\phi) \geq 0$.

Itô's lemma applied to $F(X(t))$ leads us to a consideration of the second order differential operator

(3.9) $$G\,F(\mu) := (A + B(\mu))\delta F/\delta\mu(.) + \tfrac{1}{2}\,\Omega(\mu:\delta^2F(\mu)/\delta\mu(.)\delta\mu(.)) ,$$

where

$$\delta F(\mu)/\delta\mu(x) := \lim_{\varepsilon\downarrow 0} (\ F(\mu + \varepsilon\delta_0) - F(\mu)\ )/\ \varepsilon ,$$ if the limit exists,

and δ_0 denotes the Dirac delta function.

In order to make sense of the differential operator (3.9) we introduce some appropriate subspaces of $C(M(S))$ consisting of differentiable functions. Let

(3.10) $$\mathcal{F}(M(S)) = \{F_{f,\phi}(\mu) := f(\langle\mu,\phi\rangle);\ f \in C^{\infty}(R^1),\ \phi \in C_c(S)\}$$

where $C_c(S)$ denotes the space of continuous functions with compact support on S. Let $\mathcal{P}(M(S))$ denote the smallest algebra containing all functions of the form:

(3.11) $$F_{f_n}(\mu) := \int_S \cdots \int_S f_n(x_1,\ldots,x_n)d\mu^{\times n}$$

where $f_n \in C(S^n)$ and $\mu^{\times n}(d\underline{x}) = \mu(dx_1)\ldots\mu(dx_n)$.

LEMMA 3.1. (i) $\mathcal{P}(M(S))$ is dense in $C(E)$ if E is a compact subset of $M(S)$.

(ii) $\mathcal{F}(M(S))$ is measure-determining on $M(S)$.

(iii) Functions in $\mathcal{P}(M(S))$ and $\mathcal{F}(M(S))$ are differentiable, for example,

$$\delta F_{f_f}(\mu)/\delta\mu(x) = \sum_{j=1}^{n} \underbrace{\int_S \cdots \int_S}_{j\ \text{free}} f(x_1,\ldots,x_{j-1},x,x_{j+1},\ldots,x_n)\mu^{\times(n-1)}(d\underline{x}) ,$$

where by "j free" we mean that integration over the jth variable is omitted, and

$$\delta^2F_{f_n}(\mu)/\delta\mu(x)\delta\mu(y)$$
$$= \sum_{j=1}^{n}\sum_{k=1}^{n} \underbrace{\int_S \cdots \int_S}_{j,k\,\text{free}} f(x_1,\ldots,x_{j-1},x,x_{j+1},\ldots,x_{k-1},y,x_{k+1},\ldots,x_n) .\mu^{\times(n-2)}(d\underline{x}) .$$

Let E be a compact subset of $M(S)$ and consider the pair $(\mathcal{P}(E),G)$ where G is an operator of the form (3.9). If we can verify the hypotheses of the Stroock Varadhan theorem for this pair, then we will have succeeded in constructing an E-valued Markov process. To prove the existence of solutions to the martingale problem there are a number of possible methods including the following:

(i) proof of strong existence for the SPDE and Itô's lemma,

(ii) discretization (or some other approximation method) followed by the proof of tightness for the approximating sequences and verification that a weak limit satisfies the martingale problem,

(iii) proof of the convergence of the approximating semigroups, conditional moments or characteristic functional.

Methods to prove the uniqueness condition (3.5) include:

(i) strong uniqueness for the SPDE and the Yamada-Watanabe theorem,

(ii) duality which is described in detail below, or

(iii) verify that for the approximating semigroups G_n, $G_n F \to GF$ for $F \in D$, a dense subset of $C(M(S))$ and that there exists λ_0 for which $(\lambda_0 I - G)\,D$ is also dense in $C(M(S))$.

Each of these methods works in certain cases; unfortunately there are other cases in which none of them works. In the next section we describe the method of duality which has proved to be useful for a number of measure-valued processes including the examples introduced in Section 2.

(3.4) THE METHOD OF DUALITY.

This method was originally introduced by Spitzer and used by Holley and Liggett in the context of infinite particle systems such as the voter model. We present a simplified version of the method of duality applied to measure-valued processes as introduced in Dawson and Kurtz (1982).

Let E_1 and E_2 be metric spaces and let

(3.12) $G_1 = \{(f(.,y),g(.,y)) : y \in E_2\} \subset C(E_1)\times C(E_1)$,

$G_2 = \{(f(x,.),h(x,.)) : x \in E_1\} \subset C(E_2)\times C(E_2)$,

$\alpha \in C(E_1)$, $\beta \in C(E_2)$.

THEOREM (DAWSON AND KURTZ (1982))

Assume that

(i) for $y \in E_2$,

$$f(X(t),y) - \int_0^t g(X(s),y)\,ds \quad \text{is a } P_{X(0)}\text{-martingale},$$

(ii) for $x \in E_1$,

$$f(x,Y(t)) - \int_0^t h(x,Y(s))\,ds \quad \text{is a } P_{Y(0)}\text{-martingale},$$

(iii) $g(x,y) + \alpha(x)f(x,y) = h(x,y) + \beta(y)f(x,y)$, and

(iv) a uniform integrability condition (cf. Dawson and Kurtz (1982)) which is satisfied if for example f,g,h,α,β, are all bounded functions.

Then,

$$(3.13)\quad E_{X(0)}[f(X(t),Y(0)).\exp(\int_0^t \alpha(X(u))du)]$$
$$= E_{Y(0)}[f(X(0),Y(t)).\exp(\int_0^t \beta(Y(u))du)] .$$

APPLICATION TO MODEL I.

Let X^ε denote the measure-valued process described in Model I, that is, the population of branching random walkers. We now consider the diffusion limit, $\varepsilon \to 0$, of the space-time model. Recall that in this limiting régime the total mass process converges to the Feller diffusion limit.

Consider the subspace of the space $\mathcal{F}(M(s))$ consisting of functions of the form:

$$(3.14)\quad F_\phi(\mu) := \exp(-<\phi,\mu>) \quad \text{where} \quad \phi \in (C^3(R^d))^+ \text{ (nonnegative).}$$

The generator of the process X^ε applied to such a function is given by

$$(3.15)\quad G^\varepsilon F_\phi(\mu) = [<-(\Delta+a)\phi,\mu> + \tfrac{1}{2}\gamma^2<\phi^2,\mu>]\exp(-<\phi,\mu>) + O(\varepsilon^{\frac{1}{2}})$$
$$= H\ F_\mu(\phi) + O(\varepsilon^{\frac{1}{2}})$$

where $F_\mu(\phi) = F_\phi(\mu)$ and H is the generator of the (deterministic) flow:

$$(3.16)\quad \partial u/\partial t = (\Delta+a)u - \tfrac{1}{2}\gamma^2 u^2 .$$

Then $G^\varepsilon \to G$ as $\varepsilon \to 0$ where $G\ F_\phi(\mu) = H\ F_\mu(\phi)$.
The probabilities laws P^ε of X^ε form a uniformly tight family (this can be verified by the standard techniques e.g. Ethier and Kurtz (1984)). It thus remains to prove that there is a unique limit point and to identify the latter. The Stroock Varadhan theorem then guarantees the existence of a limiting measure-valued diffusion process (the continuity of the limit process is proved using Kolmogorov's criterion).

The uniqueness is established by the method of duality. Let P^ε denote a limit point of the family P^ε . Then P^ε is a solution of the martingale problem associated with G and the class of functions described by (3.14). Since H is the generator of the flow (3.16) and $G\ F_\phi(\mu) = H\ F_\mu(\phi)$, then the above theorem (Dawson and Kurtz) yields:

(3.17) $E_\mu(\exp(-<\phi,X(t)>) \;=\; \exp(-<u(t),\mu>)$
where $u(.)$ satisfies the nonlinear partial differential equation (3.16) with initial condition $u(0) = \phi$. Equation (3.17) gives the Laplace functional for the probability transition kernel of the measure-valued Markov process X. The fact that this Laplace functional can be identified is the key to a study of the stochastic geometry of this process which is described in Section 4 .

APPLICATION TO MODEL II.

Let X_N denote the $M_1(R^d)$-valued Markov process defined by (2.8), that is, the rescaled stepwise mutation model of Ohta and Kimura. The pregenerator G_N of this process acting on a function $F_{f_n} \in P(M(S))$ is given by:

(3.18) $G_N F_{f_n}(\mu) = G\, F_{f_n}(\mu) + R_N(f_n)$

where $R_N(\ )$ is a remainder term of the order of $N^{-\frac{1}{2}}$,

(3.19) $G\, F_{f_n}(\mu) = H\, F_\mu(f_n)$ where $F_\mu(f_n) = F_{f_n}(\mu)$, and

$$H\, F_\mu(f_n) = D \sum_{j=1} (\Delta_j f_n) + \gamma^2 \sum_{j\neq k}\sum [F_\mu(\Phi_{jk} f_n) - F_\mu(f)]$$

where

$\Phi_{jk}: C(S^n) \to C(S^{n-1})$,

$\Phi_{jk} f_n(x_1,\ldots,x_n) = f_n(x_1,\ldots,x_j,\ldots,x_{k-1},x_j,x_{k+1},\ldots,x_n)$, and

Δ_j denotes the Laplacian acting on the jth variable.

H is the generator of a dual process, Y, which is a $P(M(S))$-valued Markov process with dynamics described by:

(i) at rate $\gamma^2 n(n-1)$ jumps $f \to \Phi_{jk} f$, and

(ii) between jumps of the form (i) Y evolves according to the Brownian motion semigroup on $(R^d)^n$, U_t, that is, $f \to U_t f$.

Fleming and Viot (1979) verfied that the probability laws P^N of the processes X_N form a uniformly tight family that that any limit point of the P_N is a solution of the martingale problem associated with $(G,P(M(S)))$. Then (3.19) together with the duality theorem yields the uniqueness condition for the $(G,P(M(S)))$ martingale problem. In addition we have the duality relation

(3.20) $E_\mu(F_{f_n}(X(t))) \;=\; E_{f_n}(F_\mu(Y(t)))$.

The duality relation (3.20) is the key to the study of the qualitative behavior and stochastic geometry of the limiting Fleming-Viot process X which will be described below.

(3.5) A SHORT LIST OF OTHER PREGENERATORS.

In this section we provide a short list of pregenerators which arise in other applications.

(a) Weighted Sampling. For $F_{f_n} \in P(M(S))$,

(3.5.1) $G F_{f_n}(\mu) = H F_\mu(f_n)$ where

(3.5.2) $H F_\mu(f_n) = \sum_{j \neq k}\sum \theta(x_j,x_k)[\Phi_{jk} f_n - f_n]$.

(b) Growth in a Random Environment. (Shimizu (1977, Mizuno (1978), Dawson and Salehi (1980))

Consider the bilinear stochastic evolution equation:

(3.5.3) $dX(t,x) = X(t,x)d_t W(t,x)$

where W is a function-valued Brownian motion with bounded continuous covariance density $q(.,.)$. The associated pregenerator is given by:

(3.5.4) $G F_{f,\phi}(\mu) = f''(\langle\mu,\phi\rangle).\iint q(x,y)\phi(x)\phi(y)\mu(dx)\mu(dy)$

for $F_{f,\phi} \in (M(S))$.

(c) Unnormalized Filtering Equation. Consider the nonlinear filtering problem (Pardoux (1979)) with signal process $x(.)$ and observation process given by

$dy(t) = h(x(t)) + dw(t)$

where w is a Brownian motion independent of x. The conditional distribution (unnormalized) of $x(.)$ conditioned on the observation process leads to the pregenerator:

(3.5.5) $G F_{f,\phi}(\mu) = f''(\langle\mu,\phi\rangle)\iint h(x)h(y)\mu(dx)\mu(dy)$.

(d) Stochastic flows in R^d. This example involves a generalization of the stochastic flow introduced in Model III and was originally motivated by turbulence models (Chow (1978)). Let ρ be a continuous positive definite function. Then the associated stochastic flow pregenerator in R^1 is given by:

(3.5.6) $G F_{f_n}(\mu) = F_\mu(Af_n)$ where

$$Af_n = \sum_{j \neq k} \rho(x_j - x_k)\, \partial^2 f_n(x_1,\dots,x_n)/\partial x_j \partial x_k + \kappa \sum_{j=1}^{n} \Delta_j f_n ,$$

where $\kappa \geq 0$. Stochastic flows in $d > 1$ are given by analogous pregenerators. For example, the Brownian wandering motion in R^d is given by

(3.5.7) $G F_{f_n}(\mu) = F_\mu(Af_n)$ with $A = \sum_{j=1}^{n}\sum_{k=1}^{n} (\nabla_j \cdot \nabla_k)$,

where ∇ denotes the gradient operator.

REMARK. Duality methods can be applied to pregenerators of the form (a) or (d). On the other hand for cases (b) and (c) strong methods have been successful.

(3.6) ADDING INTERACTIONS: THE CAMERON-MARTIN-GIRSANOV FORMULA.

Let $\{P_\mu : \mu \in M(S)\}$ denote the unique solution to the martingale problem associated with the pregenerator

(3.6.1) $$G\,F(\mu) = A\delta F/\delta\mu + Q(\mu:\delta^2 F/\delta\mu\delta\mu).$$

Assume that

(3.6.2) $$B(\mu)(dx) = \int\int b(\mu;y)Q(\mu:dx\times dy)$$

where b is a nice function from $M(S)$ to $C(S)$. Now define:

(3.6.3) $$R_B(t) := \exp\Big[\int_0^t b(X(s);y)M(ds\times dy) - \tfrac{1}{2}\int_0^t b(X(s);x)b(X(s);y)Q(X(s):dx\times dy)ds\Big]$$

$$\text{where } M(t) := X(t) - \int_0^t A^*X(s)ds\ .$$

Then (Dawson (1978)),

(3.6.4) $$P^B|_{F_t} := R_B(t)P|_{F_t}$$

is the unique solution to the local martingale problem for

(3.6.5) $$G_B F(\mu) = (A + B(\mu))\delta F/\delta\mu + Q(\mu:\delta^2/\delta\mu\delta\mu)\ .$$

EXAMPLE 3.6.1. Adding selection to the Fleming-Viot Model.

Let $m(x,y)$, a symmetric function of x and y, denote the "fitness" of the genotype (x,y). Define

(3.6.6) $$B(\mu)(dx) := \left(\int m(x,y)\mu(dy) - \int\int m(z,y)\mu(dy)\mu(dz)\right)\mu(dx)\ ,$$
$$= b(\mu;y)Q(\mu:dx\times dy)$$

where

$$b(\mu;y) := \int m(y,z)\mu(dz),\quad Q(\mu:dx\times dy) = \mu(dx)\delta_x(dy) - \mu(dx)\mu(dy).$$

We can then apply the Cameron-Martin-Girsanov transformation (3.6.3, 3.6.4) to obtain the unique solution to the martingale problem for the Fleming-Viot model with selection. (Refer to Fleming and Viot (1979) and Dawson for details.)

An alternative approach to the construction of the Fleming-Viot process with selection is to use the duality method - this has been carried out in Dawson and Kurtz (1982).

REMARK: Measure-valued processes related by the Cameron-Martin-Girsanov transformation have equivalent probability laws on F_t for $t < \infty$. Therefore any almost-sure sample path properties such as Hausdorff-Besicovitch dimension of support are inherited.

4. STOCHASTIC GEOMETRY OF MEASURE-VALUED PROCESSES.

For fixed t, $X(t)$ is a random measure on S. In this section we investigate the topography of this random measure. In particular we consider the following questions for finite measure-valued processes:

(i) What is the "central location" of $X(t)$?

(ii) How disperse is $X(t)$?

(iii) Is $X(t)$ locally uniform or highly clustered ?

(iv) What happens to (i) - (iii) when $t \to \infty$?

(4.1) TOPOLOGICAL AND PROPER SUPPORTS.

Let $F(S)$ denote the collection of closed subsets of the locally compact separable metric space S. The compact topology on $F(S)$ is given by the basis consisting of sets of the form

$$U^K_{G_1,\ldots,G_n} := \{F \in F(S): F\cap K = \emptyset,\ F\cap G_i \neq \emptyset,\ i=1,\ldots,n\}$$

where K is a compact subset and G_i are open subsets of S. $F(S)$ can also be metrized by the Hausdorff metric defined by:

$$d(A,B) := \max\{\sup_{x\in A} \rho(x,B), \sup_{x\in B} \rho(x,A)\} .$$

The following properties are well known (cf. Matheron (1975)):

(i) ($F(S)$,compact topology) is second countable, Hausdorff and compact.

(ii) The compact topology is coarser than the Hausdorff topology.

(iii) If S is compact, then the compact and Hausdorff topologies coincide.

Given a measure μ on S, the topological support of μ ,supp(μ), is defined to be the smallest closed set F such that $\mu(S\backslash F) = 0$. Given a reference measure ν a proper support $S_\nu(\mu)$ with respect to ν is a Borel set such that $\mu(S_\nu(\mu)^C) = 0$ and $\nu(S_\nu(\mu) \setminus B) = 0$ for any other carrying set B.

PROPOSITION 4.1.1(a) (Cutler (1984)) The mapping $\mu \to \text{supp}(\mu)$ is measurable with respect to the Borel σ-algebras on $M(S)$ and $F(S)$.

(b) Let X and Y be random measures. Then there exists $S_{X|Y} \in \beta(M\times M)\times\beta(S)$ such that for each $\omega \in M\times M$, the section $(S_{X|Y})_\omega$ is a proper support of $X(\omega)$ with respect to $Y(\omega)$. (Horowitz (1984)).

(4.2) THE LEBESGUE DECOMPOSITION OF A RANDOM MEASURE.

Lemma 4.2.1. (Cutler (1984)). Let ν $M(S)$. Then both

$$C_\nu := \{\mu:\mu << \nu\} \quad \text{and} \quad S_\nu := \{\mu:\mu\perp\nu\}$$

belong the the Borel σ-algebra $\beta(M(S))$.

PROPOSITION 4.2.1. (Yonglong (1982), Cutler (1984)). If X and Y are random measures, then $X = X_C + X_S$ with $X_C << Y$, $X_S \perp Y$, and X_C, X_S are random measures.

(4.3) STOCHASTIC GEOMETRY AT THE MICROSCOPIC LEVEL.

Let E be a subset of R^d. The Hausdorff outer measure is defined by

$$H^\alpha(E) := \lim_{\delta\downarrow 0} \inf_{\substack{\cup B_i \supset E \\ d(B_i)\le\delta}} \sum (d(B_i))^\alpha$$

where $d(B_i)$ denotes the diameter of B_i and each B_i is a closed sphere. There exists a unique $0 \le \beta \le d$ such that

$\alpha < \beta \Rightarrow H^\alpha(E) = \infty$, $\alpha > \beta \Rightarrow H^\alpha(E) = 0$.

β is called the Hausdorff-Besicovitch dimension of E, dim(E). If E is finite, then dim(E) = 0 whereas if the d-dimensional Lebesgue measure of E is positive, then dim(E) = d.

PROPOSITION 4.3.1. (Cutler (1984)) The mapping $\dim: F(S) \to [0,d]$ is measurable.

Given a measure μ the singular dimension of μ is defined as $\dim_S(\mu) := \dim(\mathrm{supp}(\mu_S))$. The Hausdorff dimension of proper support of μ, $\dim(\mu)$ is α if there exists a Borel set B of dimension α with $\mu(B) = \mu(S)$ and $\mu(B) < \mu(S)$ for any Borel set B of dimension less than α. Then $\dim(\mu) \le \dim(\mathrm{supp}(\mu))$.

DIMENSIONAL DISINTEGRATION THEOREM (Cutler (1984)).

Let $\mu \in M(R^d)$. There exists a measure $\hat{\mu}$ on $[0,d]$ and a mapping $\Phi: \beta(R^d)\times[0,d] \to R^+$, such that

(a) for $0 \le \alpha \le d$, $\Phi(.,\alpha)$ is a probability measure on $\beta(R^d)$,

(b) for $B \in \beta(R^d)$, $\Phi(B,\alpha)$ is a measurable function of α,

(c) for $B \in \beta(R^d)$, $\Phi(B,\alpha) = 0$ for $\dim(B) < \alpha \le d$, $\hat{\mu}$-almost surely, and

(d) μ has the integral representation

(4.3.1) $$\mu(B) = \int_0^d \Phi(B,\alpha)\hat{\mu}(d\alpha), \text{ and}$$

(e) $\int_0^\beta \Phi(.,\alpha)\, \hat{\mu}(d\alpha)$ is supported on a set of dimension β.

The measure $\hat{\mu}$ is called the distribution of dimension of support of μ.

REMARK. There is a related but simpler notion, namely, fractal dimension of support. To define this consider a cube V in R^d which is divided into Γ^d subcubes. Given a measure μ on V, let $N_\Gamma(\mu)$ denote the number of these subcubes having strictly positive μ-measure. Then the fractal dimension of support of μ is defined by

$$\mathrm{Fdim}(\mu) := \limsup_{N\to\infty} \log(N_\Gamma(\mu))/\log(\Gamma).$$

Note that $\mathrm{Fdim}(\mu) \ge \dim(\mu)$.

Let X be a random measure on a cube V in R^d. For $\varepsilon > 0$, let $N_\Gamma^\varepsilon(X)$ denote the minimum number of subcubes of side Γ^{-1} whose union has X-measure of at least $X(V) - \varepsilon$.

LEMMA 4.3.1. (Dawson and Hochberg (1979).

Assume that there exist sequences $\varepsilon_n, \eta_n, \varepsilon'_n$ all converging to zero as $n \to \infty$ and $\Gamma_n \to \infty$ as $n \to \infty$. Assume that

(4.3.2) $P(\log N_{\Gamma_n}^{\varepsilon_n}(X) / \log(\Gamma_n) \le D(1 + \eta_n)) \ge 1 - \varepsilon'_n$ for all n.

Then there exists a random closed set $B(\cdot)$ such that

$$X(\omega, B(\omega)) = X(\omega, V),$$

and

$$\dim B(\omega) \le D \quad \text{for all} \quad \omega.$$

LEMMA 4.3.2. (Zähle (1984)).

Let Ξ be a random closed subset of R^d and $0 \le D \le d$. Assume that Ξ supports a random measure X whose second moment measure

(4.3.3) $E(X(B)X(B')) = \int_B \kappa(x,B')EX(dx)$.

Assume that there is a measurable function $r: S \to (0,\infty)$ such that for EX-almost every x in R^d,

(4.3.4) $\int_{S(x,r(x))} |x-z|^{-D}\kappa(x,dz) < \infty,$

where $S(x,r(x))$ denotes a sphere with centre x and radius $r(x)$. Then

(4.3.5) $P(\{X(B) > 0\} \setminus \{\dim(\Xi \cap B) \geq D\}) = 0.$

APPLICATION TO THE MEASURE-VALUED DIFFUSION PROCESS

In this application we consider a generalization of the measure-valued diffusion process constructed in Section 3. Let X be a measure-valued process with Laplace functional:

(4.3.6) $E_\mu(\exp(-\langle X(t),\phi\rangle)) = \exp(-\langle u(t),\mu\rangle)$,

where u satisfies the nonlinear initial value problem

(4.3.7) $\partial u/\partial t = \Delta_\alpha u - \frac{1}{2}\gamma^2 u^2$, $u(0) = \phi$, and

Δ_α denotes the infinitesimal generator of the symmetric stable process on R^d with index $0 \le \alpha \le 2$.

THEOREM. (a) There exists a random closed set B_t for fixed $t > 0$ such that $X(t,\omega,B_t(\omega))$ K) $= X(t,K)$ a.s. for every compact set K and $\dim(B_t(\omega)) = \alpha$ for every ω.

(b) The distribution of the dimension of support $\hat{X}(t) = \delta_\alpha$.

Proof. (a) was proved in the case $\alpha = 2$ in Dawson and Hochberg (1979) and extended to the case $0 \le \alpha \le 2$ in Roelly-Coppoleta (1984).

The proof of (a) is based on Lemma 4.3.1. The proof of (b) is based on Lemma 4.3.2.

COROLLARY. For $t > 0$, $X(t)$ is almost surely singular if $d \geq 3$ or if $d = 2$ and $\alpha < 2$ or if $d = 1$ and $\alpha < 1$.

THEOREM. For $d = \alpha = 1$ or 2, $X(t.,)$ is almost surely singular for $t > 0$.

Proof. This is proved in Dawson and Hochberg (1979) for the case $\alpha = 2$ and in the case $\alpha < 2$ in Roelly-Coppoleta.

APPLICATION TO THE FLEMING-VIOT MODEL.

Let X denote the Fleming-Viot model constructed in Section 3.

THEOREM (Dawson and Hochberg (1982))

If $d \geq 2$, then the Hausdorff Besicovitch dimension of the topological support of $X(t)$ with $t > 0$ is equal to 2.

COROLLARY. If $t > 0$ and $d \geq 3$, then $X(t)$ is almost surely singular.

REMARK. The proof of the upper bound on the Hausdorff-Besicovitch dimension of the topological support of $X(t)$ is based on the "genealogical structure" of the Fleming-Viot process. In particular the population mass can classified into a hierarchy of subpopulations. In particular for each integer K it can be regarded as the sum of K subpopulations such that all members of a subpopulation can be regarded as descendents of a single "individual". Moreover the distribution of mass among the K subpopulations is given by the Dirichlet distribution. Finally the spatial range (diameter) of the subpopulation is of the order of $O(1/K^{\frac{1}{2}})$ for sufficiently large K. The precise formulation of these ideas is contained in Dawson and Hochberg (1982).

5. QUALITATIVE DESCRIPTION OF FINITE MEASURE-VALUED PROCESSES.

Consider a finite measure-valued process whose sample paths are right continuous and have left limit, that is, the sample paths belong to $D([0,\infty),M(R^d))$.

(5.1) DESCRIPTIVE STATISTICS.

Consider the following processes:

(i) total mass process $m(t) := X(t,R^d)$,

(ii) empirical mean process $\bar{x}(t) := \int xX(t,dx)/m(t)$,

(iii) empirical variance process $v(t) := \int |x - \bar{x}(t)|^2 X(t,dx)/m(t)$,

(iv) empirical support process $\Xi(t) := \mathrm{supp}(X(t))$.

The processes $m(t), \bar{x}(t)$ and $v(t)$ are real-valued progressively measurable processes and $\Xi(t)$ is a $F(R^d)$-valued progressively measurable process.

(5.2) ERGODIC CLASSIFICATION OF BEHAVIOR.

The process $X(.)$ is said to be subcritical if $\lim_{t\downarrow 0} m(t) = 0$. A subcritical process is also said to suffer extinction. The next two definitions deal with the notion of local extinction.

The process X is said to be strongly transient if for every compact set K there exists a stopping time T_K with $P(T_K < \infty) = 1$ and such that $X(t,\omega,K) = 0$ for $t > T_K(\omega)$.

The process X is said to be weakly transient if for every compact set K, $X(t,K) \to 0$ a.s. as $t \to \infty$. It is said to be weakly recurrent if for every compact set K, $X(t,K) \to 0$ in probability but for which there exists a compact set K_0 such that $P(X(t,K_0) \not\to 0) > 0$. The process X is said to be stably recurrent if it has a (not identically zero) invariant probability distribution. Finally the process is said to be supercritical if $\limsup_{t\to\infty} m(t) = \infty$.

(5.3) SCALING BEHAVIOR OF PROBABILITY MEASURE-VALUED PROCESSES.

Let X be a probability-measure-valued (conservative) process. Consider the space-time rescaling group: for $0 < \alpha \le 2$, let

$$(RS^\alpha) \qquad t \to K^\alpha t, \qquad x \to Kx \;,$$

$$X_K^\alpha(t,A) := X(K^\alpha t, A_K) \quad \text{where} \quad A_K := \{x\colon x/K \in A\}\;.$$

There are two measure-valued processes which are invariant under RS^α. These are:

(i) Pure Diffusion: X_D^α is the solution of the evolution equation

$$(5.3.1) \qquad \partial X_D^\alpha/\partial t = \Delta_\alpha {}^* X_D$$

where Δ_α is the infinitesimal generator of the symmetric stable semigroup of index α (or the Brownian motion semigroup in the case of $\alpha = 2$).

(ii) Symmetric stable or Brownian Wandering Atom: $X_W^\alpha(t) := \delta_{w(t)}$ where w denotes the symmetric stable process of index α (or Brownian motion in the case of $\alpha = 2$).

The process X is said to be ultimately diffusive of index α if

$$(5.3.2) \qquad X_K^\alpha \Longrightarrow X_D^\alpha \quad \text{as } K \to \infty, \text{ in the sense of weak convergence of } M(R^d)\text{-valued processes.}$$

The process X is said to be ultimately wandering of index α if

$$(5.3.3) \qquad X_K^\alpha \Longrightarrow X_W^\alpha \quad \text{as } K \to \infty, \text{ in the same sense.}$$

(5.4) QUALITATIVE BEHAVIOR OF THE FLEMING-VIOT MODEL.

In this section we present a brief review of the qualitative behavior of the Fleming Viot model in order to illustrate some of these concepts as well as to indicate some of the ideas involved in their proofs. Most of these results are proved in Dawson and Hochberg (1982).

(5.4.1) Let $X(0)$ have compact support. Then the support process $\Xi(t)$ is compact with probability one for each $t \geq 0$. Note that this is in striking contrast to the well-known behavior of the pure diffusion equation.

(5.4.2) Ultimately Wandering Behavior.

Consider the rescaling of the Fleming-Viot process:

(5.4.1) $$X_\varepsilon(t,dx) := X(t/\varepsilon^2, dx/\varepsilon).$$

The pregenerator of the rescaling process X_ε is given by:

(5.4.2) $$G_\varepsilon F_{f_n}(\mu) = K_\varepsilon F_\mu(f_n), \text{ where}$$

$$K_\varepsilon F_\mu(f_n) = D\sum_{j=1}^{n} F_\mu(\Delta_j f_n) + (\gamma^2/\varepsilon^2)\sum_{j\neq k}\sum [F_\mu(\Phi_{jk}f_n) - F_\mu(f_n)] .$$

Then using an "averaging limit argument" for the dual process with generator K_ε we obtain:

(5.4.3) $$G_\varepsilon \to G \text{ where } G F_{f_n}(\mu) = K F_\mu(f_n), \text{ where}$$

$$K F_\mu(f_n) = F_\mu(\Delta f_n^*) \text{ and } f_n^* \in C(R^d) \text{ is defined by}$$

$$f_n^*(x) := f_n(x,x,\ldots,x).$$

But K is the dual generator for the Brownian wandering atom and therefore we conclude that the Fleming-Viot process is ultimately wandering of index 2.

This result has an interesting extension to the Fleming-Viot process with weak selection which has been obtained with T. Kurtz. The appropriate pregenerator is:

(5.4.4) $$G_\varepsilon F_{f_n}(\mu) = F_\mu(\sum\Delta_j f_n) + \varepsilon^{-1}(F_\mu(\sum_{j=1}^{n} M_j f_n) - F_\mu(f_n))$$
$$+ (1/4\varepsilon)\sum_{j\neq k}\sum (F_\mu(2\Phi_{jk}f_n) - F_\mu(f_n))$$
$$+ \varepsilon^{-1}(n - \tfrac{1}{2}n(n-1))F_\mu(f_n) ,$$

where

$$M_j f_n(x_1,\ldots,x_n) = [m(x_j,x_{n+1}) - m(x_{n+1},x_{n+2})]f(x_1,\ldots,x_n).$$

We assume that $m(.,.)$ is a symmetric function having bounded second partial derivatives and consider the associated empirical mean processes $\bar{x}_\varepsilon(t)$. Then $\bar{x}_\varepsilon(.) \to x(.)$ where x is a R^d-valued diffusion with generator $\Delta f + b.\nabla f$ where $b(x) := (\nabla_1 m)(x,x)$, that is, a wandering atom with a drift term.

(5.4.3) Ergodic Theorem.

Consider the centered empirical process

$X^*(t,A) := X(t, \theta_{-\bar{x}(t)}A)$ where θ_y denotes translation by y,

in other words, we consider the process which is centered so as to have a zero empirical mean at all times. Then (Dawson Hochberg (1982), also see Shiga (1982)) have established the ergodic theorem:

(5.4.3.1) $\lim_{T\to\infty} T^{-1}\int_0^T X^*(t,A)\,dt = \nu(A)$ exists with probability one.

The proof is based on a duality argument. It suffices to show that for functions f_n satisfying the invariance property

$$f_n(x_1+z,\ldots,x_n+z) = f_n(x_1,\ldots,x_n),$$

that $\lim_{t\to\infty} E(F_{f_n}(X(t))$ exists. But this is easy to prove since the dual process $Y(.)$ preserves the invariance property and also

$$\lim_{t\to\infty} Y(t) \in C(R^d) \quad \text{almost surely.}$$

The result follows by observing that the only functions on R^d satisfying the invariance property are the constant functions.

(5.4.4) From Wandering to diffusive behavior.

Recall that the Fleming-Viot pregenerator is given by

$$G_\gamma F(\mu) = \Delta\delta F/\delta\mu + \gamma^2 Q(\mu:\delta^2 F/\delta\mu\delta\mu).$$

From the result of Section 5.4.2 we know that as $\gamma^2 \to \infty$, G_γ converges to the generator of the wandering Brownian motion. On the other hand as $\gamma^2 \to 0$, G_γ converges to the generator of the pure diffusion process, so that this process can exhibit a range of behavior from wandering to diffusive. We can also view G_γ for small γ as a small random perturbation of the pure diffusion process and look for Freidlin-Wentzell estimates. In joint work with J. Gärtner we have identified the action functional:

(5.4.4.1) $S_{0T}(\mu(.)) = \frac{1}{2}\int_0^T L(\mu(t), d\mu(t)/dt)\,dt$, where

$$L(\mu,\nu) := \sup_{\phi\in D} (\langle \nu - \Delta\mu, \phi\rangle)^2/(\langle\mu,\phi^2\rangle - \langle\mu,\phi\rangle^2)$$

where D denotes the space of C^∞ functions on R^d with compact support. The full Friedlin-Wentzell estimates have been obtained for a simple ensemble of interacting particles (Dawson and Gärtner (1984)). These results suggest that the analogy between the second order differential operators in infinite dimensions and those in finite dimensions can be exploited in the study of large deviations.

(5.5) QUALITATIVE BEHAVIOR OF THE FINITE MEASURE-VALUED BRANCHING PROCESS.

Let X denote the finite measure-valued branching process with generator given by (3.15,3.16). This process is subcritical if $a < 0$ and is supercritical if $a > 0$. If $X(0)$ has compact support, then the support process $\Xi(t)$ is compact with probability one for all t (Dawson (1978)).

(5.5.1) The Weighted Occupation time process. I. Iscoe (1981) has investigated the weighted occupation time process defined by

$$\langle Y(t),\phi\rangle := \int_0^t \langle \phi, X(s)\rangle\, ds \quad \text{for } \phi \in C_c(R^d) \text{ (functions with compact supports)}$$

He obtained the following characterization of the Laplace functional for Y in the critical case $a = 0$:

(5.5.1.1) $$E_\mu(\exp(-\langle Y(t),\phi\rangle)) = \exp(-\langle u(t),\mu\rangle)$$

where u satisfies the nonlinear initial value problem

(5.5.1.2) $$\partial u/\partial t = \Delta u - (\gamma^2/2)u^2 + \phi, \quad u(0) = 0.$$

Because the process is critical the limit $Y(\infty) := \lim_{t\to\infty} Y(t)$ exists. Its Laplace functional is given by:

(5.5.1.3) $$E_\mu(\exp(-\langle Y(\infty),\phi\rangle) = \exp(-\langle u,\mu\rangle)$$

where u is a bounded solution of the nonlinear elliptic equation

(5.5.1.4) $$\Delta u - (\gamma^2/2)u^2 + \phi = 0 .$$

Special cases of this equation have been studied by a number of authors including Fowler (1931), Sawyer and Fleischman (1979) and Iscoe (1981). Assuming spherical symmetry bounded solutions of (5.5.1.4) have the following asymptotic behavior for large $|x|$:

(5.5.1.5) $$u(x) \sim \begin{cases} 6/x^2, & d = 1,\\ 4/|x|^2, & d = 2,\\ 2/|x|^2, & d = 3,\\ 2/|x|^2 \log |x|, & d = 4,\\ c_d/|x|^{d-2}, & d > 4.\end{cases}$$

(5.5.2) Application to the Range of the Critical Branching Process.

Using the above ideas Iscoe (1982) has studied the range of the critical measure-valued branching process. Let

$$R_0 := \inf\{r : X(t,S_{0,r}) = X(t,R^d) \text{ for all } t \geq 0\},$$

$$r(x) := P(Y(\infty,S_{0,1}) > 0 \mid X(0)) = \delta_x ,$$

where $S_{0,r}$ denotes a sphere of radius r centered at the origin in R^d.

THEOREM (Iscoe (1982)).

(i) $P(R_0 \le r | X(0) = \delta_0) = \exp(-c/r^2)$ for $r \ge 0$,

where c is a constant.

(ii) $r(x) \sim c\, u(x)$ for large $|x|$

where c is a constant and $u(x)$ is given by (5.5.1.5).

N. El Karoui (1984) has also studied the weighted occupation time process using a Girsanov and Feynman-Kac type theorem.

6. QUALITATIVE BEHAVIOR OF THE SPATIALLY HOMOGENEOUS MEASURE-VALUED CRITICAL BRANCHING PROCESS.

I. Iscoe (1983) has shown that the measure-valued branching process with pregenerator given by (3.15,3.16) can be extended to a Markov process on the locally compact $M_p(\overline{R}^d)$ with sample paths in $D([0,\infty), M_p(R^d))$ for $d < p \le d+2$. Note that this permits us to allow $X(0)$ to be any spatially homogeneous ergodic random measure, for example, Lebesgue measure.

Consider the rescaled process:

(6.1) $$X_K(t,A) := K^{-d} X(K^2 t, A_K).$$

If $X(0)$ is Lebesgue, then the Laplace functional of X_K is given by

(6.2) $$E(\exp(-\langle X_K(t),\phi\rangle)) = \exp(-\int u_K(t,x)\,dx),$$

where u_K satisfies the initial value problem:

(6.3) $$\partial u_K/\partial t = \Delta u_K - \gamma^2 K^{2-d} u_K^2 \quad , \quad u_K(0,x) = \phi(x).$$

Equation (6.3) suggests that the behavior of the spatially homogeneous system is dimension dependent and that $d = 2$ is the critical dimension. The following theorem summarizes the main results concerning these questions.

THEROEM. Let X denote the critical measure-valued branching process in R^d with Lebesgue initial measure.

(a) $d = 1$. $X(t,K) \to 0$, a.s. as $t \to \infty$ for any compact set K, in fact it is strongly transient. (Dawson (1977)).

(b) $d = 2$. $X(t,K) \to 0$ in probability for any compact set K. (Dawson (1977)). In addition X satisfies the self-similarity property: $X_K(t)$ has the same probability law as $X(t)$. (Dawson and Hochberg (1979)). Finally Iscoe (1984) and Fleischmann and Gärtner (1984) have shown that

$$Z(t) := t^{-1}\int_0^t \langle X(s),\phi\rangle ds$$

converges in distribution as $t \to \infty$ to a positive infinitely divisible random variable whose distribution has no atom at 0.

(c) <u>$d \geq 3$.</u> (i) In this case there exists a non degenerate stationary random measure with covariance kernel $q(x,y) = |x-y|^{2-d}$. (Dawson(1977)).

(ii) Consider the fluctuation process:

$$Y_K(t,A) := K^{-(2+d)/2}[X(K^2t,A_K) - E(X(K^2t,A_K))].$$

Holley and Stroock (1978) proved that as $K \to \infty$,

$Y_K \to Y$ in the sense of weak convergence of measures on $C([0,\infty), S'(R^d))$

where $S'(R^d)$ dnotes the Schwartz space of tempered measures on R^d and Y is a generalized Ornstein-Uhlenbeck process which satisfies the linear stochastic evolution equation

$$dY(t) = Y(t)dt + dW(t)$$

where $W(.)$ is a space-time white noise.

(iii) Ergodic theorem for the occupation time process (Iscoe (1981)), (Fleischmann and Gärtner (1984)).

$T^{-1}\int_0^T X(t,dx)\, dt \to$ Lebesgue measure in the vague topology with probability one.

(iv) Central Limit theorem for the occupation time process (Iscoe (1981,1983)). For $\phi \in C_c(R^d)$, let

$$Z(T) := a_T^{\frac{1}{2}} (T^{-1}\int_0^T \langle X(t),\phi\rangle dt - \langle 1,\phi\rangle), \text{ where}$$

$$\begin{aligned} a_T &= T^{-\frac{1}{2}} \quad \text{for } d = 3, \\ &= (\log T)/T \quad \text{for } d = 4, \\ &= T^{-1} \quad \text{for } d \geq 5. \end{aligned}$$

Then $Z(T)$ converges in distribution as $T \to \infty$ to a mean zero normal random variable.

(d) <u>$d > 4$</u>. Let $Y(.)$ denote the occupation time process defined in (5.5.1) and define

$Z(T) := (Y(T) - T\Lambda)/T^{\frac{1}{2}}$ where Λ denotes Lebesgue measure.

Then

$$Z(T) \to Z(\infty) \quad \text{as} \quad T \to \infty,$$

in the sense of weak convergence of $S'(R^d)$-valued random fields, where $Z(\infty)$ is a generalized Gaussian random field with covariance kernel $q(x,y) = |x - y|^{4-d}$. (Iscoe (1981), (1983)).

<u>REMARK</u>. Fluctuation theorems for branching diffusion processes under various scalings have been studied by Gorostiza (1983 and his contribution to this volume). Recently work has begun on extending these studies to models of branching in a random spatial environment.

REFERENCES.

1. P.L. Chow. Stochastic partial differential equations in turbulence related problems, in "Probabilistic Analysis and Related Topics", Vol. 1, Academic Press, 1978.

2. C. Cutler. Some topological and measure-theoretic results for measure-valued and set-valued stochastic processes, Ph.D. thesis, Carleton University, Ottawa, 1984.

3. D.A. Dawson. Stochastic evolution equations and related measure processes, J. Mult. Anal. 5 (1975), 1-52.

4. D.A. Dawson. Geostochastic calculus, Canad. J. of Stat. 6 (1978), 143-168.

5. D.A. Dawson. Limit theorems for interaction free geostochastic systems, Coll. Math. Soc. J. Bolyai 24 (1978), 27-47.

6. D.A. Dawson and K.J. Hochberg. The carrying dimension of a stochastic measure diffusion, Ann. Prob. 7, 693-703.

7. D.A. Dawson and H. Salehi. Spatially homogeneous random evolutions, J. Mult. Anal. 10(1980), 141-180.

8. D.A. Dawson and T.G. Kurtz. Application of duality to measure-valued processes, Lecture Notes in Control and Information Sci. 42 (1982), ed. W. Fleming and L.G. Gorostiza, Springer-Verlag, 91-105.

9. D.A. Dawson and K.J. Hochberg. Wandering random measures in the Fleming Viot model, Ann. Prob. 10(1982), 554-580.

10. D.A. Dawson and G.C. Papanicolaou. A random wave process, J. Appl. Math. Opt., 1984.

11. D.A. Dawson. Asymptotic analysis of multilevel stochastic systems, Tech. Rep. 30 (1984), Laboratory for Research in Statistics and Probability, Ottawa.

12. D.A. Dawson and K.J. Hochberg. Qualitative behavior of a selectively neutral allelic model, Theor. Pop. Biol. 23 (1983), 1-18.

13. N. El Karoui. Non-linear evolution equations and functionals of measure-valued branching processes, preprint, 1984.

14. S.N. Ethier and T.G. Kurtz. The infinitely-many-neutral-alleles diffusion model, Adv. Appl. Prob. 13 (1981), 429-452.

15. S.N. Ethier and T.G. Kurtz. Markov processes: characterization and convergence, Wiley, 1984.

16. W. Feller. Diffusion processes in genetics. Proc. Second Berkeley Symp., Univ. of Calif. Press (1951), 227-246.

17. K.Fleischmann and J. Gärtner. Occupation time processes at a critical point, preprint, 1984.

18. W.H. Fleming and M. Viot. Some measure-valued Markov processes in population genetics theory, Indiana Univ. Math. J. 28 (1979), 817-843.

19. L.G. Gorostiza. High density limit theorems for infinite systems of unscaled branching Brownian motions, Ann. Prob. 11 (1983), 374-392.

20. R.A. Holley and D.W. Stroock. Generalized Ornstein-Uhlenbeck processes and infinite particle branching Brownian motions, Publ. R.I.M.S., Kyoto Univ. 14 (1978), 741-788.

21. J. Horowitz. Measure-valued random processes, preprint, 1984.

22. I. Iscoe. The man-hour process associated with measure-valued branching random motions in R^d, Ph.D. thesis, Carleton University, Ottawa, 1980.

23. I. Iscoe. Personal communication, 1982.

24. I. Iscoe. A weighted occupation time for a class of measure-valued branching processes, Carleton U. - U. Ottawa Res. Lab. Prob. Stat. Tech. Rep. 27 (1983).

25. N.V. Krylov and B.L. Rozovskii. Stochastic evolution equations, J. Soviet Math. (Itogi Nauki i Techniki 14 (1981)), 1233-1277.

26. T.G. Kurtz. Approximation of population processes, S.I.A.M. (1981).

27. B. Mandelbrot. Fractals: form, chance and dimension, Freeman, San Francisco, 1977.

28. G. Matheron. Random sets and integral geometry, John Wiley, New York, 1975.

29. S. Mizuno. On some infinite dimensional martingale problems and related stochastic evolution equations, Ph.D. thesis, Carleton University, Ottawa, 1978.

30. E. Pardoux. Stochastic partial differential equations and filtering of diffusion processes, Stochastics 3 (1979), 127-167.

31. S. Roelly-Coppoleta. Processus de diffusion a valeurs mesures multiplicatifs, Thèse de 3e cycle, Univ. Paris 6, 1984.

32. S. Sawyer and J. Fleischman. The maximum geographical range of a mutant allele considered as a subtype of a Brownian branching random field, Proc. N.A.S., U.S.A. 76 (1978), 872-875.

33. A. Shimizu. Construction of a solution of a certain evolution equation I, Nagoya Math. J. 66(1977) 23-36; II 71(1978) 181-198.

34. D.W. Stroock and S.R.S. Varadhan. Multidimensional diffusion processes, Springer-Verlag, New York, 1979.

35. T. Shiga. Wandering phenomena in infinite allelic diffusion models, Adv. Appl. Prob.14 (1982), 457-483.

36. M. Viot. Methodes de compacité et de monotonie compacité pour les equations aux dérivees partielles stochastiques, Thèse, Univ. de Paris, 1975.

37. D. Yonglong. On absolute continuity and singularity of random measures, Chinese Annals of Math. 3 (1982), 241-248.

38. U. Zähle. Random fractals generated by random cutouts, preprint, Friedrich-Schiller Univ., Jena, 1982.

DUAL PROCESSES IN POPULATION GENETICS

Peter Donnelly
Department of Mathematics
and Computer Science
University College of Swansea
Swansea SA2 8PP UK.

1. Introduction

A major tool in the study of interactive particle systems has been the use of dual processes. In examining the evolution in time of a complicated stochastic process, the basic strategy is to find an associated process (the dual process) which has a simpler structure, and by exploiting the relationship between the processes, to use information about the dual in studying the behaviour of the original process. Within population genetics, dual processes arise naturally as processes related to the genealogy of the population. As well as providing a relatively painless means of analyzing the genetic evolution of a population, these genealogical processes are often of interest in their own right.

In this paper we concentrate on the use of duality in studying the transient distribution of certain stochastic processes describing the genetic composition of a population. Throughout, our aim is to give some feeling for the methods involved and consequently we will stress the methodology at the expense of the details and explicit formulae. Applications of genealogical processes in a number of other contexts may be found in [21],[22],[23] and [7]. We note that there are at least two other approaches to finding transient distributions. The first is to do so directly by solving the appropriate differential equations, a procedure which has been carried out successfully, though not without considerable complication, for some reproductive schemes, notably the Moran model, in [13] and [14]. The more common approach is to approximate the genetic processes by diffusions whose behaviour may be analyzed explicitly. See for example [9].

For the most part we will take as our model of the reproductive mechanism the one originally due to Moran [19]. In this case an exact treatment is possible. It will be noted in the final section that many of the results obtained apply at least approximately to a large class of other models. For a population consisting of M haploid individuals, the Moran model assumes that for each individual, independently of the others and all past events, there is a probability $(\lambda/M)h + o(h)$ that it will die in the interval $(t,t+h)$. When an individual dies it is replaced by the offspring of an individual chosen at random from the members of the population present immediately before the death occurs. In this setting, for reasons which will become clear later, it is natural to choose the parameter λ to be $\frac{1}{2}M^2$. We shall adopt this choice throughout the sequel.

Direct attention to one locus at which we assume there are K neutral alleles. Thus with each individual we may associate one of the labels $A_1, A_2, \ldots, A_K$. To introduce mutation, suppose that when an individual of type A_i reproduces, the offspring will be of type A_j $(j \neq i)$ with probability u_{ij} and of type A_i with probability $1-(u_{i1}+\ldots+u_{i,i-1}+u_{i,i+1}+\ldots+u_{iK})$. In the interests of tractability, we follow other authors (for example [14]) and assume that the mutation rates u_{ij} do not depend on i, so that $u_{ij}=u_j$ $i=1,\ldots,K$ $i\neq j$. These mutation rates are typically very small and we further assume that $u = u_1+u_2+\ldots+u_K < 1$.

The composition of the population at time t may be described by the stochastic process $X(t) = \{X_1(t), X_2(t), \ldots, X_K(t)\}$ where $X_i(t)$ is the number of individuals of type i at time t. Clearly X(t) is Markovian with state space the simplex $\{(x_1,x_2,\ldots,x_K): x_i \in \{0,1,\ldots,M\}$ $i=1,2,\ldots,K$ and $x_1+x_2+\ldots+x_K = M\}$, and its infinitesimal transition probabilities follow immediately from the model.

It is convenient to introduce an alternative though equivalent model to describe the evolution of the population. As before, we assume that individuals die at rate M/2 independently of all other events. When an individual dies

i) with probability 1-u it is replaced by the non mutant offspring of an individual chosen at random from the members of the population present immediately before the death occurs, or

ii) with probability u_i, $i=1,2,\ldots,K$ it is replaced by an individual of type A_i.

This effectively just introduces the device of a spurious mutation to one's own type, and it is straightforward to check for example that the process describing the genetic composition of a population evolving according to this model has the same infinitesimal transition probabilities as the process X(t) described above. In the sequel we shall refer to the two models interchangeably.

2. Two Alleles

When there are only two possible allelic types at the locus in question the analysis is considerably simpler and, at least as far as duality is concerned, qualitatively different from the more general case. In the absence of mutation the two allele Moran model is very similar to a voter model on a complete graph (see for example [5]) and the introduction of mutation (in the equivalent formulation) corresponds to the addition of spontaneous births and deaths. In order to motivate the more general treatment of the next section, we briefly review the use of dual processes in this context. For full details the reader is referred to [4].

Either of the processes $X_1(t)$ or $X_2(t)$ now contain all the useful information about the composition of the population. Instead of finding their distribution directly, we consider certain sampling probabilities. Let $p(n,t)$ be the probability that a random sample of size n taken without replacement from the population at time t consists entirely of individuals of type A_2. An analogous "A_1" sampling probability may also be defined. Sampling probabilities of this kind are informative not only because the data available to the experimental geneticist often arise from studying a sample from the population, but because many quantities of interest, for example, heterozygosity and the distribution of absorption times, are more conveniently expressed in terms of the sampling probabilities than the transition probabilities. Furthermore, if we arbitrarily label the individuals in the population and write $X_2(t) = \chi_1(t) + \chi_2(t) + \ldots + \chi_M(t)$, where

$$\chi_i(t) = \begin{cases} 1 \text{ if individual } i \text{ is of type } A_2 \text{ at time } t \\ 0 \text{ otherwise,} \end{cases}$$

it is clear that the moments of $X_2(t)$ may be written as sums of expectations of products of the χ_i. These expectations though are just the $p(n,t)$. Thus, since the sampling probabilities uniquely determine the moments of $X_2(t)$, they also uniquely determine its distribution (see for example [2]), and in this sense contain as much information as the transition probabilities. The relationship between the sampling probabilities and the transition probabilities is given explicitly in [4].

For a voter model on a complete graph, the appropriate dual process is a system of coalescing random walks. Specifically, let $\hat{\xi}^n(t)$ be the Markov process with state space the set of subsets of $\{1,2,\ldots,M\}$ together with an additional state labelled ω, which describes the following particle process. Choose a subset of n vertices at random from the complete graph on M vertices (that is the graph with an edge between every pair of vertices) and initially place one particle at each chosen vertex. Particles behave independently; at rate $u_2M/2$ a particle is removed from the system, and at rate $(1-u)M/2$ a particle chooses one of the vertices of the graph at random (possibly the one it currently occupies) and jumps to it. Should two particles try to occupy the same vertex they coalesce. The process $\hat{\xi}^n(t)$ describes the locations of the particles present at the time t with the added proviso that if there are ℓ particles present in $\hat{\xi}^n(t)$ then at rate $\ell u_1 M/2$ the process $\hat{\xi}^n(t)$ jumps to the state ω and remains there.

Suppose that initially each individual is assigned one of the labels from the set $\{1,2,\ldots,M\}$, or equivalently associated with one of the vertices of the complete graph and thereafter new individuals adopt the label of the individual they replace, and that exactly those individuals in some set $A\epsilon\{1,2,\ldots,M\}$ are of type A_1 at time 0. The following theorem relates the behaviour of the genetic process to that of its dual process.

Theorem 1. For the two processes defined above

$$p(n,t) = P\left(\hat{\xi}^n(t) \neq \omega \text{ and } \hat{\xi}^n(t) \cap A = \phi\right) \tag{2.1}$$

Proof. This is effectively just a restatement of the standard duality relationship for additive particle systems, see for example Theorem 1.8 of [10]. For the construction of the dual process in this setting see [4].

There is a nice genealogical interpretation of this duality relationship in the case of no mutation. Choose a sample of size n at time s and trace their ancestry. A birth-death event in the forward process corresponds to a potential change in the genealogy of the sample if the event involves one of the ancestors of the individuals in the sample. Occasionally such an event will mean that two or more members of the sample will share a common ancestor: their ancestors "coalesce". It should now be evident that the dual process $\hat{\xi}^n(t)$ simply keeps track of the individuals at time s-t who are ancestors of the individuals in the sample. In order for the individuals in the sample to be of type A_2, it is necessary and sufficient that their ancestors at time 0 are of type A_2. This is exactly the duality relationship (2.1).

In the presence of mutation, for a sample to be of type A_2, it is essential that none of its ancestors arise as spontaneous A_1 births (such events cause $\hat{\xi}^n(t)$ to jump to ω), while ancestors who arise as spontaneous A_2 births may be ignored (the appropriate particle is removed). The remaining ancestors at time 0 must also be of type A_2.

A word about the dual process itself. Since the initial locations for the particles are random and, when jumping, particles move at random, it is evident that conditional on the number of particles present in the dual process, these particles will be located randomly on the complete graph. Thus (2.1) becomes

$$p(n,t) = \sum_{i=0}^{n} P(|\hat{\xi}^n(t)| = i) \binom{M-|A|}{i} \binom{M}{i}^{-1}, \qquad (2.2)$$

where, as usual, $|\cdot|$ denotes cardinality, and if $\hat{\xi}^n(t)=\omega$ we interpret $|\hat{\xi}^n(t)|$ as ω. Now the process $|\hat{\xi}^n(t)|$ is effectively just a death process and it is routine to calculate its distribution exactly (see [4]). Thus (2.2) may be used to find exact expressions for the p(n,t) and, if desired, the distribution of $X_2(t)$.

3. K Alleles

Some questions about multiple allele models, for example the distribution of absorption times, may be answered by appealing to results from the two allele case. If one is interested in a particular allele, A_1 say, it is often possible to agglomerate the other alleles into one type, "not A_1", and proceed as in the previous section. This approach is limited however, and, for example, cannot be used to find the distribution of X(t).

In this general setting, the sampling probabilities of the last section (i.e. the probability that a sample consists entirely of one type) are no longer informative enough, and we are forced to consider the composition of samples in more detail. Let $p(n; n_1, n_2, \ldots, n_K; t)$ be the probability that a sample of size n taken without replacement from the population at time t, consists of n_1 individuals of type A_1, n_2 individuals of type $A_2, \ldots,$ and n_K of type A_K (where $n_1+n_2+\ldots+n_K=n$). Once again the moments of X(t), and hence in principle its distribution, may be written as a function of these sampling probabilities.

Simply keeping track of ancestors is now no longer sufficient. To find the new sampling probabilities one not only needs to know the types of the ancestors of the sample, but also the number of descendants, in the sample, of each such ancestor. In the language of the particle system, it is now necessary to know the joint distribution of the locations and types of the particles in the dual process together with the number of coalescences to have occured in each particle.

We now introduce an appropriate dual process, called a coalescent. Choose a sample from the population at some time, s, say, and consider its genealogy (in the equivalent formulation of the model) with respect to the ancestral population present at time s-t(t<s). Individuals in the sample will be divided into K+1 types of equivalence class. First,

randomly label the individuals in the sample 1,2,...,n. We say that individuals i and j are in the same old equivalence class if they share a common ancestor at time s-t and no intervening mutation has occured between time s-t and time s. On the other hand, say that i and j are in the same new equivalence class of type ℓ if for some $r \leq t$, individuals i and j have the same ancestor at time s-r, this ancestor itself being a mutant (spontaneous birth) of type ℓ, with no intervening mutation between time s-r and time s. Thus at time s, individuals in new equivalence classes will be of the same type as that of the class while those in old equivalence classes will be of the same type as that of the appropriate ancestor at time s-t.

Denote by $D(t)$ the number of old equivalence classes present, and by $F_1(t), F_2(t), \ldots, F_K(t)$, the number of new equivalence classes of types 1,2,...,K respectively, present, after tracing back the genealogy of the sample for t units of time. Clearly

$$D(0) = n, F_1(0) = F_2(0) = \ldots = F_K(0),$$

and

$$D(t) + F_1(t) + \ldots + F_K(t) \leq n.$$

If we now denote the equivalence classes themselves by $\xi_1, \ldots, \xi_{D(t)}$; $\eta_{11}, \ldots, \eta_{1F_1(t)}, \ldots; \eta_{K1}, \ldots, \eta_{KF_K(t)}$, (respectively old, new of type 1, ..., new of type K), then the genealogy of the sample may be described by the K+1 type equivalence relation

$$R_t = (\xi_1, \ldots, \xi_{D(t)}; \eta_{11}, \ldots, \eta_{1F_1(t)}; \ldots; \eta_{K1}, \ldots, \eta_{KF_K(t)}).$$

Of course, in R_o the equivalence classes are all old, and each is a singleton. As t increases, the (stochastic) process $\{R_t, t \geq 0\}$ describes the genetic history of the sample. (In fact our construction only defines R_t for $t \leq s$ but its distributions do not depend on s, and clearly enjoy the Markov property, so are therefore those of a Markov chain $\{R_t, t \geq 0\}$ and we identify the process we have constructed with this latter chain throughout the sequel.)

The process $\{R_t, t \geq 0\}$ is called a coalescent. Coalescents were first introduced by Kingman [16],[17], and [18] as a means of representing the genealogy of a sample. Watterson [22] analyzed a related process in the context of the infinite alleles model (where it is assumed that every mutation gives rise to a completely novel allelic type). The coalescent given here represents an adaptation of Watterson's process to the K allele case.

It should be evident that each old equivalence class corresponds to a particular individual at t time units into the past: the individual whose descendants make up the equivalence class. Old equivalence classes are lost either because two of them coalesce (when a birth-death event occurs in which the ancestor of one of the classes is itself the nonmutant offspring of the ancestor of another class), or because one of them is reclassified as a new equivalence class (corresponding to a birth-death event in which the ancestor of the class is a mutant). The process $\{D(t),\ t\geq 0\}$ is thus a death process on the state space $\{n,n-1,\ldots,1,0\}$ with transition rates $\frac{1}{2}\ell(\ell-1)(1-u)+\frac{1}{2}\ell Mu$ from state ℓ. Note that (for obvious reasons) it is very similar to the process $|\hat{\xi}^n(t)|$ of the last section.

The coalescent R_t changes state exactly when the process $D(t)$ jumps. Furthermore it is straightforward to check that the embedded chain of the coalescent is independent of the death process (see for example [17]). This simple structure greatly facilitates calculation of the distribution of coalescent. Its stochastic behaviour is summarized in the following theorems in which we use the notation

$$x_{(n)} = x(x+1)\ldots.(x+n-1), \qquad x_{(0)} = 1$$

and write

$$\theta = Mu/(1-u), \quad \text{and } \theta_i = Mu_i/(1-u) \qquad i = 1,2,\ldots,K.$$

Theorem 2. The distribution of the death process is given by

$$P\big(D(t) = i\big) = \sum_{j=i}^{n} h_{ij} \exp\big(-j(j+\theta-1)Mt/2(M+\theta)\big)$$

where $h_{ij} = (-1)^{j-i}(2j-1+\theta)\dfrac{n!\,(1+\theta)_{(j-1)}}{i!\,(j-i)!\,(n-j)!\,(n+\theta)_{(j)}}$,

unless $i=j=0$, in which case $h_{00}= 1$.

Proof. The result follows on applying standard techniques to the appropriate forward equations. See [4] for details.

Theorem 3. The distribution of the jump chain of the coalescent is given by

$$P\left[R_t = (\xi_1,\dots,\xi_\ell;\ \eta_{11},\dots,\eta_{1f_1};\dots;\ \eta_{K1},\dots,\eta_{Kf_K})\,|\,D(t)=\ell\right]$$

$$= \frac{(n-\ell)!\,\ell!}{n!\,(\ell+\theta)_{(n-\ell)}} \prod_{i=1}^{\ell} \lambda_i! \prod_{i=1}^{K} \prod_{j=1}^{f_i} \theta_i (\mu_{ij}-1)!, \qquad (3.1)$$

where we have written $\lambda_i = |\xi_i| \quad i=1,2,\dots,\ell$

and $\mu_{ij}=|\eta_{ij}| \quad i=1,2,\dots,K, \quad j=1,2,\dots,f_i$

<u>Proof</u>. Watterson ([22], Theorem 1) calculates the distribution of a related coalescent in the infinite alleles case. In our setting, there is a probability $u_i/u = \theta_i/\theta$ that a new equivalence class is in fact of type i, and (3.1) follows.

The way in which the individuals in the sample are labelled is often immaterial. What is of interest is the collection of class sizes, which we call the <u>description</u> of the sample. To find the distribution of sample descriptions, it is necessary to sum (3.1) over the various ways in which a particular description may arise. Denoting by α_j the number of old classes of size j and by β_{ij} the number of new classes of type i which are of size j, we must multiply (3.1) by $n!\left(\prod_{i=1}^{\ell} \lambda_i! \prod_{i=1}^{n} \alpha_i! \prod_{i=1}^{K} \left(\prod_{j=1}^{f_i} \mu_{ij}! \prod_{j=1}^{n} \beta_{ij}!\right)\right)^{-1}$.

Thus, in an obvious notation,

$$P(\lambda_1,\dots,\lambda_\ell;\ \mu_{11},\dots,\mu_{1f_1};\dots;\mu_{K1},\dots,\mu_{Kf_K}\,|\,D(t)=\ell)$$

$$= \frac{(n-\ell)!\,\ell!}{(\ell+\theta)_{(n-\ell)}} \; \frac{\theta_1^{f_1}\cdots\theta_K^{f_K}}{\alpha_1!\cdots\alpha_n! \prod_{i=1}^{K}\left(\prod_{j=1}^{f_i}\mu_{ij}! \prod_{j=1}^{n}\beta_{ij}!\right)} \qquad (3.2)$$

This gives the transient distribution of class sizes in a K allele Moran model.

Letting $t\to\infty$ corresponds to sampling from the model at stationarity. In this case there are no old equivalence classes (almost surely) and (3.2) becomes

$$P(\mu_{11},\dots\mu_{1f_1};\dots;\ \mu_{K1},\dots,\ \mu_{Kf_K}\,|\,D(t)=0)$$

$$= \frac{n!}{\theta_{(n)}} \; \frac{\theta_1^{f_1}\cdots\theta_K^{f_K}}{\prod_{i=1}^{K}\left(\prod_{j=1}^{f_i}\mu_{ij} \prod_{j=1}^{n}\beta_{ij}!\right)} \qquad (3.3)$$

The distribution (3.3) is the analogue of the celebrated Ewens sampling formula (Ewens [8]) in this K allele setting.

Watterson ([22] equation 3.4.1) gives the marginal distribution of old class sizes for his coalescent. Since our coalescent effectively just reclassifies (multinomially) his new equivalence classes, it follows that in our case also, writing $z=\lambda_1+\ldots+\lambda_\ell$ for the total number of old genes in the sample,

$$P(\alpha_1,\ldots,\alpha_n \mid D(t)=\ell) = \frac{\ell!}{\alpha_1!\ldots\alpha_n!} \; \frac{(n-\ell)!\,\theta_{(n-z)}}{(n-z)!\,(\ell+\theta)_{(n-\ell)}} \,. \tag{3.4}$$

Thus

$$P\left(\mu_{11},\ldots,\mu_{1f_1};\ \ldots;\ \mu_{K1},\ldots,\ \mu_{Kf_K} \mid D(t)=\ell,\ \alpha_1,\ \ldots,\alpha_n\right)$$

$$= \frac{(n-z)!}{\theta_{(n-z)}} \; \frac{\theta_1^{f_i}\ \ldots\ \theta_K^{f_K}}{\prod_{i=1}^{K}\left\{\prod_{j=1}^{f_i}\mu_{ij}\ \prod_{j=1}^{n}\beta_{ij}!\right\}} \tag{3.5}$$

Comparison of (3.5) with (3.3) shows that the new class sizes, conditional on there being n-z new genes and irrespective of the old class sizes, behave as a stationary sample of size n-z. This result is known for the diffusion approximation to the K allele Wright-Fisher model with a certain mutation structure (Griffiths [12]), and for the infinite alleles Moran model (Watterson [22]). A caveat should be added here however. The introduction of spurious mutations to one's own type means that classes which we classify as new may in fact be descendants, without a change of type, from some founder gene and hence not really "new" at all.

Once the distribution of class sizes is known (and in particular on its factorization as the product of (3.4) and (3.5)), calculation of the sampling probabilities simply involves the combinatorial grouping of classes. Sadly the elegance of the approach ends here. While perfectly straightforward, the ensuing combinatorics is far from pretty; the summations involved appear resistant to simplification and closed expressions for the sampling probabilities seem difficult to find. This should not be too surprising. The exact formulae for the transition probabilities given in Karlin and McGregor [14] together with those for the diffusion approximation in, for example, Griffiths [11], take an extremely complicated form, as do expressions for sampling probabilities in the diffusion approximation (obtained only in the case of equal mutation rates) in Perlow [20].

Rather than reproduce expressions for particular transient sampling probabilities, we content ourselves with a special case.

Theorem 4. At equilibrium the sampling probabilities are given by

$$p(n;\ n_1,n_2,\ldots,\ n_K;\ \infty) = \frac{n!}{\theta_{(n)}} \prod_{i=1}^{K} \left(\theta_{i\,(n_i)} / n_i! \right)$$

Proof. To find the sampling probability in question, we must sum (3.3) over all possible configurations of class sizes, and numbers of classes, such that $\mu_{i1}+\ldots+\mu_{if_i} = n_i$. First fix the values $f_1,\ldots f_K$.

$$\sum_{\mu_{11}+\ldots+\mu_{1f_1}=n_1} \cdots\cdots \sum_{\mu_{K1}+\ldots+\mu_{Kf_K}=n_K} \frac{1}{\mu_{11}\cdots\mu_{1f_1}\beta_{11}!\cdots\beta_{1n}!} \cdots\cdots$$

$$\frac{1}{\mu_{K1}\cdots\mu_{Kf_K}\ \beta_{K1}!\cdots\beta_{Kn}!} = \frac{\left|S_{n_1}^{(f_1)}\right|}{n_1!} \frac{\left|S_{n_2}^{(f2)}\right|}{n_2!} \cdots\cdots \frac{\left|S_{n_K}^{(f_K)}\right|}{n_K!} \tag{3.6}$$

where $S_n^{(m)}$ represents the Stirling number of the first kind (see [1] page 823). Furthermore, ([1] page 824 for example), for $n_i \neq 0$

$$\sum_{i=1}^{n_i} \theta_j^i \left|S_{n_i}^{(i)}\right| = \theta_{j\,(n_i)} \tag{3.7}$$

The theorem follows on combining (3.6) and (3.7).

This result is not new, see for example Kelly [15].

4. Robustness

An invariance result underlies much of this work. Kingman [18] shows that his coalescent arises as the appropriate process for describing the geneaology of a sample of fixed size, in the limit as the population size tends to ∞, for any one of a large class of reproductive models; most of the so-called exchangeable models of Cannings [3], including the Wright-Fisher model. The same result remains true for coalescents with mutation when $M \to \infty$ with θ fixed [7], and in particular applies here. The limiting process is almost exactly that of section 3, the only difference is that the death process has death rate $\frac{1}{2}\ell(\ell+\theta-1)$ from ℓ to $\ell-1$. (The choice $\lambda = \frac{1}{2}M^2$ for the Moran model means that $D(t)$ converges to this latter process as $M \to \infty$.) Thus, for example, for samples whose size is small compared to the population size, the sampling probabilities for any of these models will be approximately the same as those for the Moran model.

Since the distribution of X(t) may be written in terms of sampling probabilities (via its moments), the invariance result for dual processes may also be used to derive the diffusion approximations used in this context, (see [5] for the two allele case). It is not surprising then that the results on lines of descent derived by Griffiths [12] for these diffusion approximations closely parallel those derived here, and elsewhere, using dual processes. Griffiths [11] also studies sampling probabilities in diffusion approximations but he finds these using the transition density of the diffusion, in contrast to our direct approach.

References

[1] Abramowitz, M. and Stegun, I.A. *Handbook of Mathematical Functions.* Dover, New York (1972)

[2] Billingsley, P. *Probability and Measure.* Wiley, New York (1979)

[3] Cannings, C. The latent roots of certain Markov chains arising in genetics: a new approach 1. Haploid models. *Adv. Appl. Prob.* 6, 260-290 (1974)

[4] Donnelly, P. The transient behaviour of the Moran model in population genetics. *Math. Proc. Camb. Phil. Soc.* 95, 349-358 (1984)

[5] Donnelly, P. Dual processes and an invariance result for exchangeable models in population genetics. To appear.

[6] Donnelly, P. and Welsh, D. Finite particle systems and infection models. *Math. Proc. Camb. Phil. Soc.* 94, 167-182 (1983)

[7] Donnelly, P. and Tavaré, S. The ages of alleles and a coalescent. To appear.

[8] Ewens, W.J. The sampling theory of selectivity neutral alleles. *Theor. Pop. Biol.* 3, 87-112 (1972)

[9] Ewens, W.J. *Mathematical Population Genetics.* Springer, Berlin (1979)

[10] Griffeath, D. *Additive and Cancellative Interacting Particle Systems.* Springer, Lecture Notes in Mathematics, vol. 724, (1979)

[11] Griffiths, R.C. Exact sampling distributions from the infinite neutral alleles model. *Adv. Appl. Prob.* 11, 326-354 (1979)

[12] Griffiths, R.C. Lines of descent in the diffusion approximation of neutral Wright-Fisher models. *Theor. Pop. Biol.* 17, 37-50 (1980)

[13] Karlin, S. and McGregor, J. On a genetics model of Moran. *Proc. Camb. Phil. Soc.* 58, 299-311 (1962)

[14] Karlin, S. and McGregor, J. Linear growth models with many types and multidimensional Hahn polynomials. In *Theory and Applications of Special Functions.* Academic Press, New York, (1975)

[15] Kelly, F.P. *Reversibility and Stochastic Networks.* Wiley, New York (1979).

[16] Kingman, J.F.C. On the genealogy of large populations. *J. Appl. Prob.* 19A, 27-43 (1982).
[17] Kingman, J.F.C. The coalescent. *Stoch. Proc. Appl.* 13, 235-248 (1982).
[18] Kingman, J.F.C. Exchangeability and the evolution of large populations. In *Exchangeability in Probability and Statistics*. G.Koch, F. Spizzichino (Eds.). North Holland, Amsterdam, 97-112 (1982).
[19] Moran, P.A.P. Random processes in genetics. *Proc. Camb. Phil. Soc.* 54, 60-71 (1958).
[20] Perlow, J. The transition density for multiple neutral alleles. *Theor. Pop. Biol.* 16, 223-232 (1979).
[21] Tavaré, S. Lines of descent and genealogical processes and their applications in population genetics models. *Theor. Pop. Biol.* To appear.
[22] Watterson, G.A. Lines of descent and the coalescent. *Theor. Pop. Biol.* 26, 77-92 (1984).
[23] Watterson, G.A. The genetic divergence of two species. To appear.

SOME PECULIAR PROPERTIES OF A PARTICLE SYSTEM WITH SEXUAL REPRODUCTION

Richard Durrett
U.C.L.A.

The purpose of this paper is to describe a two dimensional growth model with sexual reproduction (i.e. two particles are needed to produce a new one) and contrast its properties with those of a similar model with asexual reproduction (i.e. an additive process in the sense of Harris (1978) and Griffeath (1979)). The results reported here were obtained in collaboration with Larry Gray. Detailed proofs are given in Durrett and Gray (1985), a source which will be referred to as DG(1985) below.

In both models the state of the system at time t is ξ_t a subset of Z^2, and particles die at rate one, that is, if $x \in \xi_t$ then $P(x \notin \xi_{t+\delta} | \mathfrak{F}_t) = \delta + O(\delta)$ as $\delta \to 0$, where $\mathfrak{F}_t = \sigma(\xi_\delta : s \leq t)$ = the σ-field generated by the process up to time t. The two models then are distinguished by their birth rates which can be described as follows:

Example 1. (asexual reproduction).
If $x \notin \xi_t$ and $x + (1,0)$ OR $x + (0,1) \in \xi_t$ then

$$P(x \in \xi_{t+\delta} | \mathfrak{F}_t) = \lambda\delta + O(\delta).$$

Example 2. (sexual reproduction)
If $x \notin \xi_t$ and $x + (1,0)$ AND $x + (0,1) \in \xi_t$ then

$$P(x \in \xi_{t+\delta} | \mathfrak{F}_t) = \lambda\delta + O(\delta).$$

The reasons for the names and the difference between the models can be seen in the two capitalized words. The rest of this paper is devoted to explaining how this simple change results in drastic differences in the behavior of the two processes. We will look at four aspects of their behavior below.

I. If we consider what happens when we start from $\xi_0^A = A$, A a finite set, then immediately we see differences between the two models.

Ex. 1. If $\lambda \geq 4$ then $P(\xi_t^A \neq \emptyset$ for all $t) > 0$.

Proof. If we restrict the process to $\{(x,0): x \in Z\}$ then it is a one-sided contact process so the conclusion follows from a result of Holley and Liggett (1978).

Ex. 2. If A is finite then $P(\xi_t^A \neq \emptyset$ for all $t) = 0$ for all λ.

Proof. If $[-L,L]^2$ contains A then no births can ever occur outside $[-L,L]^2$, so a simple argument shows there is an $\varepsilon_L > 0$ so that

$$P(\xi_n^A \neq \emptyset) \leq (1-\varepsilon_L)^n.$$

To sum things up if we let

$$\lambda_f = \inf\{y: P(\xi_t^A \neq \emptyset \text{ for all } t) > 0 \text{ for some finite set } A\}$$

then we have

Ex. 1 $\lambda_f \leq 4$

Ex. 2 $\lambda_f = \infty$.

Remark. Projecting onto $\{(x,-x): x \in Z\}$ and comparing with the two sided contact process shows $\lambda_f \leq 2$ in Ex. 1. It is trivial that $\lambda_f \geq \frac{1}{2}$ and as usual practically impossible to figure out exactly what λ_f is.

II. Having started from a finite set the next thing we want to contemplate is what happens when we start from $\xi_0^1 = Z^2$. In both cases it follows from general results about attractive spin systems (see Liggett (1985), Chapter 3) that we have:

(a) As $t \to \infty$ $\xi_t^1 \Rightarrow \xi_\infty^1$ a stationary distribution.

(b) If $\xi_\infty^1 = \delta_\emptyset$ (the point mass on the empty set) then there are not other stationary distributions.

From (a) and (b) it follows that if we let $\lambda_c = \inf\{\lambda: \xi_\infty^1 \neq \delta_\emptyset\}$ then the stationary distribution is $\delta_\emptyset$ for $\lambda<\lambda_c$ and is not unique for $\lambda>\lambda_c$. Comparing with the contact process again shows that in Ex. 1 $\lambda_c\le 4$. The corresponding result for the other example is Theorem 1 in DG(1985) Ex. 2 $\lambda_c\le 110$. The last result gives a ridiculous upper bound for λ_c (try to find a better one!). Simulations done by Tom Liggett suggest $\lambda_c\ge 12$ and a look at the over estimates in the proof in DG(1985) suggests $\lambda_c\le 20$ but beyond this we have no idea what λ_c is.

Comparing the last result with the one in paragraph I. shows that in Ex. 2

$$\lambda_c\le 110 \qquad \lambda_f=\infty$$

and hence $\lambda_c\ne\lambda_f$. (This result is somewhat surprising in view of the fact that $\lambda_c=\lambda_f$ in one dimensional attractive nearest neighbor (see Gray (1985)) and reversible nearest particle systems (see Liggett (1985), Chapter 7) and it was conjectured that $\lambda_c=\lambda_f$ for the contact process in any dimension (see Durrett and Griffeath (1982)).

<u>Note</u>: For the reader who thinks we have cheated by putting $\lambda_c<\infty$, $\lambda_f=\infty$ we would like to observe that if we let Ex. 3 = $(1-\varepsilon)$Ex.2 + εEx.1 and ε is small then $\lambda_c<\lambda_f<\infty$. This is Theorem 3 in DG(1985).

<u>Problem</u>. Once you see the proof of Theorem 1 it is easy to construct lots of examples with $\lambda_c<\infty,\lambda_f=\infty$. Since it is believed that for many models $\lambda_c=\lambda_f$ this brings up the problem of finding sufficient conditions for this to occur. It seems likely that this is true in Ex. 1 (although we do not know how to prove this) and we conjecture that this holds for all additive growth models but the latter question even in one dimension seems a very difficult problem.

III. Having started last time from $\xi_0^1=Z^2$ our next step is consider what happens starting from other simple initial distributions: ξ_0^p = product measure with density p, i.e. the events $\{x\in\xi_0^p\}$ are independent and each has probability p. With these initial distributions, we have

Ex. 1 $\quad \xi_t^p \Rightarrow \xi_\infty^1$ for any $p>0$.

Ex. 2 If $p<p^*\equiv 1-$ (the critical probability for two dimensional oriented percolation) then

$$\xi_t^p \not\Rightarrow \delta_\emptyset.$$

Proof. Suppose there is an infinite sequence $x_n \in (\xi_0)^c$ so that for each $n \geq 0$, $x_{n+1} \in \{x_n + (1,0),\ x_n + (0,1)\}$ then $\{x_n : n \geq 0\} \subset (\xi_t)^c$ for all t and an easy argument shows $\xi_t \Rightarrow \delta_\emptyset$.

Looking back at the last proof we see that the only property of the product measures we needed was the existence of the x_n's. This suggests that we cannot have an equilibrium distribution in which the density is too low, so we

Conjecture: $\rho(\lambda) \equiv P(0 \in \xi^1_\infty)$ is discontinuous at λ_c. In support of the conjecture we would like to observe that if Z^2 is replaced by the binary tree a simple argument shows that $\lambda\rho(\lambda)(1-\rho(\lambda)) \geq 1$ (see DG(1985) Section 1 for details), so $\rho(\lambda)$ cannot $\to 0$ as $\lambda \downarrow \lambda_c$. The reader should note that "general nonsense" (see Griffeath (1981)) implies $\rho(\lambda)$ is right continuous and hence $\rho(\lambda_c) > 0$.

IV. Last but not least we want to consider what happens when we add spontaneous births at rate β, i.e. the new birth rates = old rates + β. Again there is a drastic difference in the two results.

Ex. 1. There is a unique stationary distribution and convergence to equilibrium occurs exponentially fast.

Ex. 2. If λ is such that $\xi^\infty_1 \neq \delta_\emptyset$ when $\beta = 0$, and β is chosen so that $6\beta^{1/4}\lambda^{3/4} < 1$ then the process with parameters β and λ has two translation invariant stationary distributions.

In view of the results in III i.e.

Ex. 1 $\quad \xi^p_t \Rightarrow \xi^1_\infty$ for any $p > 0$

Ex. 2 $\quad \xi^p_t \Rightarrow \delta_\emptyset$ for $p < p^* \approx .345$

these conclusions should not be surprising. In Ex. 1 $\delta_\emptyset$ is an unstable equilibrium because points artibtarily close to it converge to another fixed point. In Ex. 2 it is an attracting fixed point and hence a stable equilibrium.

It would be nice to prove results which make the last two sentences precise, but failing this we are satisfied with the result given above because outside of the Ising model and some examples due to Toom, there are very few (if any) examples in which all the flip rates are positive and there are two stationary distributions, and furthermore the new example has two new properties the old ones don't. (i) There is an open set of nonergodic examples in the set of nearest neighbor translation invariant growth models and (ii) as $\lambda\downarrow\lambda_c$ the lower bound on the allowed β's approaches $1/6^4\lambda_c^3>0$ a phenomenon which suggests that $\xi_1^\infty\neq\delta_\emptyset$ at λ_c.

The statements of the results above are already quite lengthy so there is no time or space to do anything but to make some simple general remarks about the proofs.

A. The proofs for Ex. 1 are all based on the fact that it is an additive process and has a set valued dual process $\tilde{\xi}_t \subset Z^2$. For Ex. 2 no such dual exists but we can define a new type of dual process (which can be defined for any attractive process and reduces to the usual notion for additive processes) where the state at time t χ_t is a collection of subsets of Z^2 with the interpretation that $\{0\}$ is in ξ_t if and only if some $A \in \chi_t$ is completely occupied in ξ_0.

B. Theorems 1, 2, and 3 all concern the behavior when some parametric (i.e. $1/\lambda$, β, or ε respectively) is small so it should not come as a surprise that the results are proved by "perturbation arguments." We write an expansion for the quantity of interest in terms of the small parameter and prove that the series has a positive radius of convergence.

Proofs of the type referred to above are "contour arguments." A good example of an argument of this type is the proof given in Gray and Griffeath (1982) and in fact with hindsight our proofs of Theorems 1, 2, and 3 are straightforward generalizations of the argument obtained by rewriting the easy case of Gray and Griffeath's proof in terms of the dual process. Further details are left to the reader or see DG(1985).

References

Durrett, R. (1984). Oriented percolation in two dimensions. Ann. Probab.12,

Durrett, R. and Gray, L. (1985). Some peculiar properties of a particle system with sexual reproduction.

Durrett, R. and Griffeath, D. (1982). Contact processes in several dimension. Z. Wahrsch. Verw. Gebiete, 59, 535-552.

Gray, L. (1985). Duality for general attractive spin systems with applications.

Griffeath, D. (1979). Additive and Cancellative Interacting Particle Systems.

Griffeath, D. (1981). The basic contact process. Stoch. Proc. Appl. 11, 151-186.

Harris, T. (1978). Additive set-valued Markov processes and graphical methods. Ann. Probab.6, 355-378.

Holley, R. and Liggett, T. (1978). The survival of contact processes. Ann. Probab. 6, 198-206.

Liggett, T. (1985). Interacting Particle Systems. Springer-Verlag, New York.

COMPUTER SIMULATION OF DEVELOPMENTAL PROCESSES IN BIOLOGY: MODELS FOR THE DEVELOPING LIMB

D. A. Ede
Department of Zoology,
University of Glasgow,
Scotland, U.K.

Of all biological events in which stochastic models are potentially applicable, the processes whereby a fertilized egg cell develops into an embryo by a process of cell proliferation and cell differentiation are among the most challenging. Among these processes, those termed morphogenetic, in which the specific forms of the organism and its component structures are produced, are particularly interesting. The ultimate control of these changes must be genetic, and problems at this level will concern the molecular events determining which genes will be active and which inactive at different times and in different regions of the embryo. These events will result in the transcription of programmatic rules controlling the activity of each of the cells of which the developing embryo is made up, each cell following the rules appropriate to its location within the whole embryo, its past history, and its relation to its neighbours. In this respect, the developing embryo may be likened to a population of individual organisms, as for example ants in an ant colony, whose activities are so integrated as to form a unified whole. The cells in the once case, and the individuals in the other, are presented perpetually with behavioural alternatives, the choice being governed by rules which have greater or lesser probabilities of imposing themselves according to the particular situation. In this way, while organisms develop according to the manner of their species, even where there is genetic identity, as in monozygotic twins, development unfolds with some variation and absolute identity of structure is never attained. Stochastic models relating to population dynamics may therefore find application in analyzing developmental processes. Much less attention has been given to construction of models specifically designed to elucidate morphogenetic processes at the level of cellular behaviour. This is partly due to the difficulty of obtaining information about the activity of cells which, for the most part, are embedded in the interior of an embryo. Here I shall describe an attempt to approach these problems by combining the following approaches:

1. Use of mutants affecting embryonic development, in which either the

genetic rules or the cells' responses to these rules are altered.

2. Analysis of normal and mutant embryonic cell behaviour _in vitro_, where the activity of individual cells and interaction with their neighbours may be directly observed.
3. Construction of a single model - essentially descriptive of cell social behaviour in space with cells spread on a lattice - to explain how the overall developmental abnormalities may be explained by disturbances in this behaviour in the mutant embryonic cells.

Development of the embryonic limb

Early in embryonic development, a small hillock of cells arises from the flank to form the limb bud which grows out to give an elongated stem region, then expands and flattens distally to give a paddle shape. The mesenchyme cells of the bud undergo a condensation process whereby they produce a pattern of more densely packed cells, which differentiate to form the cartilage rudiments which prefigure the limb skeleton (Fig. 1). The cartilage pattern of the wing and leg is the pentadactyl limb pattern characteristic of land living vertebrates, with some reduction of digits (Fig. 2A,C). In the _talpid_^{3} mutant of the fowl the outgrowth of the limb bud is modified, being fan-shaped rather than paddle-like, and in the wings and legs there are fusions of cartilage resulting from a disurbance of the condensation process, and polydactyly (Fig.2B,D).

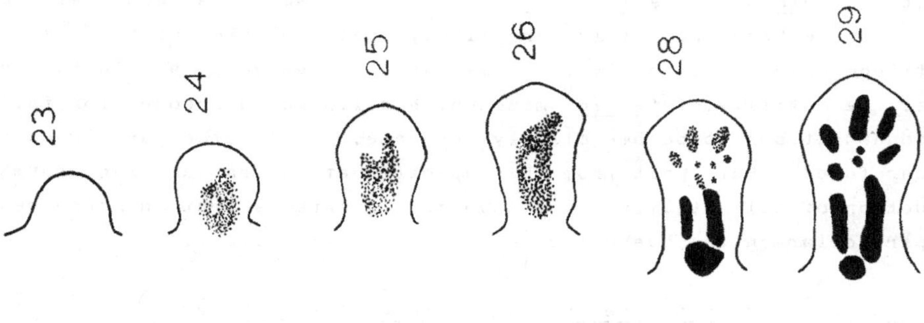

Fig. 1. Development of the wing bud in the fowl embryo, showing condensations of mesenchyme cells (stippled) which are superseded by the cartilage rudiments of the limb skeleton (black).

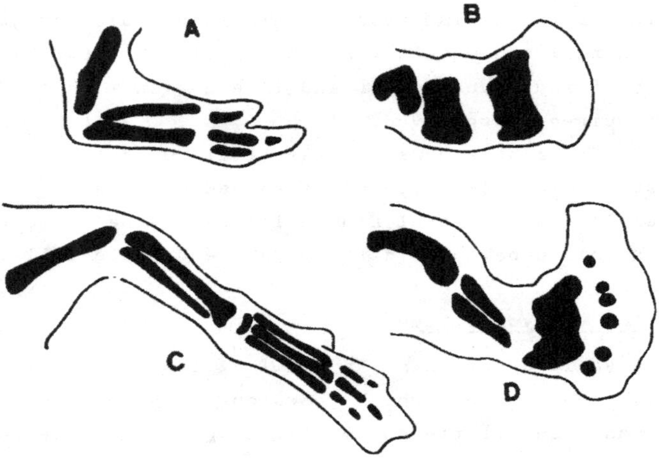

Fig. 2. Cartilage patterns in the wing (A) and leg (B) in the fowl embryo, and the corresponding patterns in a *talpid*[3] mutant embryo.

Cellular activity in limb development

The failure to achieve proper condensation in *talpid*[3] embryos led to a series of investigations of the behaviour of limb mesenchyme cells in *in vitro* culture (reviewed Ede 1976, 1982). Measures of cell adhesion showed that mutant cells were more adhesive to each other than normal cells. This suggested that their motility might be reduced, and time-lapse ciné studies confirmed that this was the case. This reduced motility of *talpid*[3] cells *in vitro* does not necessarily affect their activity in the embryo, but it is very likely that formation of the pre-cartilage condensations involves small-scale movements, and in this case defective motility of *talpid*[3] mesenchyme cells would account for failure of condensations to become clearly separated and for the cartilages to appear fused. But apart from this specific effect on cartilage formation, reduction of all motility might have more general effect on morphogenesis, in particular on the overall shape of the growing limbs.

Computer simulation of the changing form of the limb bud

In order to test this possibility Ede and Law (1969) devised a very simple model in which cells are represented by numbers stored in the computer and their genetic instructions are represented by parts of the computer program. A two-dimensional array of numbers is allocated to represent the available cell sites, 1 indicating that a particular space is occupied by a cell, 0 an empty site. At each iteration the array is scanned, some of the cells are made to reproduce and each daughter cell is made to occupy a vacant site in the array. Rules ('genetic instructions') incorporated in the program determine whether a cell is to divide or not, and what happens to its daughter cell if it should do so, according to its position in the total cell mass. When the array is read for print-out a convenient symbol, an asterisk, is used to represent each cell. A typical print-out displays the patterns formed by the cell mass at each iteration in sequence so that the developing form is shown graphically.

The main principles on which our program was designed are as follows:

1. Cell number is determined by the array size, 80 x 80. The model starts from a base-line of 40 cells, representing flank cells, along one side of the array.
2. Cell proliferaton is obtained by having each *n*th cell produce a daughter cell. In reality the daughter cells would displace other cells, but in our model they are positioned in the nearest available vacant site in the array; since all cells are uniform in this model, this has an equivalent effect on the cell mass.
3. Cell position (strictly, the local conditions consequent on the position of the cell in the cell mass) is important in determining a cell's activity, and in this study it was assumed that the position of the cell in a gradient of some kind, for example, of a diffusible substance, extending from the base to the apex of the limb, is important in relation to its rate of proliferation. This is incorporated in the program in a simple way by instructing that cells within a specified number of rows from the distal tip of the cell mass shall reproduce while more proximal cells shall not.
4. Cell movement is provided for in one direction by making each daughter cell move distally in the array by a specified number of places before being positioned in the nearest vacant site.

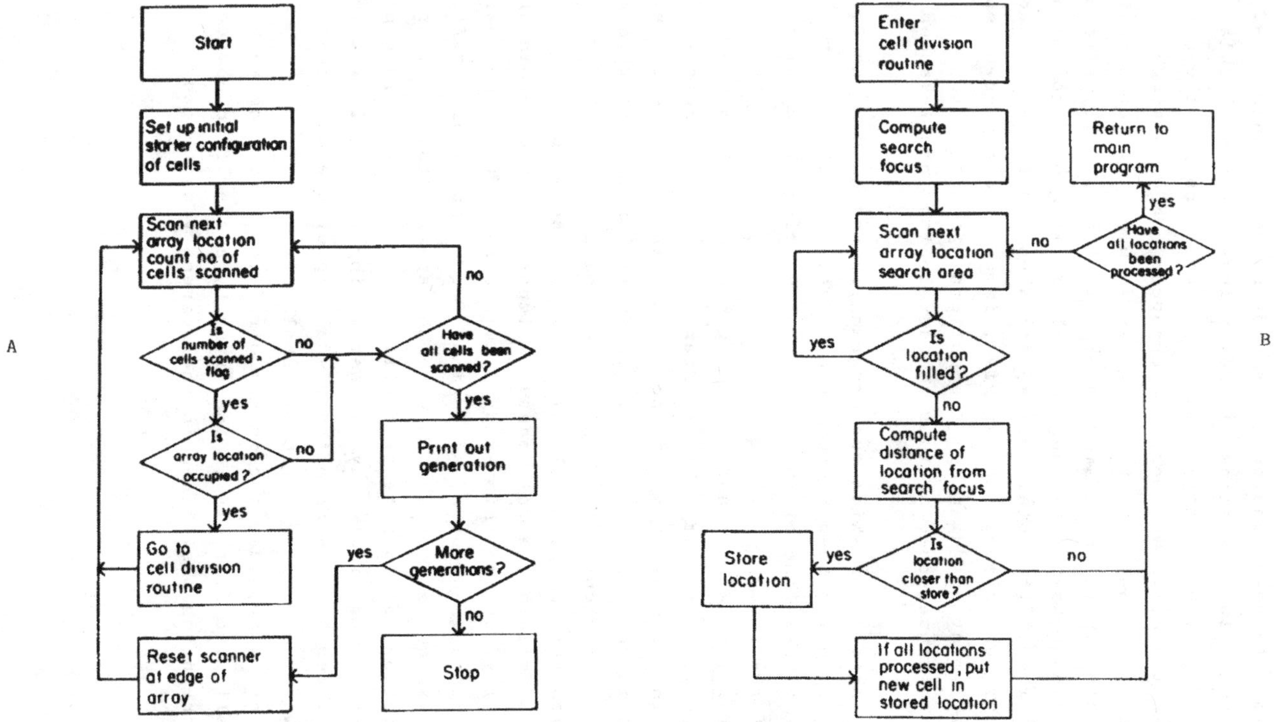

Fig. 3. A. Flow chart for main programme in Ede and Law model.
B. Flow chart for cell division routine.

The program is extremely simple and of course there are many restraints imposed on the model which are not biologically meaningful; nevertheless it is sufficiently comprehensive to test some simple models of limb morphogenesis and in particular to test the hypothesis that cell movement plays an important part in producing the form of the normal limb bud and that reduction of this movement would account for the distortion of this form in the talpid^3 mutant (Fig. 4).

Starting from our line of 'flank cells', simulation of simple cell proliferation such as is believed to occur in the earliest stages produces a hillock of cells which does resemble the form of the early limb bud. One of the chief problems of limb bud development is to explain how elongation of the stem region is produced. There is a gradient of cell division at this stage, with basal cells proliferating more than apical ones. Incorporating this in our model produces a broadening of the apex of the bud but no significant elongation. The elongation can however be obtained by incorporating a slight cell movement in a distal direction. In combination with the gradient of cell division, and reducing the amount of movement slightly at the apex, this produces a shape which rather closely resembles the elongate stem and expanded distal paddle of a late limb bud. The model does in fact suggest that a crucial part is played by cell movements; these are not supposed to be of a large order but are envisaged rather as a distalward drift of mesenchyme cells, jostling forwards on each other in the general direction of the apex. Using the same program, but omitting the distalward movement, a close approximation to the form of the talpid^3 limb bud is produced. Since we have shown that talpid^3 limb mesenchyme cells are in fact much less motile than normal cells we may accept this as supporting our hypothesis that not only the defect in mesenchymal precartilage condensation but also the distortion of the total shape of the talpid^3 limb arises from this abnormality in cell behaviour.

As a model there are of course many serious defects in this program, and in particular the means of achieving cell proliferation. This entails placing daughter cells at a distance from the dividing cell, which, while this is unimportant where changes of cell state are not considered, becomes fatal when cell differentiation is included, since here it is essential for cells of a particular lineage to stay together. This requires that when a cell on the lattice divides, neighbouring cells should be displaced to make room for the additional cell produced, and a program which does this was devised by Wilby and Ede (1976).

In this, assuming that control of cell division is effected by a localised region in the limb, acting to regulate the number of cells entering the division cycle in any part of the limb, the following

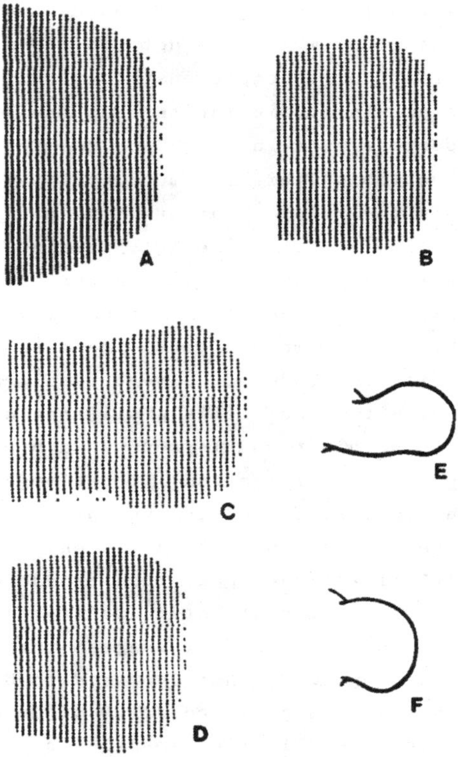

Fig. 4. A–D. Computer simulated models of limb growth: A, with simple cell proliferation; B, with a gradient of cell proliferation, limiting reproduction to the more distal cells; C, with a gradient of cell proliferation as in B, but daughter cells move distalwards by a small amount which is reduced after initial growth; D, as in model C but with reduced distalward movement. E–F. Outlines of (E) normal, (F) <u>talpid</u>[3] mutant limb buds.

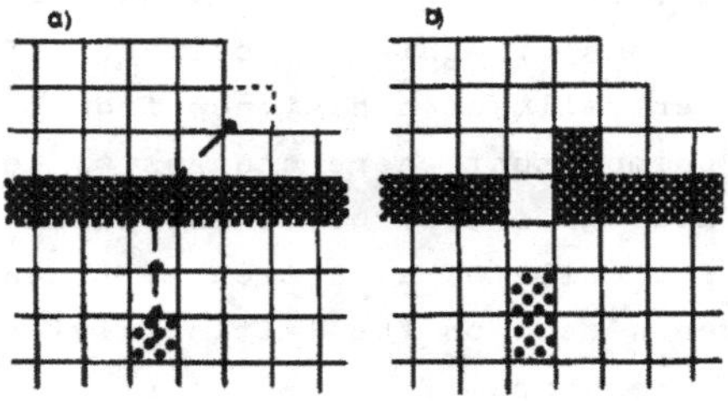

Fig. 5. Cell displacement in the Wilby and Ede (1976) program for cell division.

algorithmic process is used to determine which cells should divide. The limb is scanned sequentially cell by cell, anterior to posterior, proximal to distal; for each cell the distance from the control region is found and an appropriate 'control level' is calculated as a function of this distance. A pseudo-random number generator is then used to decide if the cell's 'internal level' is greater than that of the control, in which case it divides, or less than the control, in which case the program skips to the next cell. Consider, for example, a cell 50 units (cell diameters) from the control region and a control pattern that specifies a linear decrease in mitotic index from 0.1 at the control region in steps of 0.001 per cell diameter. The appropriate control level for this cell will be 0.95, giving a mitotic index of 0.05 since 5% of all random numbers generated in the range 0.0 to 1.0 will equal or exceed 0.95. In this way cells are picked at random but in the correct proportion for each region.

The next step is to simulate the effect of each division before proceeding to the next cell. Firstly the program finds the nearest free space (NFS), that is the nearest empty array element, to the dividing cell by scanning the whole array. Then it calculates the shortest path by which rows and files of cells may be displaced, starting with one of the eight nearest neighbours of the dividing cell, such that a cell adjacent to the NFS comes to occupy that space. For simplicity the present program uses only eight possible displacement paths, two along rows, two along files, four along diagonals, since any NFS can be reached by the appropriate combination of two of these paths intersecting at an intermediate 'bend point' cell as indicated in Figure 4, giving a fair approximation to uniform growth.

If in subsequent steps of a given scan, a daughter cell of a previous division in that scan is found, the program proceeds directly to the next cell. At the end of a scan, all cells are again competent to divide in the next scan. This sequential scanning may introduce slight artifacts into the growth pattern but we believe they are insignificant compared to the increase in time and complexity needed to select cells completely at random.

This program has been used to stimulate the effect of various gradients of cell division within the developing limb bud (Wilby and Ede, 1976). The results achieved (Fig. 6A) show that in order to achieve a limb-like elongation of the model a steep gradient is required, which becomes limited to the tip as the limb grows, and this is not realistic. Studies by Ede, Flint and Teague (1975) showed that in both normal and *talpid*^{3} embryos cell division occurs throughout the developing bud, with a slight peak just behind the apex. Simulating this using realistic

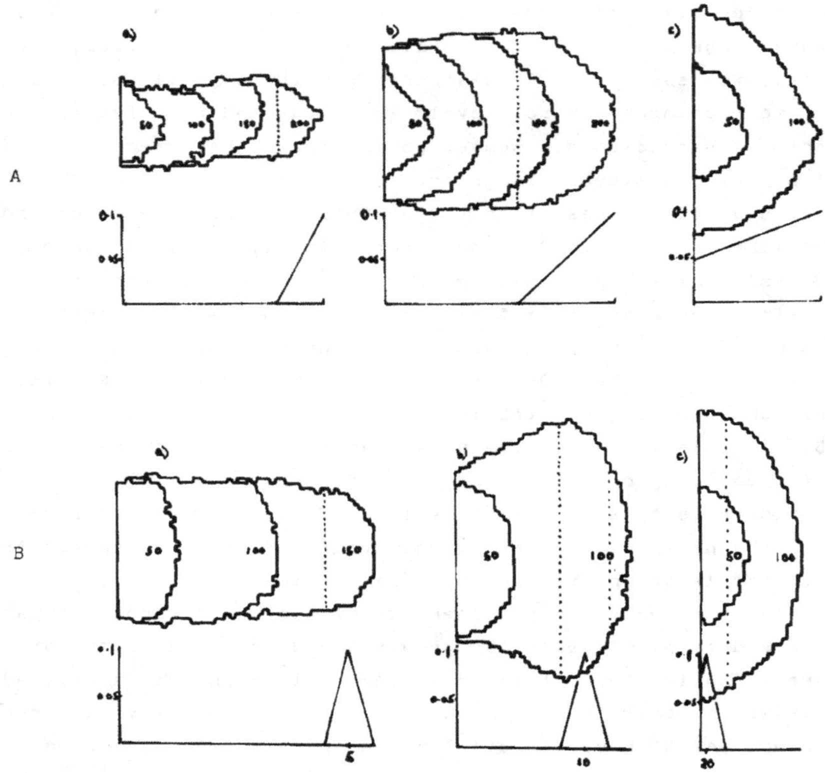

Fig. 6. Mitotic gradients in the Wilby and Ede (1976) model. Outline of print-outs for 50, 100, etc. scans is shown above. Mitotic index scale, below, shows fraction of cells dividing.
A. a) steep, b) medium, c) shallow disto-proximal gradients.
B. Mitotic gradients with a medial peak, a) displaced 5 rows from apex, b) 10 rows, c) 20 rows.

shallow gradients produces excessive lateral growth, and so steep gradients were used, with the control region displaced proximal to the apex. This produced a fairly normal limb outgrowth with the control region close to the apex, but produced a dramatic transformation to a talpid-like shape merely by displacing the control region five cell rows proximally (Fig. 6B). No cell movement component has been incorporated in these simulations but they indicate that regional cell proliferation control as it appears to exist in the real embryonic limb bud is not sufficient to produce the normal or $talpid^3$ limb, though it clearly plays an important part. The essential role of cell movement is confirmed.

Models for the development of the developing cartilage skeleton pattern

The models described above do not include changes of cell state (differentiation) and therefore do not define the formation of the cartilage skeletal pattern. Specification of this pattern may be imposed upon (though it has not yet been incorporated into) models of these types. One model (Tickle, Summerbell and Wolpert, 1975) proposes that there is a monotonic gradient of some chemical morphogen set up across the limb bud, with multiple thresholds defining where cells should differentiate as cartilages of digit 1, digit 2, etc. (Fig. 7A).

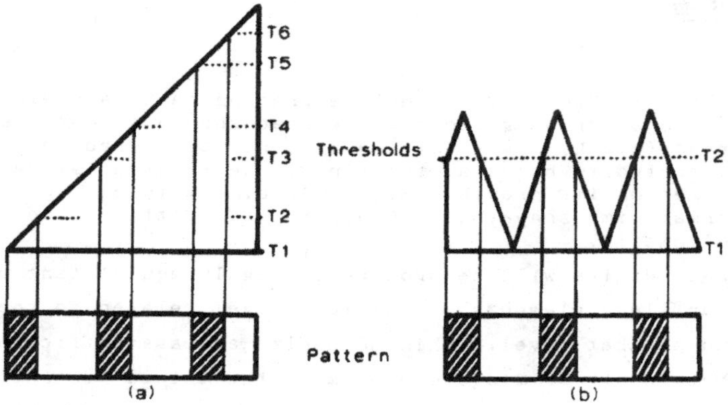

Fig. 7. Alternative types of model for generating a pattern of limb cartilages. A. Monotonic gradient with multiple thresholds; B. Periodic gradient with a single threshold.

Wilby and Ede (1975) proposed a periodic gradient, with only a single threshold (Fig. 7B), and suggested a molecular mechanism by which this might be generated, incorporating this mechanism in a computer program (Fig. 8). The chief difference between the two models is that whereas with a monotonci gradient, if the limb bud is expanded laterally the number of digits should remain constant, since the threshold levels remain the same, with an expanded periodic gradient of fixed periodiating the number of digits should increase, as illustrated in the computer simulation for the talpid-like limb. That such polydactyly does in fact occur in the talpid[3] embryo limb indicates that there is some such periodic control mechanism operating in the limb. Even if the cells of the limb bud are dissociated in experimental conditions, repacked in a limb ectodermal jacket, and cultured by grafting on another embryo (Ede, 1982),

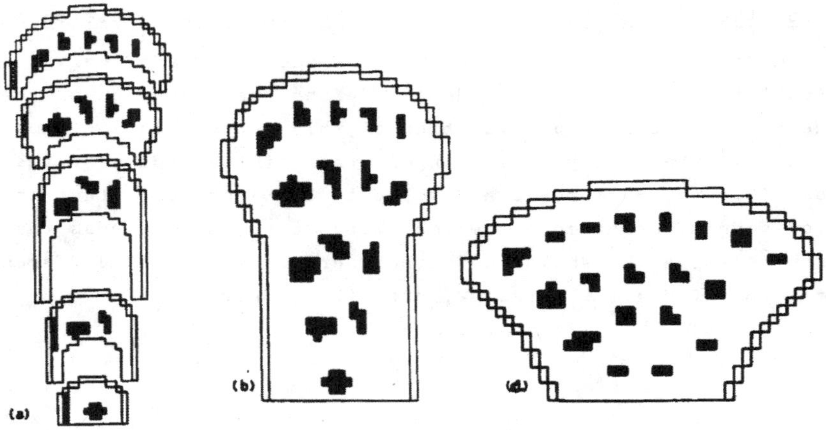

Fig. 8. Wilby and Ede (1975) model generating cartilage elements within outgrowing limbs of defined shape. A. Cartilage areas (black) are specified by a periodic gradient generated in bands at the tip of the limb bud at various stages of its growth. B. The same, combined to give a normal limb shape, with 5 digits. C. Pattern producing polydactly.

still separate digits will be produced, roughly equidistant and with the number of cartilage elements at any limb level related to the width of the limb bud at that level. This mechanism appears sufficient to explain the distribution of cartilage elements in the talpid[3] mutant limb, but also in the limb-like fin base of the crossopterygian fossil fish from which land-living vertebrates may have evolved (Fig. 9A). However,

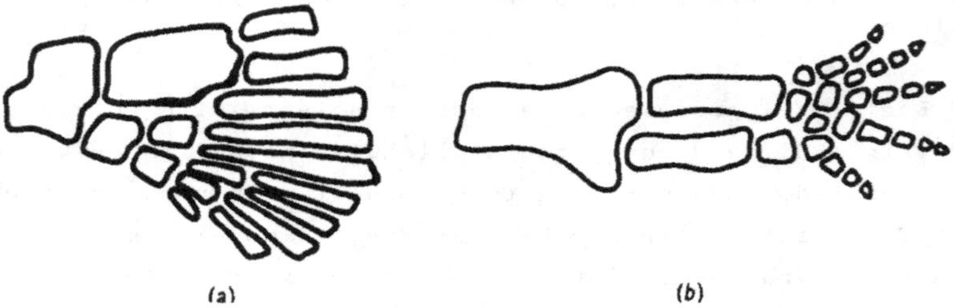

Fig. 9. A. Fin of a fossil fish, Sauripterns.
B. Limb of a primitive amphibian.

some further "fine tuning" mechanism, such as that proposed in the alternative model, evolved with the land-living vertebrates (Fig. 9B) to produce the differences between individual digits and the remarkably stable pentadactyl limb pattern which persists in animals whose fully-developed limbs are amazingly diverse.

References

Ede, D.A. 1976. Cell interaction in vertebrate limb development. In "The Cell Surface in Animal Embryogenesis and Development." (G. Poste and G.L. Nicolson, eds), Elsevier, Amsterdam, pp. 495-543.

Ede, D.A. 1982. Levels of complexity in limb-mesoderm cell culture systems. In "Differentiation *in vitro*." (M.M. Yeoman and D.E.S. Truman, eds), Cambridge University Press, pp. 207-229.

Ede, D.A., Flint, O.P. and Teague, P. 1975. Cell proliferation in the developing wing-bud of normal and *talpid*^{3} mutant chick embryos. J. Embryol. exp. Morph. **34**, 589-607.

Ede, D.A. and Law, J.T. 1969. Computer simulation of vertebrate limb morphogenesis. Nature, Lond. **221**, 244-248.

Tickle, C., Summerbell, D. and Wolpert, L. 1975. Positional signalling and specification of digits in chick limb morphogenesis. Nature, Lond. **254**, 199-202.

Wilby, O.K. and Ede, D.A. 1975. A model generating the pattern of cartilage skeletal elements in the embryonic chick limb. J. Theoret. Biol. **52**, 199-217.

Wilby, O.K. and Ede, D.A. 1976. Computer simulation of vertebrate limb development: the effect of cell division control patterns. In "Automata, Languages, Development: at the Crossroads of Biological, Mathematical and Computer Science." (A. Lindenmayer and G.Rozenberg, eds), North-Holland, Amsterdam, pp. 25-37.

ASYMPTOTICS AND SPATIAL GROWTH OF BRANCHING RANDOM FIELDS[1]

Luis G. Gorostiza
Centro de Investigación y de Estudios Avanzados
México 07000 D.F.
and SEGICyT-UPIICSA-IPN, México

1. INTRODUCTION

Systems consisting of particles in a d-dimensional Euclidean space R^d which undergo migration, reproduction, transformation of species and interaction arise in many applications as mathematical models for the evolution of spatially distributed populations (see e.g. Dawson [5]). In this paper we will survey some recent results on the asymptotic behavior of systems of particles of a single species, subject to migration and reproduction. These systems may consist of a finite or an infinite number of particles, and they are composed of two subpopulations: the particles whose first ancestors appeared at an initial time and those whose first ancestors immigrated later on. Different types of asymptotics may be studied; here we will consider large spatial scale, large space-time scales, and scaling of system parameters (density of particles, mean particle lifetime, branching law).

The process usually studied concerning these systems is the measure-valued process $N \equiv \{N_t, t \geq 0\}$, where N_t is a random point measure whose atoms are the locations of the particles present at time t, all of them having weight 1. We will restrict ourselves for simplicity to the case where the particle migration process is Brownian motion and the particle lifetime distribution is exponential, and the results we shall give for the process N are laws of large numbers, fluctuation limits, Langevin equations for the fluctuation limits, and behavior of fluctuation limits for large time (other more technical aspects of the fluctuation limits are omitted). For brevity we will not write the distributions of the fluctuation limits, but we will indicate how they can be obtained from the corresponding Langevin equations. In this setting our results contain as special cases results of Dawson and Ivanoff [3], [4], [7], [15], [16], Holley and Stroock [17], and Gorostiza [10]. We refer the reader to the works of Dawson and Ivanoff for results on existence of steady state random fields for the process N and central limit theorems for the steady states.

The fluctuation limits and their Langevin equations given previously by several authors have been obtained case by case using somewhat different techniques. General theorems proved recently by Bojdecki and Gorostiza [1], [2] allow a unified and simpler treatment of problems of this type. We refer the reader to [2] concerning the proofs of the results presented here for the infinite particle case.

In the case of finite particle systems with supercritical reproduction and no

[1]Research partially supported by CONACyT grant PCCBBNA 002042

immigration we will give results that are useful for the study of the asymptotic space-time growth of the systems, and we will consider also other types of migrations and more general lifetime distributions. In this case, rather than looking at the process N described above it is more convenient to work with measures on a space of trajectories. These results rely on an almost-sure invariance theorem for the empirical distribution of the trajectories of the lines of descent of the system generated by a single initial particle (Gorostiza and Kaplan [13]).

We will refer only to papers directly connected to the results given here. There are many interesting studies of branching random fields with other properties (e.g. transport-type migration, boundaries, particles of several species, varying environments, interactions, controls) which we shall not discuss.

The precise formulations of spaces, definitions, etc., we shall use can be found in the references or in their bibliographies.

2. THE BRANCHING RANDOM FIELD WITH IMMIGRATION (BRFI)

We will consider a special type of BRFI which corresponds to a system of particles in R^d evolving in the following way. At time $t=0$ the particles are distributed according to a homogeneous Poisson random field with intensity $\gamma \geq 0$. As time elapses, each particle independently migrates as a standard Brownian motion and independently reproduces subject to a branching law $\{p_n\}_{n=0,1,\ldots}$ after an exponentially distributed lifetime with parameter V; the offspring particles appear at the same location where their parent branched, and they migrate and reproduce in the same way; the branching law is assumed to have finite variance, and its mean and its second and third factorial moments are denoted m_1, m_2 and m_3, respectively. In addition, particles from an external source immigrate into R^d according to a homogeneous space-time Poisson random field with intensity $\beta \geq 0$, and each immigrant particle generates an independent branching Brownian motion as described above. The initial Poisson field and the position component of the immigration Poisson field can both be restricted to bounded Borel sets of R^d; in this case the particle system is finite at each time, otherwise it is infinite. The Malthusian parameter of the underlying branching process is given by $\alpha = V(m_1-1)$; the cases $\alpha = 0, >0$ and <0 are called *critical*, *supercritical* and *subcritical*, respectively. In the case without immigration $(\beta = 0)$ the system is called a branching random field (BRF).

We consider the process $N \equiv \{N_t, t \geq 0\}$, where for each t N_t is the random point measure determined by the locations of the particles at time t, i.e. $N_t(A)$ is the number of particles lying in the Borel set A of R^d at time t. The process N can be realized in the Skorohod space $D([0,\infty), S'(R^d))$, where $S'(R^d)$ is the space of tempered Schwartz distributions on R^d.

We denote $N^K \equiv \{N_t^K, t \geq 0\}$ the scaled process, where K (≥ 1) is the scaling parameter. For N^K the quantities defined above are designated $\gamma^K, \beta^K, \{p_n^K\}_{n=0,1,\ldots}$, m_1^K, m_2^K, m_3^K, V^K, α^K; scale changes in terms of K may be introduced in the space and time coordinates. In section 4 we will give the asymptotic behavior of N^K as $K\to\infty$ for several different scalings.

3. GENERALIZED WIENER PROCESSES AND LANGEVIN EQUATIONS

This section is needed for expressing the fluctuation limits of the BRFI.

Let $\langle\cdot,\cdot\rangle$ denote the canonical bilinear form on $\mathcal{S}'(R^d)\times\mathcal{S}(R^d)$, where $\mathcal{S}(R^d)$ is the space of C^∞ rapidly decreasing functions on R^d, $\mathcal{S}'(R^d)$ being its topological dual.

The *standard spatial Gaussian white noise* J *on* R^d is the random element of $\mathcal{S}'(R^d)$ whose characteristic functional is given by

$$Ee^{i\langle J,\phi\rangle} = \exp\{-\frac{1}{2}\int_{R^d}\phi^2(x)dx\},\quad \phi\in\mathcal{S}(R^d)$$

(see [8]).

The $\mathcal{S}'(R^d)$*-valued Wiener process* $W\equiv\{W_t,\ t\geq 0\}$ *associated to* $Q\equiv\{Q_t,\ t\geq 0\}$ is a continuous centered Gaussian process whose covariance functional $K_W(s,\phi;t,\psi) = E\langle W_s,\phi\rangle\langle W_t,\psi\rangle$ has the form

$$K_W(s,\phi;t,\psi) = \int_0^{s\wedge t}\langle Q_u\phi,\psi\rangle du,\quad s,t\geq 0,\quad \phi,\psi\in\mathcal{S}(R^d),$$

where for each u $Q_u:\mathcal{S}(R^d)\to\mathcal{S}'(R^d)$ is a continuous linear operator, symmetric and positive, and for each $\phi,\psi\in\mathcal{S}(R^d)$ the function $u\to\langle Q_u\phi,\psi\rangle$ is right-continuous with left-limits (see [2]).

A *generalized Langevin equation* for an $\mathcal{S}'(R^d)$-valued Gaussian process $X\equiv\{X_t,\ t\in[0,T]\}$ is a stochastic evolution equation of the form

$$dX_t = A^*X_t dt + dW_t,\quad t\in[0,T],$$

where A^* is the adjoint of a continuous linear operator A of $\mathcal{S}(R^d)$ into itself and W is an $\mathcal{S}'(R^d)$-valued Wiener process. We consider here solutions of Langevin equations in the "mild" sense, i.e. X satisfies

$$\langle X_t,\phi\rangle = \langle X_0,\phi\rangle + \int_0^t\langle X_s,A\phi\rangle ds + \langle W_t,\phi\rangle,\quad t\in[0,T],$$

for each $\phi\in\mathcal{S}(R^d)$. Solutions of generalized Langevin equations are called *generalized Ornstein-Uhlenbeck processes*.

If A generates a strongly continuous semigroup of continuous linear operators $\{T_t,\ t\geq 0\}$ on $\mathcal{S}(R^d)$, and if $X\equiv\{X_t,\ t\in[0,T]\}$ is an $\mathcal{S}'(R^d)$-valued centered Gaussian process whose covariance functional $K_X(s,\phi;t,\psi) = E\langle X_s,\phi\rangle\langle X_t,\psi\rangle$ satisfies the condition

$$K_X(s,\phi;t,\psi) = K_X(s,\phi;s,T_{t-s}\psi),\quad s\leq t,\quad \phi,\psi\in\mathcal{S}(R^d),$$

then X is a Markov process and obeys the generalized Langevin equation above, where the Q determining the S'-Wiener process W is given by

$$\langle Q_u\phi,\psi\rangle = \frac{d}{du} K(u,\phi;u,\psi) - K(u,A\phi;u,\psi) - K(u,\phi;u,A\psi),$$

assuming the function $u \to K(u,\phi;u,\phi)$ is continuously differentiable (see [2]).

Given a semigroup $\{T_t\}$ with generator A and W as above, the $S'(R^d)$-valued centered Gaussian process X defined by

$$\langle X_t,\phi\rangle = \langle X_0,T_t\phi\rangle + \int_0^t \langle dW_s,T_{t-s}\phi\rangle, \quad t \in [0,T],$$

for each $\phi \in S(R^d)$, also satisfies the generalized Langevin equation and its distribution coincides with that of a "mild" solution (the latter is called a "mild evolution" solution). The covariance functional of X can be obtained from the expression of the mild evolution solution, knowing the covariance of the initial condition X_0 and the S'-Wiener process W (in our examples X_0 and W are independent).

4. ASYMPTOTICS OF THE INFINITE PARTICLE BRFI

In the following, $\Longrightarrow$ denotes convergence in distribution, and for the fluctuation processes $\Longrightarrow$ refers to convergence in the Skorohod space $D([0,\infty), S'(R^d))$ (see [1]).

1) <u>Space scaling</u>: Let $\gamma^K = \gamma$, $\beta^K = \beta$, $V^K = V$, $\{p_n^K\} = \{p_n\}$, assume $\{p_n\}$ has finite third moment, and introduce the space scaling $x \to Kx$, i.e. $\langle N_t^K,\phi\rangle = \langle N_t,\phi(\cdot/K)\rangle$.

Law of large numbers: For each $t \geq 0$,

$$K^{-d}\langle N_t^K,\phi\rangle \to e^{\alpha t}[\gamma + \beta(1-e^{-\alpha t})/\alpha] \int_{R^d} \phi(x)dx \quad \text{in } L^2 \text{ as } K\to\infty.$$

Fluctuation limit and Langevin equation: Let $X_t^K = K^{-d/2}(N_t^K - EN_t^K)$. Then $X^K \Longrightarrow X$ as $K\to\infty$, where X is an $S'(R^d)$-valued centered Gauss-Markov process satisfying

$$dX_t = \alpha X_t dt + dW_t, \quad t \geq 0,$$

$$X_0 = \gamma^{1/2} J,$$

and the S'-Wiener process W is associated to

$$\langle Q_t\phi,\psi\rangle = \{\gamma e^{\alpha t}(m_2V-\alpha) + \beta[2-e^{\alpha t} + m_2V(e^{\alpha t}-1)/\alpha]\} \int_{R^d} \phi(x)\psi(x)dx.$$

Fluctuation limit for large time: In the subcritical case with possible immigration $(\alpha < 0,\ \beta \geq 0)$, $X_t \Longrightarrow X_\infty$ as $t\to\infty$, where

$$X_\infty = (-\beta/\alpha + m_2V\beta/2\alpha^2)^{1/2} J.$$

Observe that this limit depends only on the immigration; the effect of the initial particles disappears as $t \to \infty$ due to subcriticality.

We note that these results contain information from the branching process but not from the particle migration process; this is because the space scaling annihilates the motion and preserves the branching in the limit; in fact, if the migration is any other time-homogeneous Markov process which preserves the spatial homogeneity of the system, then the results are exactly the same. Moreover, these results coincide with those of the model where the particles do not migrate at all; therefore the migration has no effect in the limit under this scaling.

Our results for the space scaling appear in [2], [11]. This scaling was first studied by Dawson and Ivanoff [7], [15], [16]; they obtained fluctuation limits for fixed t, including $t = \infty$ when a steady-state random field exists.

2) Space-time scaling in the critical case with low immigration density: Let $\gamma^K = \gamma$, $\beta^K = \beta/K^2$, $V^K = V$, $\{p_n^K\} = \{p_n\}$, assume that $\{p_n\}$ is critical $(\alpha = 0)$ and has finite third moment, and introduce the space-time scaling $(x,t) \to (Kx, K^2 t)$, i.e. $\langle N_t^K, \phi \rangle = \langle N_{K^2 t}, \phi(\cdot/K) \rangle$. In this case the results depend on the dimension of the space; we need $d \geq 3$.

Law of large numbers: For each $t \geq 0$,

$$K^{-d} \langle N_t^K, \phi \rangle \to (\gamma + \beta t) \int_{R^d} \phi(x) dx \quad \text{in } L^2 \quad \text{as } K \to \infty.$$

Fluctuation limit and Langevin equation: Let $X_t^K = K^{-d/2-1}(N_t^K - EN_t^K)$. Then $X^K \Rightarrow X$ as $K \to \infty$, where X is an $S'(R^d)$-valued centered Gauss-Markov process satisfying

$$dX_t = \frac{1}{2} \Delta X_t dt + dW_t, \quad t \geq 0,$$
$$X_0 = 0,$$

where the S'-Wiener process W is associated to

$$\langle Q_t \phi, \psi \rangle = m_2 V(\gamma + \beta t) \int_{R^d} \phi(x) \psi(x) dx.$$

The Laplacian in the Langevin equation reflects the fact that this scaling preserves the migration process (Brownian motion).

This fluctuation limit appears in [2]. The case $\beta = 0$ was first studied by Dawson [3], [4], and by Holley and Stroock [17]. Dawson has also obtained results for $d \leq 2$.

Fluctuation limit for large time: In the case without immigration $(\beta = 0)$, $X_t \Rightarrow X_\infty$ as $t \to \infty$, where X_∞ is a centered Gaussian random element of $S'(R^d)$ with covariance functional

$$\mathrm{Cov}(\langle X_\infty, \phi \rangle, \langle X_\infty, \psi \rangle) = m_2 V \int_{R^d} \int_{R^d} \phi(x) \psi(y) k(x,y) dx dy,$$

where $k(x,y)=\Gamma(d/2-1)/4\pi\|x-y\|^{d-2}$ (this is the potential kernel of Brownian motion for $d\geq 3$) [3], [4], [17].

3. <u>High density scaling</u>: Let $\gamma^K=K\gamma$, $\beta^K=K\beta$, $V^K=V$, $\{p_n^K\}=\{p_n\}$.

Law of large numbers: For each $t\geq 0$,

$$K^{-1}\langle N_t^K,\phi\rangle \to e^{\alpha t}[\gamma+\beta(1-e^{-\alpha t})/\alpha]\int_{R^d}\phi(x)dx \quad\text{in } L^2 \text{ as } K\to\infty.$$

Fluctuation limit and Langevin equation: Let $X_t^K=K^{-1/2}(N_t^K-EN_t^K)$. Then $X^K\Rightarrow X$ as $K\to\infty$, where X is an $S'(R^d)$-valued centered Gauss-Markov process satisfying

$$dX_t=(\tfrac{1}{2}\Delta+\alpha)X_t dt+dW_t,\quad t\geq 0,$$

$$X_0=\gamma^{1/2}J,$$

where the S'-Wiener process W is determined by

$$\langle Q_t\phi,\psi\rangle=\gamma e^{\alpha t}[(m_2V-\alpha)\int_{R^d}\phi(x)\psi(x)dx+\int_{R^d}\nabla\phi(x)\cdot\nabla\psi(x)dx]$$

$$+\beta\{[m_2V(e^{\alpha t}-1)/\alpha+2-e^{\alpha t}]\int_{R^d}\phi(x)\psi(x)dx+(e^{\alpha t}-1)/\alpha\int_{R^d}\nabla\phi(x)\cdot\nabla\psi(x)dx\},$$

with $\cdot$ denoting the usual scalar product in R^d. We note that in this scaling both the motion and the branching are preserved in the limit, as the Langevin equation shows.

Fluctuation limit for large time: In the subcritical case with immigration ($\alpha<0$, $\beta>0$), $X_t\Rightarrow X_\infty$ as $t\to\infty$, where X_∞ is a centered Gaussian random element of $S'(R^d)$ with covariance functional

$$\mathrm{Cov}(\langle X_\infty,\phi\rangle,\langle X_\infty,\psi\rangle)=-(\beta/\alpha)\{\int_{R^d}\phi(x)\psi(x)dx$$

$$+m_2V\int_{R^d}\int_{R^d}\phi(x)\psi(y)k(x,y)dxdy\},$$

where

$$k(x,y)=\begin{cases} e^{-(-2\alpha)^{1/2}\|x-y\|}/2(-2\alpha)^{1/2}, & d=1,\\ (-2)^{d/4-1/2}K_{d/2-1}((-2\alpha)^{1/2}\|x-y\|)/(2\pi)^{d/2}\|x-y\|^{d/2-1}, & d\geq 2,\end{cases}$$

and $K_{d/2-1}$ is the usual modified Bessel function. Note that $k(x,y)$ is one half of the potential kernel of Brownian motion killed at an independent exponentially distributed time with parameter $-\alpha$. We observe that this limit depends only on the immigration; the progenie of the initial particles disappears by subcriticality.

These results are in [2]. The case $\beta = 0$ was studied in [10].

4) High density, small mean lifetime and branching law scaling: A different type of limit is a diffusion approximation obtained by Dawson and Ivanoff [4], [7]: Let $\gamma^K = K\gamma$, $\beta^K = K\beta$, $V^K = KV$, $m_1^K = 1 + \mu/K$ ($\mu \in R$), $\lim_{K\to\infty} m_2^K = m_2 \geq 0$ and $\sup_K m_3^K < \infty$. Assume each particle has a mass equal to $1/K$, and let $\hat{N}_t^K$ denote the mass measure determined by the locations of the particles at time t, i.e. $\hat{N}_t^K = N_t^K/K$. Then $\hat{N}^K \Rightarrow Y$ as $K \to \infty$ in the Skorohod space $D([0,\infty), M(R^d))$, $M(R^d)$ being the space of Borel measures on R^d, where Y is a continuous process such that $Y_0 = \gamma \cdot$ Lebesgue measure, and the Laplace functional of Y_t is given by

$$E \exp\{-\int_{R^d} \phi(x) Y_t(dx)\} = \exp\{-\gamma \int_{R^d} H_t(\phi,x)dx - \beta\int_0^t \int_{R^d} H_{t-s}(\phi,x)dxds\},$$

$\phi \in C(R^d)$ with compact support and $0 \leq \phi \leq 1$, where $H \equiv H_t(\phi,x)$ satisfies the nonlinear equation

$$\frac{\partial}{\partial t} H = \frac{1}{2}\Delta H + \mu V H - \frac{1}{2} m_2 V H^2,$$

$$H_0 = \phi.$$

Let us consider the special case of this scaling where $m_2^K = \upsilon/K$, $\upsilon > 0$ (necessarily $\upsilon > \mu$). Hence $m_2 = 0$ and the function H above is given by $H_t(\phi,x) = e^{\mu V t} T_t \phi(x)$, where $\{T_t\}$ is the Brownian semigroup. In this situation the process Y becomes deterministic and one can find a law of large numbers and a fluctuation limit for the process N^K; these limits are the same as for the high density scaling 3), with α replaced by μV ($= \alpha^K$) and m_2 replaced by υ (since $m_2^K V^K = \upsilon V$). In particular, the law of large numbers implies the result $\hat{N}^K \Rightarrow Y$ of [4], and Y is the deterministic process $Y_t = e^{\mu V t}[\gamma + \beta(1 - e^{-\mu V t})/\mu V] \cdot$ Lebesgue measure. The fluctuation limit can be obtained by the methods of [1], [2].

5. GROWTH OF THE FINITE PARTICLE SUPERCRITICAL BRF

Here we are interested in analyzing the growth of the supercritical BRF in terms of the empirical distribution in function space of the trajectories of the lines of descent. We assume that the initial Poisson field is restricted to a bounded Borel set A of R^d; hence the number of particles of the BRF is finite at each time with probability one. We assume $p_0 = 0$ for simplicity.

Let us consider the BRF up to time T. We introduce the space-time scaling $(x,t) \to (T^{1/2}x, Tt)$ as in example 2) of the previous section (with $K = T^{1/2}$). Therefore the trajectories of the descent lines of the scaled BRF are in the space $C \equiv C([0,1], R^d)$ of continuous functions from $[0,1]$ into R^d. Let N^T denote the random point measure on C such that $N^T(\mathcal{B})$ is the number of descent line trajectories contained in the Borel set $\mathcal{B}$ of C.

In the following, W^x stands for the standard Wiener measure on C with initial point $x \in R^d$.

The asymptotic behavior of N^T as $T \to \infty$ is given by the following results.

Law of large numbers: For each Borel set B of C,

$$T^{-d/2} e^{-\alpha T} N^T(B) \to \gamma \int_A W^x(B)dx \quad \text{in} \quad L^2 \quad \text{as} \quad T \to \infty.$$

Central limit theorem: Let $X^T = T^{-d/4} e^{-\alpha T}(N^T - EN^T)$. Then for each Borel set B of C, $X^T(B) \Rightarrow X(B)$ as $T \to \infty$, where $X(B)$ is a Gaussian random variable with mean 0 and variance

$$\sigma^2(B) = \gamma \frac{m_2}{m_1 - 1} \int_A (W^x(B))^2 dx.$$

We remark that for each $T < \infty$ $X^T(\cdot)$ is a random signed measure on C, but $X(\cdot)$ is not; a suitable state space for X should be a space of "distributions" on C.

These results follow from an almost-sure invariance theorem (Gorostiza and Kaplan [13], see also Gorostiza and Ruíz Moncayo [14] for the case of Brownian motion). They can be used to study the asymptotic space-time growth of the supercritical BRF by making appropriate choices of Borel sets B. For example, let $a > 0$, $s \in (0,1]$, and $B = \{f \in C : \sup_{0 \leq t \leq s} \|f(t)\| > a\}$, $\| \ \|$ being the Euclidean norm on R^d. Then

$N^T(B)$ = number of descent line trajectories (starting from the bounded set $T^{1/2}A$) which at some instant in the time interval $[0,Ts]$ exceed the distance $T^{1/2}a$ from the origin.

Assuming that the set A is very small and contains the origin and approximating $W^x(B)$ by $W^0(B)$, from the previous limit theorems we have the following <u>approximate</u> results:

$$T^{-d/2} e^{-\alpha T} N^T(B) \to \gamma |A| \, P[\tau_a < s] \quad \text{in} \quad L^2 \quad \text{as} \quad T \to \infty,$$

and $X^T(B) \Rightarrow X(B)$ as $T \to \infty$, where $X(B)$ is a Gaussian random variable with mean 0 and variance

$$\sigma^2(B) = \gamma \frac{m_2}{m_1 - 1} |A| (P[\tau_a < s])^2;$$

here $|A|$ is the Lebesgue measure of A and τ_a is the first passage time through the level a of a Bessel process starting from 0 of index $(d-2)/2$; the Laplace transform of τ_a is

$$Ee^{-u\tau_a} = [a(2u)^{1/2}]^b / 2^b \Gamma(b+1) I_b(a(2u)^{1/2}), \quad u \geq 0,$$

where $b = (d-2)/2$, Γ is the gamma function, and I_b is the usual modified Bessel function (see [9]).

For other particle migration processes and lifetime distributions the law of large numbers and the central limit theorem are the same as above with W^x replaced by other measures on C; [13] shows how to obtain these measures. We will give two examples.

Suppose the particle lifetime distribution G is non-lattice and has no atom at 0; denote $\tilde{G}$ the distribution $\tilde{G}(dt) = m_1 e^{-\alpha t} G(dt)$, where α is the Malthusian parameter, i.e. the unique (positive) constant such that $\tilde{G}$ is a probability distribution (supercritical corresponds to $m_1 > 1$); denote $\tilde{\mu}$ and $\tilde{\sigma}^2$ the mean and variance of $\tilde{G}$.

1) Branching random walk: The particles jump instantaneously at birth independently and remain at their locations. Assume the jump distribution has mean 0 and finite, strictly positive-definite covariance matrix C. Then W^x is replaced by Wiener measure on C with covariance matrix $\tilde{\mu}^{-1}C$ and initial point x.

2) Branching transport process: The motion takes place in R^d, $d \geq 3$. Assume G has finite moment or order $3+\delta$ for some $\delta > 0$. Suppose the particles travel along straight lines at constant speed 1, and the offspring particles independently choose directions with radial symmetry forming a random angle θ (not supported on $\{0,\pi\}$) with respect to their parent particle's direction. Then W^x is replaced by Wiener measure on C with variance parameter

$$\tilde{\mu}^{-1}[\tilde{\sigma}^2 + \tilde{\mu}^2(1 + E\cos\theta)/(1 - E\cos\theta)]$$

and initial point x. This model is relevant in neutron transport theory and polymer chemistry.

Remark. If $\mathcal{B}$ is a Borel set of C with base B , a Borel set of R^d, at time $t \in [0,1]$, then $N^T(\mathcal{B}) \equiv N^T_t(B)$ defines a measure-valued process $N^T \equiv \{N^T_t, t \in [0,1]\}$ The asymptotic behavior of this process as $T \to \infty$ is studied in Dawson and Gorostiza [6], [12]. Although the results are similar to those of Section 4, they are qualitatively different in that in Section 4 the migrations along the descent lines have a central limit effect, whereas in this case the migrations enter through a law of large numbers (the theorem of [13]), and the central limit effect is due to the branching structure of the system. This difference is discussed in [6].

6. SKETCH OF PROOFS

The asymptotic results for the infinite particle BRFI are proved the same way for all the various scalings. We sketch here the main ideas of the proofs.

The characteristic functional of N_t is given by

$$E \exp iu\langle N_t,\phi\rangle = \exp\{\gamma\int_{R^d} E[\exp iu\langle N^x_t,\phi\rangle - 1]dx$$
$$+ \beta\int_0^t \int_{R^d} E[\exp iu\langle N^x_{t-s},\phi\rangle - 1]dxds\},\ u \in R,\ \phi \in \mathcal{S}(R^d),$$

where $\{N^x_t, t \geq 0\}$ is the BRF generated by a single particle located initially at

$x \in R^d$. This can be obtained by the method of [7], as well as an analogous expression for the joint characteristic functional of $N_{t_1},\dots,N_{t_m}$, $t_1 < \dots < t_m$. By differentiating the characteristic functionals in the usual way one finds the mean and the covariance functionals:

$$E\langle N_t,\phi\rangle = \gamma\int_{R^d} E\langle N_t^x,\phi\rangle dx + \beta\int_0^t \int_{R^d} E\langle N_{t-s}^x,\phi\rangle dxds$$

and

$$\mathrm{Cov}(\langle N_s,\phi\rangle,\langle N_t,\psi\rangle) = \gamma\int_{R^d} E\langle N_s^x,\phi\rangle\langle N_t^x,\psi\rangle dx$$

$$+ \beta\int_0^s \int_{R^d} E\langle N_{s-r}^x,\phi\rangle\langle N_{t-r}^x,\psi\rangle dxdr, \qquad s \le t.$$

The integrals on the right hand sides are computed by the renewal argument in [10], and we obtain

$$E\langle N_t,\phi\rangle = [\gamma e^{\alpha t} + \beta(e^{\alpha t}-1)/\alpha]\int_{R^d}\phi(x)dx, \qquad t \ge 0,$$

and

$$\mathrm{Cov}(\langle N_s,\phi\rangle,\langle N_t,\psi\rangle) = e^{\alpha t}[\gamma + \beta(1-e^{-\alpha s})/\alpha]\int_{R^d}\phi(x)T_{t-s}\psi(x)dx$$

$$+ e^{\alpha t}\beta m_2 V\int_0^s e^{\alpha(s-r)}(1-e^{-\alpha r})/\alpha\int_{R^d}\phi(x)T_{t+s-2r}\psi(x)dxdr, \quad s \le t,$$

where $\{T_t\}$ is the standard Brownian semigroup.

We may now introduce the different scalings in the expressions above and write the characteristic, the mean and the covariance functionals for the process N^k in each case.

The laws of large numbers for the N_t^k follow immediately from the corresponding mean and covariance functionals.

We now consider the fluctuation process $X^K \equiv \{X_t^K, t \ge 0\}$ in each case. From the previous results we may write the joint characteristic functional of $X_{t_1}^K,\dots,X_{t_m}^K$, $t_1<\dots<t_m$, and take the limit as $K\to\infty$. Using Lévy's continuity theorem we then identify the finite-dimensional distributions of the Gaussian limit process X in each case. Tightness of the families $\{X^K\}_{K\ge 1}$ is established by the martingale method of [17] and the result on tightness of $\mathcal{S}'$-valued processes of [18]; this procedure is described in [11] for the space scaling case.

The different conditions on the branching law and the space dimension for the various scalings are necessary to obtain the laws of large numbers and the asymptotic fluctuations in each case.

To show how the Langevin equations for the fluctuation limits are obtained let us take the space-time scaling in the critical case with low immigration intensity. The covariance functional of the process X, which is the limit of the covariance functional of N^K under this scaling and normalization, is

$$K_X(s,\phi;t,\psi) = m_2 V\{\gamma \int_{R^d} \phi(x) \int_0^s T_{t+s-2r}\,\psi(x)drdx$$

$$+ \beta \int_{R^d} \phi(x) \int_0^s r\, T_{t+s-2r}\psi(x)drdx\}, \quad s \leq t.$$

We observe that the condition $K_X(s,\phi;t,\psi) = K_X(s,\phi;s,T_{t-s}\psi)$ is satisfied, and therefore the Markov property of X and the Langevin equation follow from the result quoted in Section 3.

The results of Section 5 are direct consequences of the almost-sure invariance theorem in [13].

REFERENCES

1. Bojdecki, T., Gorostiza, L.G., and Ramaswamy, S. Convergence of S'-valued processes and space-time random fields. J. Functional Analysis (to appear).
2. Bojdecki, T. and Gorostiza, L.G. Langevin equations for S'-valued Gaussian processes and fluctuation limits of infinite particle systems (to appear).
3. Dawson, D.A. The critical measure diffusion process. Z. Wahrsch. verw. Geb., Vol. 40, (1977), 125-145.
4. Dawson, D.A. Limit theorems for interaction free geostochastic systems. Seria Coll. Math. Soc. Janos Bolyai, Vol. 24. North Holland, Amsterdam (1980), 27-47.
5. Dawson, D.A. Stochastic measure processes. In: Stochastic Nonlinear Systems in Physics, Chemistry, and Biology (L. Arnold and R. Lefever, eds.) Springer-Verlag. Berlin-Heidelberg-New York (1981), 185-199.
6. Dawson, D.A. and Gorostiza, L.G. Limit theorems for supercritical branching random fields. Math. Nachr., Vol. 119 (1984), 19-46.
7. Dawson, D.A. and Ivanoff, B.G. Branching diffusions and random measures. In: Advances in Probability (A. Joffe and P. Ney, eds.), Vol. 5, 61-104. New York. Dekker, (1978).
8. Gelfand, I.M. and Vilenkin, N.J. Generalized Functions, Vol. 4. Academic Press. New York, (1966).
9. Getoor, R.K. and Sharpe, M.J. Excursions of Brownian motion and Bessel processes. Z. Wahrsch. verw. Geb., Vol. 47, No.1, (1979), 83-106.
10. Gorostiza, L.G. High density limit theorems for infinite systems of unscaled branching Brownian motions. Ann. Probab., Vol.11, No.2, (1983), 374-392.
11. Gorostiza, L.G. Space scaling limit theorems for infinite particle branching Brownian motions with immigration. In: Stochastic Differential Systems (M. Metivier and E. Pardoux, eds.). Lect. Notes in Control and Inf. Sci. Vol. 69. Berlin. Springer-Verlag. (1985), 91-99.
12. Gorostiza, L.G. Limit theorems for supercritical branching random fields with immigration. Tech. Report no. 58, Lab. for Research in Statistics and Probability, Carleton University (1985).
13. Gorostiza, L. G. and Kaplan, N. Invariance principle for branching random motions, Bol. Soc. Mat. Mexicana, Vol. 25, No. 2, (1980), 63-86.
14. Gorostiza, L.G. and Ruiz Moncayo, A. Almost-sure invariance principle for branching Brownian motion. In: Studies in Probability and Ergodic Theory (G.C. Rota, ed.). Academic Press, New York, (1978), 95-111.

15. Ivanoff, B.G. The Branching random field. Adv. in Probab., Vol. 12, (1980), 825-847.

16. Ivanoff, B.G. The branching diffusion with immigration. J. Appl. Probab., Vol. 17, (1981), 1-15.

17. Holley, R. and Stroock, D.W. Generalized Ornstein-Uhlenbeck processes and infinite particle branching Brownian motions. Publ. RIMS, Kyoto Univ., Vol. 14, (1981), 741-788.

18. Mitoma, I. Tightness of probabilities in $C([0,1],S')$ and $D([0,1],S')$. Ann. Probab., Vol. 11 (4), (1983), 989-999.

GENERATION-DEPENDENT BRANCHING PROCESSES WITH IMMIGRATION: CONVERGENCE OF DISTRIBUTIONS

Thomas Götz
Universität Heidelberg, SFB123
Im Neuenheimer Feld 294
D-6900 Heidelberg

1. INTRODUCTION

There are cell systems in the organism which have the ability to maintain a constant cell production rate. Examples are the blood forming system (Lord and Schofield, 1979), the intestinal crypts (Cheng and Leblond, 1974), the surface epithelium (Potten, 1976) or the sperm forming system (Clermont and Hermo, 1975). A common feature of these systems is the presence of undifferentiated stem cells, which give rise to new stem cells and to so called committed precursor cells, who differentiate and after further divisions evolve into the mature non-proliferating end cells. Generation-dependent Bellman-Harris processes (see Fildes(1974), Fearn(1976), Edler(1978)) with the additional property that the offspring of a generation-zero-individual does not as usual belong to generation one, but partially remains in generation zero (in order to keep the stem cell population constant) may serve as a model for such systems. Another point of view is the following: The subpopulation of generation-zero individuals remains constant (equal to one); at the points of a renewal process one generation-one individual immigrates, which is the ancestor of a Bellman-Harris process with generation-dependent lifetime and offspring distributions.

We use the following notation:

G_o:	Lifetime distribution of an individual from generation zero or waiting time distribution between two immigrations of generation-one individuals
G_k:	Lifetime distribution of an individual from generation k
g_k:	Generating function of the offspring-size
$Z_k(t)$:	Total number of individuals at time t if the process started with one individual from generation k
$F_k(s,t)$:	Generating function of $Z_k(t)$

We assume, that all individuals live and reproduce independently of each other, that there is a finite constant B such that $m_k=g_k'(1)\leq B$ and that G_k is continuous for all $k\geq 0$, and is dominated by some continuous distribution function G(t), which puts mass zero on t=0. Let

$$\mu_k=\int_o^\infty t\ dG_k(t) \quad \text{and} \quad \rho_k=\int_o^\infty t^2 dG_k(t)$$

denote the first resp. second moments of the lifetime distributions.

We assume $\rho_k \leq \rho < \infty$ for all $k \geq 1$ which implies of course $\mu_k \leq \mu_+ < \infty$.

$A*B(t) = \int_0^t A(t-u)dB(u)$ denotes the convolution of two functions A and B whenever the integral makes sense and for each real x. [x] is the integer part of x. As usual we define for k<j

$$\sum_{i=j}^{k} a_i = 0 \quad \text{and} \quad \prod_{i=j}^{k} a_i = 1.$$

Our process is defined by the following system of non-linear integral equations

$$F_o(s,t) = s(1-G_o(t)) + \int_0^t F_o(s,t-u)F_1(s,t-u)dG_o(u) \tag{1}$$

$$F_k(s,t) = s(1-G_k(t)) + \int_0^t g_k(F_{k+1}(s,t-u))dG_k(u)$$

and it is not very hard to show that this system has a unique solution and that this solution is a generating function (see e.g. Fearn(1980) for a similar result). We have shown (Götz, 1985) that under certain conditions the expected population size grows like t^γ and it is a natural question, whether under similar conditions the distribution of $Z_o(t)/t^\gamma$ converges to some limit. The answer is a bit surprising: for $0 \leq \gamma < 1$ $Z_o(t)$ has a limit without any normalization, while for $\gamma = 1$ $Z_o(t)/t$ converges weakly to a gamma distributed random variable - a result that has been obtained for non-generation dependent processes by Pakes(1972). The case $\gamma > 1$ is not cosidered in this paper. One can show however that in this case the variance of $Z_o(t)/t^\gamma$ converges to zero, while the expectation converges to some positive constant, such that $Z_o(t)/t^\gamma$ converges stochastically to some constant. These results are stated in section 3 while in section 2 the main tool is developed: the investigation of the embedded Galton-Watson process and its generating function. Some of the proofs are only sketched, more detailed versions of them may be found in Götz(1985).

2. THE EMBEDDED GALTON-WATSON PROCESS

The main tool for the investigation of the asymptotic behaviour of $Z_k(t)$ resp. $F_k(s,t)$ is the comparison of the original process with its embedded Galton-Watson process.

Let $f(k,k,s) = s$, and if $f(k,k+j,s)$ is still defined for each $k \geq 1$,

$$f(k,k+(j+1),s) = g_k(f(k+1,(k+1)+j,s)) \; ,$$

then obviously $f(k,k+j,s)$ is the generating function of the size of generation k+j of the embedded Galton-Watson process starting with one individual from generation k.

The following lemma, which is a straightforward generalization of Fearn's version of Goldstein's inequalities (Fearn, 1980), describes the connection between $F_k(s,t)$ and $f(k,k+j,s)$.

LEMMA 1: Assume $m_k \leq 1$, then for $j,k \geq 1$, $s \in [0,1]$ and $t \geq 0$

$$1-f(k,k+j,s) - (1-s)\left(\prod_{l=k}^{k+j-1} m_l\right)(G_k * \ldots * G_{k+j-1})(t)$$

$$\leq \quad 1 - F_k(s,t) \tag{2}$$

$$\leq 1-f(k,k+j,s) + (1-s)\sum_{i=0}^{j-1}\left(\prod_{l=k}^{k+i-1} m_l\right)(1-G_{k+i}) * G_{k+i-1} * \ldots * G_k(t)$$

In the following lemmas we investigate the behaviour of $f(k,k+j,s)$.

LEMMA 2: Assume $m_k \leq 1$ and $g_k''(1) < \infty$ for $k \geq 1$, then for $k,j \geq 1$

$$\frac{(1-s)\prod_{l=k}^{k+j-1} m_l}{1 + (1-s)\prod_{l=k}^{k+j-1} m_l \sum_{i=0}^{j-1}\left(\prod_{l=k}^{k+i}\frac{1}{m_l}\right)\frac{g''_{k+i}(1)}{2g'_{k+i}(s)}}$$

$$\leq 1 - f(k,k+j,s) \leq \tag{3}$$

$$\frac{(1-s)\prod_{l=k}^{k+j-1} m_l}{1 + (1-s)\prod_{l=k}^{k+j-1} m_l \sum_{i=0}^{j-1}\left(\prod_{l=k}^{k+i}\frac{1}{m_l}\right)\frac{g''_{k+i}(s)}{2g'_{k+i}(1)}}$$

PROOF: Applying the mean value theorem and substituting s by $f(k+1,k+j,s)$ yields

$$1 - f(k,k+j,s) = g_k'(\vartheta)(1 - f(k+1,k+j,s))$$

with $f(k+1,k+j,s) \leq \vartheta \leq 1$. Hence

$$g_k'(f(k+1,k+j,s))(1-f(k+1,k+j,s))$$

$$\leq 1 - f(k,k+j,s) \leq m_k(1 - f(k+1,k+j,s)) \, .$$

$m_k \leq 1$ implies

$$s \leq g_k(s) = f(k,k+1,s) \leq g_k \circ g_{k+1}(s) \tag{4}$$

$$= f(k,k+2,s) \leq \ldots \leq f(k,k+j,s)$$

and by the monotonicity of g_k

$$g_k'(s)(1-f(k+1,k+j,s) \leq 1 - f(k,k+j,s) \leq m_k(1-f(k+1,k+j,s))$$

or

$$\frac{1}{m_k} \leq \frac{1 - f(k+1,k+j,s)}{1 - f(k,k+j,s)} \leq \frac{1}{g_k'(s)} \qquad s \in [0,1) \, . \tag{5}$$

Taylor expansion of g_k and substituting s by f(k+1,k+j,s) yields

$$f(k,k+j,s) = 1 - m_k(1-f(k+1,k+j,s)) + \frac{g_k''(\eta)}{2}(1 - f(k+1,k+j,s))^2$$

with $f(k+1,k+j,s) \le \eta \le 1$. This implies

$$\frac{1}{1 - f(k,k+j,s)} = \frac{1}{m_k}\frac{1}{1 - f(k+1,k+j,s)} + \frac{g_k''(\eta)}{2m_k}\frac{1 - f(k+1,k+j,s)}{1 - f(k,k+j,s)}$$

which implies together with (4) and (5)

$$\frac{1}{m_k}\frac{1}{1 - f(k+1,k+j,s)} + \frac{1}{m_k}\frac{g_k''(s)}{2g_k'(1)}$$

$$\le \frac{1}{1 - f(k,k+j,s)}$$

$$\le \frac{1}{m_k}\frac{1}{1 - f(k+1,k+j,s)} + \frac{1}{m_k}\frac{g_k''(1)}{2g_k'(s)}$$

and by iteration

$$\frac{1}{1-s}\prod_{l=k}^{k+j-1}\frac{1}{m_l} + \sum_{i=0}^{j-1}\left[\prod_{l=k}^{k+i}\frac{1}{m_l}\right]\frac{g_{k+i}''(s)}{2g_{k+i}'(1)}$$

$$\le \frac{1}{1 - f(k,k+j,s)}$$

$$\le \frac{1}{1-s}\prod_{l=k}^{k+j-1}\frac{1}{m_l} + \sum_{i=0}^{j-1}\left[\prod_{l=k}^{k+i}\frac{1}{m_l}\right]\frac{g_{k+i}''(1)}{2g_{k+i}'(s)}$$

from which the assertion follows.

A direct consequence is the following lemma:

LEMMA 3: Let $m_k \le 1$, $g_k''(0) \ge r > 0$, and $g_k''(1) < \infty$. Assume further

$$\sum_{i=1}^{k}\left(\prod_{l=1}^{i} m_l\right)\mu_{i+1} \sim \beta k^{\gamma}, \quad k \to \infty,$$

then for each $k \ge 1$ there is a constant $B_k < \infty$ such that

$$1 - f(k,k+j,0) \le \frac{B_k}{j^{2-\gamma}}.$$

Lemma 3 is sufficient to prove Theorem 1 while for the more refined results of Theorem 2 the denominator in (4) must be estimated by the following lemma:

LEMMA 4: Assume for the sequence $\{g_k(s)\}$

a) $0 < \lim_{k\to\infty} \prod_{l=1}^{k} m_l = \beta < \infty$

b) $\bar{r} = \lim_{k\to\infty} g_k'' < \infty$ exists

c) the sequence $g_k'''(1)$ is bounded.

then for each function $k:[0,\infty)\to\mathbb{N}$ tending to infinity for $t\to\infty$ and for each $x\geq 0$

$$\lim_{t\to\infty} \frac{1}{k(t)} \sum_{j=0}^{k(t)} \left(\prod_{l=1}^{j} \frac{1}{m_l}\right) \frac{g_j''(e^{-x/t})}{2g_j'(1)} = \frac{\bar{r}}{2\beta} \tag{6}$$

$$\lim_{t\to\infty} \frac{1}{k(t)} \sum_{j=0}^{k(t)} \left(\prod_{l=1}^{j} \frac{1}{m_l}\right) \frac{g_j''(1)}{2g_j'(e^{-x/t})} = \frac{\bar{r}}{2\beta} \tag{6'}$$

PROOF: By condition c)

$|g_k''(s) - g_k''(1)| < \varepsilon \qquad$ for $k\geq 1$.

Together with the estimation

$$\left|\frac{1}{k(t)} \sum_{j=0}^{k(t)} \left(\prod_{l=1}^{j} \frac{1}{m_l}\right) \frac{g_j''(e^{-x/t})}{2g_j'(1)} - \frac{\bar{r}}{2\beta}\right|$$

$$\leq \frac{1}{k(t)} \sum_{j=0}^{k(t)} \frac{g_j''(e^{-x/t})}{2g_j'(1)} \left| \left(\prod_{l=1}^{j} \frac{1}{m_l}\right) - \frac{1}{\beta}\right|$$

$$+ \frac{1}{k(t)} \sum_{j=0}^{k(t)} \frac{1}{2\beta g_j'(1)} \left|g_j''(e^{-x/t}) - g_j''(1)\right|$$

$$+ \left|\frac{1}{k(t)} \sum_{j=0}^{k(t)} \frac{g_j''(1)}{2\beta g_j'(1)} - \frac{\bar{r}}{2\beta}\right| .$$

this proves (6) because $a_k \to a$ for $k\to\infty$ implies

$$\frac{1}{k(t)} \sum_{j=0}^{k(t)} a_j \to a \text{ für } t\to\infty .$$

The proof of (6') follows along the same lines.

3. THE ASYMPTOTIC BEHAVIOUR OF $Z_0(t)$

We will now use the preceeding lemmas to investigate the behaviour of $1-F_1(0,t)$ - the probability that at time t at least one offspring of an individual which immigrated at time 0 is alive. By a theorem of Pakes and Kaplan(1974) this function determines wether there is a limit distribution for $Z_o(t)$ or not.

THEOREM 1: Let $m_k \leq 1$, $g_k''(0) \geq r > 0$ and $g_k''(1) < \infty$. Then

$$\sum_{i=1}^{k} \Big(\prod_{l=1}^{i} m_l\Big)\mu_{i+1} \sim \beta k^{\gamma} \quad \text{for some} \quad \gamma < 1 \tag{7}$$

implies, that $Z_o(t)$ has a limit distribution for $t\to\infty$.

PROOF: By Theorem 1 of Pakes and Kaplan(1974) a necessary and sufficient condition for the existence of a limit distribution is

$$\int_0^\infty (1 - F_1(0,t))dt < \infty.$$

By Lemma 1

$$1-F_1(0,t) \leq 1-f(1,j+1,0) + \sum_{i=0}^{j-1}\Big(\prod_{l=1}^{i} m_l\Big)(1-G_{i+1})*G_i*\ldots*G_1(t).$$

Put $j(t)=\left[\frac{1}{2}\,\frac{t}{\mu_+}\right]$, then one has to show that

$$\int_0^\infty (1-f(1,j(t)+1,0))dt + \int_0^\infty \left\{\sum_{i=0}^{j(t)-1}\Big(\prod_{l=1}^{i} m_l\Big)(1-G_{i+1})*G_i*\ldots*G_1(t)\right\}dt$$

is finite. For the first term this follows directly by applying Lemma 2. to estimate the second, observe that for each $j\geq 1$

$$\sum_{i=0}^{j-1}\Big(\prod_{l=1}^{i} m_l\Big)(1-G_{i+1})*G_i*\ldots*G_1$$

$$= \sum_{i=1}^{j-1}\Big(\prod_{l=1}^{i-1} m_l\Big)(1-m_i)(1-G_i*\ldots*G_1)+\Big(\prod_{l=1}^{j-1} m_l\Big)(1-G_j*\ldots*G_1)$$

Thus one must prove the finiteness of

$$\int_0^\infty \left\{\sum_{i=1}^{j(t)-1}\Big(\prod_{l=1}^{i-1} m_l\Big)(1-m_i)(1-G_i*\ldots*G_1(t))\right\}dt \tag{8}$$

$$+ \int_0^\infty \left\{\Big(\prod_{l=1}^{j(t)-1} m_l\Big)(1-G_{j(t)}*\ldots*G_1(t))\right\}dt.$$

To estimate the first term in (8), we note that

$$\sum_{i=1}^{j(t)-1}\Big(\prod_{l=1}^{i-1} m_l\Big)(1-m_i)(1-G_i*\ldots*G_1(t)) \tag{9}$$

$$\leq \sum_{i=1}^{\infty}\Big(\prod_{l=1}^{i-1} m_l\Big)(1-m_i)(1-G_i*\ldots*G_1(t))\mathbf{1}_{[2i\mu_+,\infty)}(t) \;.$$

and the righthand side of (9) is dominated by

$$\sum_{i=1}^{\infty}\Big(\prod_{l=1}^{i-1} m_l\Big)(1-m_i).$$

It follows from the identity

$$\sum_{i=1}^{j}\Big(\prod_{l=1}^{i-1} m_l\Big)(1-m_i) = 1 - \prod_{l=1}^{j} m_l .$$

and condition a), that the last expression is equal to one. Thus the righthand side of (9) is uniformly convergent and may be integrated term by term. This yields

$$\int_0^{\infty}\left\{\sum_{i=1}^{j(t)-1}\Big(\prod_{l=1}^{i-1} m_l\Big)(1-m_i)(1-G_i*\dots*G_1(t))\right\}dt \tag{10}$$

$$\le \sum_{i=1}^{\infty}\Big(\prod_{l=1}^{i-1} m_l\Big)(1-m_i)\int_{2i\mu_+}^{\infty} (1-G_i*\dots*G_1(t))dt.$$

By Chebychev's inequality one obtains for $t\geq 2i\mu_+$

$$1-G_i*\dots*G_1(t) \le \frac{i\rho}{(t-i\mu_+)^2}$$

which implies

$$\int_{2i\mu_+}^{\infty} (1-G_i*\dots*G_1(t))dt \le \int_{2i\mu_+}^{\infty}\frac{i\rho}{(t-i\mu_+)^2}dt = \frac{\rho}{\mu_+}.$$

Substituting this into (10) one sees that the first term in (8) is finite.

It follows from (7) and because the sequence μ_k is bounded away from zero that there exists a constant $B_1<\infty$, such that

$$\prod_{l=1}^{j(t)-1} m_l \le B_1 t^{\gamma-1}.$$

One obtains by applying Chebychev's inequality again

$$1-G_{j(t)}*\dots*G_1(t) \le \frac{j(t)\rho}{(t-j(t)\mu_+)^2} \le 2\frac{\rho}{\mu_+}t^{-1} .$$

and one sees that also the second term in (8) is finite for each $\gamma<1$, which completes the proof.

EXAMPLE: In cell kinetics one often has

$$g_k(s) = p_o(k) + p_1(k)s + p_2(k)s^2.$$

where $p_i(k)$ is the probability that a cell from generation k splits into i daughters. In this case

$$m_k = g_k'(1) = p_1(k) + 2p_2(k) \qquad \text{and } g_k''(0) = 2p_2(k).$$

From (7) it follows that m_k converges to one, thus the condition $g_k''(0) \geq r > 0$ excludes the trivial case $p_1(k) \to 1$.

The assertion of the following theorem is a generalization of a result obtained by Pakes(1972). Note that in contrast to Theorem 1 in the case $\gamma=1$ a normalization of $Z_o(t)$ is necessary.

THEOREM 2: Let the moments of the offspring distributions satisfy the conditions

a) $m_k \leq 1$ for $k \geq 1$ and $\beta = \lim_{k\to\infty} (\prod_{l=1}^{k} m_l) > 0$ exists

b) $0 < \bar{r} = \lim_{k\to\infty} g_k''(1) < \infty$ exists

c) the sequence $\{g_k'''(1)\}$ is bounded.

Assume further that

$$\mu_* = \lim_{k\to\infty} \frac{1}{k}\sum_{l=1}^{k} \mu_l$$

exists and is finite, then the distribution of $Z_o(t)/t$ converges weakly to the gamma distribution with density

$$\frac{1}{\Gamma(\nu)} \lambda^{\nu} y^{\nu-1} e^{-\lambda y} \qquad y \geq 0$$

with $\nu = \dfrac{\beta}{\bar{r}/2} \dfrac{\mu_*}{\mu_o}$ and $\lambda = \dfrac{\mu_*}{\bar{r}/2}$.

PROOF: This follows along the same lines as the proof of Theorem 7.1.5 of Jagers(1975). Use Lemma 1 instead of Jager's Theorem 6.6.7, and Equation (3),(6) and (6') instead of Equation 3.1.5 (see also Pakes (1972)).

REFERENCES

Cheng,H., Leblond,C.P.(1974) Origin, differentiation and renewal of four main epithelial cell types in the mouse small intestine. V. Unitarian theory of the origin of the four epithelial cell types. Am.J.Anat.**141**,537-562.

Clermont,Y., Hermo,L.(1976) Spermatogonial stem cells and their behaviour in the seminiferous epithelium of rats and monkeys. In: Stem Cells of Renewing Cell Populations, A.B.Cairnie, P.L.Lala, D.G.Osmond (eds.). Academic Press, London.

Edler,L. (1978) Strict supercritical generation-dependent Crump-Crump-Mode-Jagers branching processes. Adv.Appl.Prob.**10**.744-763.

Fearn,D.H.(1976) Supercritical age-dependent branching processes with generation dependence. Ann.Prob.**4**.27-37.

Fearn,D.H. (1980) Extinction probability for critical age-dependent branching processes with generation dependence. J.Appl.Prob.**17**. 16-24.

Fildes,R. (1974) An age-dependent branching process with variable lifetime distributions: The generation size. Adv.Appl.Prob.**6**. 291-308.

Götz,T. (1985) Generationsabhängige Verzweigungsprozesse mit Einwanderung. Dissertation, Heidelberg.

Jagers,P.(1975) Branching Processes with Biological Applications. John Wiley and Sons, New York.

Lord,B.I., Schofield,R.(1979) Some observations on the kinetics of haemopoetic stem cells and their relationship to the spatial cellular organization of the tissue. In: Biological Growth and Spread, Mathematical Theories and Applications. W.Jäger, H.Rost, P.Tautu (eds.). Springer, Berlin.

Pakes,A.G.(1972) Limit theorems for an age-dependent branching process with immigration. Math.Biosc.**14**,221-234.

Pakes,A.G., Kaplan,N.L.(1974) On the subcritical Bellman-Harris process with immigration. J.Appl.Prob.**11**,652-668.

Potten,C.S.(1976) Identification of clonogenic cells in the epidermis and the structural arrangement of the epidermal proliferative unit (EPU). In: Stem Cells of Renewing Cell Populations. A.B.Cairnie, P.L.Lala, D.G.Osmond (eds.). Academic Press, London.

Seneta,E.(1968) On asymptotic properties of subcritical branching processes. J.Austral.Math.Soc.**8**.671-682.

ON A CLASS OF INFINITE PARTICLE SYSTEMS EVOLVING IN A RANDOM ENVIRONMENT

Andreas Greven
Universität Heidelberg
Institut für Angewandte Mathematik
Im Neuenheimer Feld 294
D-6900 Heidelberg 1

Abstract

We consider a class of processes on $(\mathbb{N})^S$ $(S=\mathbb{Z}^n)$, modelling population growth. The dynamics of the system consists of: motion of particles, birth and death of individual particles, extinction of all particles at a site and splitting of all particles at a site.

We investigate the changes in the longterm behaviour of these systems, changes, which occur if we replace parameters of the evolution (as offspring distribution or site killing rate) by collections indexed by the sites and generated by a random mechanism at time 0.

We study the system for each (a.s.) realisation of the random environment and show that the exponential growth rate of the expected number of particles per site (given the environment)depends heavily on the character of the underlying motion; the growth rate is maximal (and can be calculated explicitly) iff the underlying motion has no drift. We propose an approach for the more detailed study of the asymptotic behaviour $(t\to\infty)$ of the process and show for Branching Random Walks a law of large numbers, respectively convergence to a "Poisson limit". Furthermore we show that nontrivial equilibria for our evolutions can exist only in the case of a translation-invariant structure of the mean offspring size and the mean death rates.

1. Introduction to the problem and description of the model

a) The aim of this paper is to show how the longterm behaviour of certain infinite particle systems changes if the evolution mechanism is not anymore translation-invariant but has an interesting spatial structure.

Here in this paper we shall consider a class of infinite particle systems whose evolution mechanism is motivated by population dynamics. It provides the possibility to model: random motion of particles, birth and death of individual particles, extinction of all particles at a certain site, splitting of all particles at a site at once. Certain parameters of the evolution as offspring distribution, site killing rate are choosen at random at time t=0, they form the random environment. Our attention will always be focused on the behaviour of the system for a given realisation of the random environment (a.s. of course).

We will prove, that the longterm behaviour of the system in random environment is (different from the deterministic case!) strongly dependent on the character of the underlying motion. The influence of the other parameters (site-killing rate etc.) is on the other hand similar to the case of a translation-invariant evolution mechanism.

The motivation for studying these models and questions is twofold: for one it is desirable to know how strongly do depend results in the ergodic theory of infinite particle systems on the special form of the evolution mechanism. Results we think of are: behaviour of the particle density, existence of equilibria, phase transition. Here it is important to get a collection of examples. The other reason is that for modelling the growth of a population in space it would be very unrealistic to assume spatial homogeneity of the evolution mechanism.

D. Dawson and K. Fleischmann in [1],[2], were the first introducing a random spatial structure in the evolution mechanism of an infinite particle system. They studied Branching Random Walks in a Random Environment, but use a discrete time model and more important they keep the expected offspring size constant in space. We will see later that giving this mean offspring size a spatial structure changes the features of the model essentially.

b) Now we give an exact definition of our model. For that purpose we shall need the following ingredients:

(i) A measure μ on $(\mathbb{N})^S$ (with $S=\mathbb{Z}^n$) which has the properties:

(1) μ is translation-invariant and shiftergodic

(2) $\int \eta(x)d\mu < \infty \qquad (\eta \in \mathbb{N}^S)$

This measure μ will become the initial distribution of our process.

(ii) We have numbers $b_1, m \in \mathbb{R}^+$ and transition kernels $p(x,y), \tilde{q}(x,y)$, $\tilde{\tilde{q}}(x,y)$ from S to S, which are homogeneous and $mp(x,y)+b_1\tilde{q}(x,y)$ is assumed irreducible.

(iii) A random collection $\{P_x(\cdot)\}_{x\in\mathbb{Z}^n}$ of probability distributions on $\mathbb{N}$ (these will become our offspring distributions) satisfying the following conditions:

(3) $\sum_{n=1}^{\infty} P_x(n)n \leq K < \infty \qquad$ a.s.

(4) The $\{P_x(\cdot)\}_x$ are i.i.d. random variables.

(iv) We have independent random fields $\{b_2(x)\}_{x\in S}$, $\{d_1(x)\}_{x\in S}$, $\{d_2(x)\}_{x\in S}$ which satisfy the conditions:

$$(5)\qquad 0 \leq b_2(x) \leq K < \infty \qquad \text{a.s.}$$

$$(6)\qquad 0 \leq d_i(x) < +\infty \quad i=1,2, \qquad \text{a.s.}$$

$$(7)\qquad \{b_2(x)\}_{x\in S},\{d_1(x)\}_{x\in S},\{d_2(x)\}_{x\in S} \qquad \text{are i.i.d.}$$

We denote by ω a specific realisation of our random environment: $(\{P_x(\cdot)\}_{x\in S},\{d_1(x)\}_{x\in S},\{d_2(x)\}_{x\in S},\{b_2(x)\}_{x\in S})$.

For each ω we define now a Markov process $(\eta_t^\mu)_{t\in\mathbb{R}^+}$ with state space $(\mathbb{N})^S$, initial distribution μ and generator G. To define G we need the following notation:

(i) $\delta_x\in\mathbb{N}^S$ is defined as $\delta_x(z)=1$ for $x=z$ and 0 otherwise;

(ii) $N^n_{x,y}:=\#\{i|X_i^{\{x\}}=y\}$, where the $X_i^{\{x\}}$ are i.i.d. random variables with $\text{Prob}(X_i^{\{x\}}=y)=q(x,y)$ and i runs runs from 1 to n.

Let f be a bounded function on $(\mathbb{N})^S$ which is of the form $f(\eta)=\tilde{f}(\eta(x_1),\ldots,\eta(x_n))$ for some $x_1,\ldots,x_n\in S$.

$$(8)\qquad \begin{aligned} Gf_{(\eta)} := \sum_{x\in S} [& m\eta(x)(\sum_{y\in S} p(x,y)(f(\eta-\delta_x+\delta_y)-f(\eta))) \\ & + b_1\eta(x)(\sum_{n=0}^{\infty} P_x(n)\sum_{y\in S} Ef(\eta+N^n_{x,y}\delta_y)-f(\eta)) \\ & + \sum_{y\in S} b_2(y)q(y,x)(f(\eta+\eta(x)\delta_y)-f(\eta)) \\ & + d_1(x)\eta(x)(f(\eta-\delta_x)-f(\eta)) \\ & + d_2(x)(f(\eta-\eta(x)\delta_x)-f(\eta))] \end{aligned}$$

The transitions are interpreted as (in this order): motion of a particle, birth of particles with instantaneous independent displacement, splitting of all particles at a site and replacement, death of an individual particle, extinction of all particles at a site.

With the methods developed by Liggett and Spitzer [6] it can be shown that there exists a Markov process whose generator is the extension of the operator G defined in (8) to the class of Lipschitz functions with respect to a suitable norm on the configuration space. We leave the details to the reader.

Note that we suppress in our notation the ω-dependence of our process to keep the notation handy.

2. The asymptotic behaviour of the particle density

a) The first object of interest is the longterm behaviour of the particle density $\hat{E}(\eta_t(x)) := E(\eta_t(x)|\omega)$. In order to obtain properties of the particle density, which are independent of ω (a.s.) as well as of x, we consider the quantity:

(9) $$\overline{\lim_{t\to\infty}}\left(\frac{1}{t}(\log \hat{E}(\eta_t^{\mu}(x)))\right)$$

The behaviour of this quantity depends on the environment through the parameters defined below:

(10) $$T_x(\omega) := \left[\sum_{y\in S}\left((b_1 q(y,x)\sum_{n=0}^{\infty} P_y(n)n) + (b_2(y)\tilde{\tilde{q}}(y,x))\right)\right] - [d_1(x)+d_2(x)]$$

(11) $$\lambda := \operatorname{supess}_{\omega}(T_x(\omega)) (=\sup_x(T_x(\omega)) \text{ a.s.})$$

(12) $$\tau := E_{\omega}(T_x(\omega))$$

We determine first the possible range of the quantity in (9):

Proposition 1:

(13) $$\tau \le \overline{\lim_{t\to\infty}}\left(\frac{1}{t}\log \hat{E}(\eta_t^{\mu}(x))\right) \le \lambda \quad \text{a.s.}$$

So we obtain a possibly larger growthrate of the system, if we look at the system evolving in a random environment and compare it with the system where we replace the random parameters by their mean values. The reason for this is that spots most favorable for growth, i.e. x with $T_x(\omega) > E_{\omega}(T_x(\omega))$, can dominate the growth of the particle density. The extent to which this effect of domination occurs depends on the character of the underlying motion, namely it's drift. Precisely:

Theorem 1:

a) If $p(x,y)$, $\tilde{q}(x,y)$, $\tilde{\tilde{q}}(x,y)$ are all symmetric, then:

(14) $$\lim_{t\to\infty}\left(\frac{1}{t}\log \hat{E}(\eta_t^{\mu}(x))\right) = \lambda \text{ a.s.}$$

b) Suppose there exists an octant of $\mathbb{Z}^n$ such that the distributions $p(0,\cdot)$, $\tilde{q}(0,\cdot)$, $\tilde{\tilde{q}}(0,\cdot)$ are concentrated on this octant and $p(0,0)+\tilde{q}(0,0)+\tilde{\tilde{q}}(0,0)=0$, then for $T_x(\omega)\neq$const :

(15) $$\overline{\lim_{t\to\infty}}\left(\frac{1}{t}\log \hat{E}(\eta_t^{\mu}(x))\right) < \lambda \quad \text{a.s.}$$

c) Suppose $S=\mathbb{Z}^1$ and $\exists\, z\in\mathbb{Z}^1$ such that $p(x,y)=\tilde{q}(x,y)=\tilde{\tilde{q}}(x,y)=\delta_z(x-y)$. Then

$$(16) \qquad \tau < \overline{\lim_{t\to\infty}}\left(\frac{1}{t}\log \hat{E}(\eta_t^\mu(x))\right) := \tau' < \lambda \; (\text{for } T_x(\omega) \not\equiv const)$$

There is an important special case, where we can classify the kernels exactly with respect to the question whether the exponential growth rate is λ or smaller than λ:

Theorem 1':

Suppose that $\tilde{q}(x,y)=\tilde{\tilde{q}}(x,y)=\delta(x,y)$ and that $p(x,y)$ has the property: $\sum_y p(0,y)e^{\alpha|y|}<\infty$, $\forall\alpha\in\mathbb{R}^+$. Then we have in the case $T_x(\omega)\not\equiv$const:

$$(17) \quad \text{For } p(0,y) = p(0,-y) \; : \; \lim_{t\to\infty}\left(\frac{1}{t}\log \hat{E}(\eta_t^\mu(x))\right) = \lambda$$

$$(18) \quad \text{For } \sum_y y\,p(0,y) \neq 0 \; : \; \overline{\lim_{t\to\infty}}\left(\frac{1}{t}\log \hat{E}(\eta_t^\mu(x))\right) < \lambda$$

We can interpret these results by saying that for certain evolutions which include those with symmetric kernels $p,\tilde{q},\tilde{\tilde{q}}$ the growth rate is completely determined by the points with the most favorable conditions for growth (i.e. x with $T_x(\omega)$ very close to λ). On the other hand this effect is destroyed if we have kernels which have a drift. An important application would be to environments with offspring distributions concentrated on $\{n\}$ or $\{m\}$ with $n<m$, the inhomogeneity of the medium comes here from impurities i.e. points with smaller offspring size n compared to the "normal" one equal to m. With symmetric $p(x,y)$ we would not observe any change in the growth rate due to these impurities, but if $p(x,y)$ has a drift we would see a strictly smaller exponential growth rate.

The statement of the theorem can also be viewed as a stability property for symmetric systems: these systems are insensitive to random perturbations of $T_x(\omega)$ which are negative. On the other hand asymmetric systems of the type in theorem 1'(b) are very sensitive to perturbations of this type. For perturbations with mean zero the roles are reversed.

b) In this section we want to formulate an interesting open problem. Consider the case $S=\mathbb{Z}^1$, $\tilde{q}(x,y)=\tilde{\tilde{q}}(x,y)=\delta(x,y)$ and $p(x,y)$ of the type $p(0,1)=p$, $p(0,-1)=1-p$. Then one can show with the methods used to prove theorem 1 that for a distribution on the ω such that: $\lambda>0>\tau'$ we have

Proposition 2: There exist $\delta,\delta'>0$ such that for Branching Random walks:

$$(19') \qquad \underline{\lim}_{t\to\infty}\left(\frac{1}{t}\log \hat{E}(\eta_t^\mu(x))\right) > 0, \quad \eta_t^\mu(x)\xrightarrow[t\to\infty]{}\infty \text{ a.s. } ; \text{ for } p\in[1/2,1/2+\delta]$$

(19") $\overline{\lim_{t\to\infty}}(\frac{1}{t}\log \hat{E}(\eta_t^{\mu}(x))) < 0\ ,\ \eta_t^{\mu}(x) \underset{t\to\infty}{\longrightarrow} 0$ a.s., for $p\in[1-\delta',1]$

What one would like to show is that a <u>phase transition</u> occurs i.e.

<u>Conjecture:</u> Under the assumptions described above there exists a $p^*\in(1/2,1)$ such that:

(20) $L(\eta_t^{\mu}) \underset{t\to\infty}{\longrightarrow} \delta_{\{\eta\equiv 0\}}$ for $p\in[p^*,1]$, $L(\eta_t^{\mu}) \underset{t\to\infty}{\longrightarrow} \delta_{\{\eta\equiv\infty\}}$ for $p\in[1/2,p^*)$

c) The reader interested in applications may now ask the question what these results mean for "real world" systems which are finite, i.e. $|S|<\infty$.

To answer this question consider $S_m=[-m,m]^n$ and matrices $p_m(x,y)$ $(=\sum_{z\sim y} p(x,z)$, where $\sim\ \hat{=}\ \mathrm{mod}(m))$, $\tilde{q}_m(x,y)$, $\tilde{\tilde{q}}_m(x,y)$. Then one can show (using formula (42) in section 4) that:

(21) $\frac{1}{t}\log \hat{E}\,(\eta_t^{\mu,m}(0)) \underset{t\to\infty}{\longrightarrow} \lambda_m(\omega)$ a.s.

It is now easy to show, using the shift-ergodicity of our random environment and the arguments in section 4, that:

<u>Proposition 3:</u> Under the assumptions of theorem 1a) [1b,1'] we have:

(22) $\lambda_m(\omega) \to \lambda$ in probability

[(22') $\lim_{m\to\infty}(\mathrm{Prob}(\lambda_m(\omega)<\lambda)) = 1$]

This means that our theorem 1 is a statement for an idealized system, but is relevant already for large enough finite systems: for large finite systems we will already observe the dependence of the exponential growth rate on the character of the kernels $p,\tilde{q},\tilde{\tilde{q}}$.

d) Let us conclude this section with a statement about $E(\eta_t^{\mu}(x))=E(\eta_t^{\mu}(0))$ a quantity of interest if we want to learn how $\lim_{n\to\infty}|V_n|^{-1}\sum_{x\in V_n}\eta_t^{\mu}(x)$ behaves.

<u>Proposition 4:</u> Suppose that $T_x(\omega)\not\equiv\mathrm{const}$, $\tilde{q}(x,y)=\tilde{\tilde{q}}(x,y)=\delta(x,y)$ and $\sup_x(p(0,x))<1$. Then:

(23) $$\tau < \overline{\lim_{t\to\infty}}(\frac{1}{t}\log E(\eta_t^{\mu}(0)))\begin{cases} = \lambda & \text{for } p(x,y)=p(y,x)\\ < \lambda & \text{for } \sum_y p(0,y)y\neq 0\end{cases}$$

3. Longterm behaviour of η_t^μ.

Now one would like to know in more detail how (η_t^μ) grows respectively dies out. We focus here on the case of symmetric kernels $p,\tilde{q},\tilde{\tilde{q}}$ and we shall treat first the case $\lambda\neq 0$ and then in the second section of this chapter the case $\lambda=0$.

a) Law of large numbers $(\lambda>0)$, Poisson limit $(\lambda<0)$. The behaviour of η_t^μ for $t\to\infty$ can be studied best by looking at the following new process (ξ_t^μ) living on the state space $(\mathbb{R}^+)^S$:

$$(24)\qquad \xi_t^\mu(x) := \begin{cases} \eta_t^\mu(x)[\hat{E}(\eta_t^\mu(x))]^{-1} & \text{in the case } \hat{E}(\eta_t^\mu(x)) \underset{t\to\infty}{\longrightarrow} \infty \\ \sum\limits_{z\in I_t} \eta_t^\mu([\tfrac{1}{2}xf_t]+z) & \text{in the case } \hat{E}(\eta_t^\mu(x)) \underset{t\to\infty}{\longrightarrow} 0 \end{cases}$$

Here I_t is a cube of the form $[a,b]^n$ where I_t is chosen such that:

$$(25)\qquad \sum_{x\in I_t} \hat{E}(\eta_t^\mu(x)) \underset{t\to\infty}{\longrightarrow} \rho \quad (\rho := E^\mu(\eta(x))) \qquad f_t = \sqrt[n]{|I_t|}$$

(The fact that (25) can be achieved follows form the monotonicity of $\hat{E}(\eta_t(x))$ in t and (36)).

This process $(\xi_t^\mu)_{t\in\mathbb{R}^+}$ is very useful in the situation of a translation invariant evolution mechanism (compare [5]). Most of these models exhibit some phase transition that means that $\text{w-}\lim\limits_{t\to\infty} (\xi_t^\mu)$ exists and equals $\delta_{\{\xi\equiv 0\}}$ for certain values of the parameters of the evolution or dimension and equals ν_ρ with $\int\eta(x)d\nu_\rho = \rho$ for others. In our situation we have by construction that:

$$(26)\qquad \hat{E}(\xi_t(x)) \underset{t\to\infty}{\longrightarrow} \rho \qquad \text{for almost all } \omega .$$

The relation above means that for almost all ω the set $\{\mathcal{L}(\xi_t^\mu|\omega)\}_{t\in\mathbb{R}^+}$ has weak limit points (for $t\to\infty$) in the set of probability distributions on $\mathbb{R}^+\setminus\{\infty\}$. This leads of course to the problem to show that $\mathcal{L}(\xi_t^\mu|\omega)\underset{t\to\infty}{\longrightarrow}\nu_\rho$ for all μ with $E^\mu(\eta(x))=\rho$ and then in the next step to decide whether $\nu=\delta_{\{\xi\equiv 0\}}$ or $\int\xi(0)d\nu_\rho=\rho$.

These problems are very hard to settle in general, even in the symmetric situation. The main reason is that the different models in our class show very different behaviour as far as these questions are concerned. We treated the case of the Coupled Branching Process ($b_2=0$, $d_2>0$, $P_x(\cdot)=\delta_1$) in [4]. Let us focus in this paper on the case of a Branching Random Walk, i.e. $b_2\equiv d_2\equiv 0$. Then we have the following result:

Theorem 2

Suppose that $b_2(x)\equiv d_2(x)\equiv 0$; $\tilde{q}(x,y)=\tilde{q}(y,x)$, $p(x,y)=p(y,x)$, $\sum_n P_x(n)n^2 \leq K<\infty$ a.s.; and assume that μ discribes a Poisson system with parameter ρ.

Then the following holds:

(27) for $\lambda>0$: $L(\xi_t^\mu|\omega)\xrightarrow[t\to\infty]{w}\delta_{\{\xi\equiv 1\}}$

(28) for $\lambda<0$: $L(\xi_t^\mu(0)|\omega)$ has for $t\to\infty$ mixed Poissondistributions as weak limits.

So in this example the effect of the random environment is (in the case $\lambda\neq 0$!) only seen in the <u>rates</u> of renormalisation respectively rescaling. So the study of the longterm behaviour of this process reduces to studying the behaviour of $\hat{E}(\eta_t^\mu(x))$.

The situation in the case $d_2(x)+b_2(x)\neq 0$ is different and more complicated due to the phenomenon often called phase transition (compare [4]). The general conjecture however is, that we observe for the process $(\xi_t^\mu)_{t\in\mathbb{R}}$ qualitatively very similar behaviour in the case of an evolution in a random environment compared to the case of a translation invariant evolution mechanism; all the influence of the randomness is captured by the quantity $\hat{E}(\eta_t^\mu(x))$ and the changed critical points.

In the case where the mean offspring size does not vary one can generalize Theorem 2 to arbitrary kernels fulfilling our irreducibility condition:

<u>Theorem 2':</u>
Under the assumption $T_x(\omega)\equiv$const and $\sum_n P_x(n)n^2\leq K<\infty$, $\int\eta^2(x)d\mu<\infty$ we have (for arbitrary $p,\tilde{q}$) that in the case of a Branching Random Walk:

(28') $L(\xi_t^\mu(0)|\omega)\xrightarrow[t\to\infty]{}$(<u>mixed</u> Poissondistribution, mean ρ), for $T_x(\omega)<0$

(27') $L(\xi_t^\mu)\xrightarrow[t\to\infty]{}\delta_{\{\xi\equiv 1\}}$, for $T_x(\omega)>0$

If in addition $r(x,y)$ (as defined in (31)) is transient, then:

(27") $\xi_t^\mu(x)\xrightarrow[t\to\infty]{} 1$ a.s. for $T_x(\omega)>0$

b) Nonexistence of equilibria for $\lambda=0$. Now we shall study the longterm behaviour of our process in the case $\lambda=0$. Let us consider for a moment those processes of our class which have translation invariant evolution mechanism, i.e. $P_x(\cdot)=P_0(\cdot)$, $d_1(x)\equiv d_1$, $d_2(x)\equiv d_2$, $b_2(x)\equiv b_2$. Then of course $T_x(\omega)\equiv T$ and $E(\eta_t^\mu(x))=E^\mu(\eta(0))e^{Tt}$. For the case $T=0$ it is meaningful to ask whether there exist non-trivial equilibrium states, i.e. measures ν on $(\mathbb{N})^S$ such that:

$$L(\eta_t^\nu)=\nu \quad \forall t\in\mathbb{R}^+ \quad \text{and } \nu\neq\delta_{\{\eta\equiv 0\}}$$

It is known (compare [5]) that under the condition (29) given below our processes have a one parameter set $(\nu_\rho)_{\rho\in\mathbb{R}^+}$ of extremal translation-invariant equilibria, where $\rho = E^{\nu_\rho}(\eta(0))$. (All other possible equilibria are mixtures of the $(\nu_\rho)_{\rho\in\mathbb{R}^+}$). The condition (29) reads as follows:

$$(29)\qquad T=0;\ \left(\frac{1}{2}\,\frac{b_2+d_2}{m+b_1k+b_2}\,G_{\hat r}(0,0)\right)<1 \quad\text{and}\quad G_{\hat r}(0,0)<\infty$$

where the $r, \hat r, G_{\hat r}$ are defined as follows:

$$(30)\qquad \hat r(x,y) = 1/2(r(x,y)+r(y,x))\ ; \quad k = \textstyle\sum_n P_x(n)n$$

$$(31)\qquad r(x,y) = \left(\frac{b_1k}{m+b_1k+b_2}\,\tilde q(x,y) + \frac{b_2}{m+b_1k+b_2}\,\tilde{\tilde q}(x,y) + \frac{m}{m+b_1k+b_2}\,p(x,y)\right)$$

$$(32)\qquad G_{\hat r} = \sum_{n=0}^{\infty} \hat r^n$$

(Note that the condition (29) is sufficient but in general not necessary). The situation is very different for evolutions in a random environment. Here we have the following situation:

Theorem 3:
Suppose that $T_x(\omega) \not\equiv$ const and that $p(x,y), \tilde q(x,y)$ and $\tilde{\tilde q}(x,y)$ are symmetric.
a) Suppose we have a set $\{\nu(\omega)\}$ of probability measures, on $\mathbb{N}^S$ such that:

$$(33)\qquad \mathcal{L}(\eta_t^{\nu(\omega)}|\omega) = \nu(\omega)\ \forall t\in\mathbb{R}^+, \quad \sup_x\left(\int\eta(x)d\nu(\omega)\right)<\infty \quad \text{a.s.}$$

Then $\nu(\omega) \equiv \delta_{\{\eta\equiv 0\}}$ a.s.
b) If $\lambda \leq 0$ we have:

$$(34)\qquad \mathcal{L}(\eta_t^\mu|\omega) \underset{t\to\infty}{\overset{w}{\Longrightarrow}} \delta_{\{\eta\equiv 0\}}\ (\omega\text{- a.s.}), \left(\eta_t^\mu(x) \underset{t\to\infty}{\longrightarrow} 0 \quad \text{a.s. for } \lambda<0.\right)$$

and if $\lambda>0$ we have that for Branching Random Walks:

$$(35)\qquad \eta_t^\mu(x) \underset{t\to\infty}{\longrightarrow} +\infty \quad \text{a.s. } \forall x\in S\ ;\ \omega\text{ - a.s.}$$

c)

$$(36)\qquad \hat E(\eta_t^\mu(x)) = o(e^{\lambda t}) \quad \omega\text{ - a.s.}$$

Remark: There exist infinite particle systems evolving in a random environment having nontrivial equilibria; for example exlusion processes with a randomset of forbidden edges.

This theorem says, that in the case of an evolution in a random environment (and with symmetric kernels $p,\tilde{q},\tilde{\tilde{q}}$) for no values of the parameters interesting equilibria can exist and there is furthermore a dichotomy for the system either to die out locally or to explode locally.

The main open problem in this section is to show that the analogue of the theorem above holds also for kernels with drift. We only know:

Theorem 3':

Consider the case $S=\mathbb{Z}^1$. If for some $z\neq 0$: $p(x,y)=\tilde{q}(x,y)=\tilde{\tilde{q}}(x,y)=\delta_z(x-y)$ then there exists no equilibrium state with bounded particle density $\not\equiv 0$ as long as $T_x(\omega)\not\equiv$const. Furthermore for $T_x(\omega)\not\equiv$const and $\mu\neq\delta_{\{\eta\equiv 0\}}$

$$(37)\qquad \underline{\lim}_{t\to\infty}(e^{-\bar{\gamma}t}\hat{E}(\eta_t^\mu(x))) = 0, \quad \overline{\lim}_{t\to\infty}(e^{-\bar{\gamma}t}\hat{E}(\eta_t^\mu(x))) = +\infty$$

If $T_x(\omega)\equiv$const, but $P_x(\cdot)$, $d_1(x),\ldots$ are still random then different phenomena occur. This has been investigated in discrete time models for Branching Random Walks by Dawson and Fleischmann [1],[2].

4. Proof of theorem 1, 3 and Proposition 1-3

a) A formula for $\hat{E}(\eta_t^\mu(x))$.

Throughout this section we consider ω to be fixed and we will use from now on the following abbreviation:

$$(38)\qquad f_t(x) := \hat{E}(\eta_t^\mu(x))$$

In order to describe our representation of $f_t(x)$ we will need a Markov process $(X_s)_{s\in\mathbb{R}^+}$ on $\mathbb{Z}^n$ with jumprate $\varepsilon(x)$ and transition kernel $r(x,y)$. Furthermore we will need a function $\gamma(x)$.

$$(39)\qquad r(x,y) := (\varepsilon(x))^{-1}(mp(y,x)+b_1(\sum_{n=0}^{\infty}P_y(n)n)\tilde{q}(y,x)+b_2(y)\tilde{\tilde{q}}(y,x))$$

$$(40)\qquad \varepsilon(x) := m+\sum_{y\in S}(b_1(\sum_{n=0}^{\infty}P_y(n)n)\tilde{q}(y,x)+b_2(y)\tilde{\tilde{q}}(y,x))$$

$$(41)\qquad \gamma(x) := (\varepsilon(x)-m)-(d_1(x)+d_2(x)) \quad (=T_x(\omega))$$

Lemma 1: Define $\rho=E^\mu(\eta(0))$. Then

$$(42)\qquad f_t(x) = \rho\hat{E}\exp(\int_0^t\gamma(X_s^{(x)})ds)$$

Proof: Denote by $g_x(\eta)$ the functions $g_x(\eta):=\eta(x)$ for $\eta\in(\mathbb{N})^S$. It is meanwhile wellknown (see [6]) how to prove for models of our type that:

$$(43)\qquad \frac{d}{dt}\hat{E}(g_x(\eta_t^\mu)) = \hat{E}(Gg_x)(\eta_t^\mu)$$

Using the definition of G (see (8)) and of g_x we obtain from (43) that:

(44) $$\frac{d}{dt}f_t(x) = \varepsilon(x)\Big(\sum_{y\in S} r(x,y)f_t(y)-f_t(x)\Big)+\gamma(x)f_t(x)$$

$$f_o(x) \equiv \rho \quad (= E^{\mu}(\eta(0)))$$

An elementary calculation shows that the functions $\{\rho\hat{E}(\exp(\int_0^t\gamma(X_s^{\{x\}})ds)\}_{x\in S}$ fulfill also the system (44). Since (44) has a unique solution we have proved assertion (42),(unique in $L^{\infty}(S)$).

b) Proof of Proposition 1.

We obtain from formula (42) (Jensen's inequality) that:

(45) $$\frac{1}{t}\int_0^t \gamma(X_s^{\{x\}})ds \leq \frac{1}{t}\log f_t(x) \leq \lambda$$

Integrate this inequality over ω and note that the integrand is bounded in ω, and that the value of the integral on the left is τ. This implies immediately the assertion by a version of Fatou's Lemma. (see remark to(9)).

c) Proof of theorem 1a)

In order to derive theorem 1 from the formula (42) we consider for $\varepsilon>0$ the following set (which depends on ω of course)

(46) $$M_\varepsilon = \{x\in\mathbb{Z}^n \mid T_x(\omega) \geq \lambda-\varepsilon\}=\{x\in\mathbb{Z}^n\mid\gamma(x)\geq\lambda-\varepsilon\}$$

Now observe that:

(47) $$\exp\Big(\int_0^t\gamma(X_s^{\{x\}})ds\Big) \geq e^{(\lambda-\varepsilon)t}1_{\{X_s^{\{x\}}\in M_\varepsilon,\ \forall s\in[0,t]\}}$$

So that we can conclude:

(48) $$f_t(x) \geq \rho e^{(\lambda-\varepsilon)t}\mathrm{Prob}(\{X_s^{\{x\}}\in M_\varepsilon,\ \forall s\in[0,t]\})$$

Define now $\alpha=\sup_x(\varepsilon(x))$. With a coupling argument it is easy to show that:

(49) $$\mathrm{Prob}(X_s^{\{x\}}\in M_\varepsilon,\forall s\in[0,t]) \geq \sum_{n=0} e^{-\alpha t}\frac{(\alpha t)^n}{n!}\mathrm{Prob}(X_k^{\{x\}}\in M_\varepsilon,\ k=1,\dots,n)$$

Here $(X_n)_{n\in\mathbb{N}}$ is a Markov chain with transition kernel $r(x,y)$. In order to evaluate the right hand side of (49) we introduce:

(50) $$a(x,\varepsilon) := \overline{\lim_{n\to\infty}}(|n^{-1}\log(\mathrm{Prob}\ X_k^{\{x\}}\in M_\varepsilon,\forall\ k=1,\dots,n\})|)$$

Then we can conclude from (48) and (49) that there exist for every

$\varepsilon'>0$ a $c>0$ such that:

$$(51) \qquad f_t(x) \geq c\cdot e^{(\lambda-\varepsilon)t}\exp(\alpha(e^{-(a(\varepsilon,x)+\varepsilon')}-1)t)$$

Now let A_k^ε, B_k^ε be the following random sets:

$$(52) \qquad A_k^\varepsilon = \{x|x+[-k,k]^n \subseteq M_\varepsilon\} \quad B_k^\varepsilon=\{z|\text{Prob}(\{X_s^{\{z\}}|s\in\mathbb{R}^+\}\cap A_k^\varepsilon\neq\emptyset)>0)$$

and define:

$$(53) \qquad \hat{a}(\varepsilon,k) := \sup_x (a(\varepsilon,x)|x\in A_k^\varepsilon)$$

We obtain from (51) that:

$$(54) \qquad \forall\ z\in B_k^\varepsilon : \underline{\lim}_{t\to\infty}(\tfrac{1}{t}\log f_t(z))\geq \lambda-\varepsilon+\alpha(\exp(-a(\varepsilon,k)-\varepsilon')-1)$$

Since we have a random environment such that the $\{P_x(\cdot),d_1(x),d_2(x),b_2(x)\}_{x\in S}$ are all independent and have translation-invariant distributions we know that the points x lying in A_k^ε have positive density. Since $r(x,y)$ is irreducible this means that:

$$(55) \qquad B_k^\varepsilon = S \qquad \forall\varepsilon>0,\ k\in\mathbb{N}$$

Since ε,ε' in (54) can be made arbitrarily small we can conlcude from (54) using (55) that:

$$(56) \qquad (\inf_{\varepsilon,k}(\hat{a}(\varepsilon,k)) = 0 \text{ a.s.})\Rightarrow(\underline{\lim}_{t\to\infty}(\tfrac{1}{t}\log f_t(x)) \geq \lambda, \forall x\in S \text{ a.s.})$$

Now we introduce the following quantity (depending on ω through $r(x,y)$!)

$$(57) \qquad \tilde{a}(\varepsilon,k;\omega) = \lim_{m\to\infty}|\tfrac{1}{m}\log(\text{Prob}\{X_j^{\{0\}}\in[-k,k]^n\subseteq M_\varepsilon, \forall j\leq m\})|$$

Obviously it suffices (in view of (55), (56) and the stationarity of the process $\{T_x(\cdot)\}_{x\in S}$) to show the Lemma 2 below in order to prove theorem 1a).

Lemma 2:

$$(58) \qquad \text{infess}(\tilde{a}(\varepsilon,k;\omega)|\omega,\ \varepsilon>0,\ k\in\mathbb{N}) = 0$$

d) Proof of Lemma 2:
Crucial in this proof is the notion of a Markov chain restricted to a

set A. This restricted chain is a Markov chain with transition kernel $r_A(x,y)$ from A to A ($|A|<\infty$) defined as follows:

(59) $$r_A(x,y) := r(x,y)\ (\sum_{y\in A} r(x,y))^{-1}$$

$$e_A(x) := \sum_{y\in A} r(x,y) \quad \forall\ x\in A$$

We will be interested in sets A of the form $A=[-k,k]^n$.

Note first that for the purpose of our Lemma we can assume without loss of generality that $e_A(x)>0$ on A and that $r_A(x,y)$ is ergodic. (For purpose of large deviation theory we can replace $r(x,y)$ by a kernel $\bar{r}(x,y)$ with $\|r(x,\cdot)-\bar{r}(x,\cdot)\|\leq\delta$ such that we can make δ arbitrarily small!)

Making this assumption we have an irreducible chain on a finite state space. Therefore the restricted chain has a unique invariant measure, which we denote by π_A.

We can express the probability of $X_k^{\{0\}}$ to stay in A up to time n as $\int \prod_{i=1}^{n} e_A(x_i)dP(x_1,\ldots,x_n)$ where P is the distribution of the restricted chain. By the ergodic theorem for the chain restricted to A and Jensen's inequality we obtain:

(60) $$\overline{\lim_{n\to\infty}} \frac{1}{n}|\log(\text{Prob}(X_j^{\{x\}}\in A \text{ for } j\leq n))| \leq |\sum_{x\in A} \Pi_A(x)\log e_A(x)|$$

In order to evaluate the right hand side of this inequality we fix some $\tilde{\varepsilon}>0$ and define the $\tilde{\varepsilon}$-boundary of A with respect to $r(x,y)$:

(61) $$\delta A = \{y\in A \mid r(y,A)\leq 1-\tilde{\varepsilon}\}$$

Note first that on $A\setminus\delta A$ the quantity $|\log(e_A(x))|$ can be made small uniformly on A by making $\tilde{\varepsilon}$ small. So we have to investigate the behaviour of $e_A(x)$ on δA. Consider a sequence $A_k=[-k,k]^n$ (which we substitute for A). We shall show that for every $\tilde{\varepsilon}>0$ sufficiently small:

(62) $$\text{infess}(\sum_{x\in\delta A_k} \Pi_{A_k}(x) \mid \omega, \varepsilon>0,\ k\in\mathbb{N}) = 0$$

This proves via (61) and the remark above immediately our assertion.

Now we show (62). If $r(x,y)$ would be symmetric itself then we would have:

(63) $$\Pi_A(x) = e_A(x)(\sum_{x\in A} e_A(x))^{-1}$$

since r(x,y)=r(y,x) implies immediately the detailed balance condition:

(64) $$e_A(x)r_A(x,y) = e_A(y)r_A(y,x) \ .$$

The relation (63) would of course prove the assertion (62) since $\sum_{x\in A} e_A(x)$ is of order $|A|$ and $|\delta A|$ of order $o(|A|)$ for $|A|\to\infty$.

We sketch now how to conclude the proof in the general situation. Observe that we can achieve by making ε small that for fixed k and ω such that $A_k \subseteq M_\varepsilon$ we have: $r_{A_k}(x,y)$ is pointwise close to a kernel $\tilde{r}_{A_k}(x,y)$ which is a restriction of a symmetric kernel (this holds uniformly for all ω such that $A_k \subseteq M_\varepsilon$). The reason for this is that $b_2(y)$, $\sum_n P_y(n)n$ are almost constant for $y\in M_\varepsilon$. Then we conclude that the invariant measure of r_{A_k} is pointwise close to the invariant measure of $\tilde{r}_{A_k}$. q.e.d.

e) Proof of theorem 1b), c) Theorem 1'.

The key to all these assertions is the following Lemma relating the range of (X_s) to the growth of the particle density:

Lemma 3: Let (X_n) be the chain with transition kernel r(x,y). Assume that there exist $\beta,\delta>0$ such that with $R(n)=\{X_i | i=1,\ldots,n\}$ the following holds:

(65) $$\text{Prob}(|R(n)|n^{-1}\le\delta) \le c\cdot e^{-\beta n}, \quad T_x(\omega)\not\equiv\text{const} \ ; \ a.s.$$

Then

(66) $$\overline{\lim_{t\to\infty}} \left(\frac{1}{t} \log \hat{E}(\eta_t^\mu(x))\right) < \lambda$$

Proof of Lemma 3:

Define $L_{z,t}$ to be the time the process (X_s) spends at site z during it's first visit before time t. Denote by R(t) the range of (X_s) until time t. Then we define:

(67) $$C_t := \sum_{z\in R(t)} L_{z,t}$$

With the assumption (65) we can conclude from the fact that the jumprate $\varepsilon(x)$ of (X_s) is bounded that:

(68) $$\exists\, \tilde{\beta},\tilde{\delta} > 0 \quad \text{such that} \quad \text{Prob}(C_t t^{-1}\le\tilde{\delta}) \le ce^{-\tilde{\beta}t}$$

To the process $\{T_x(\cdot)\}_{x\in S}$ we can apply the ergodic theorem to obtain after some analysis that:

$$(69)\qquad |R(t)|^{-1}\sum_{z\in R(t)}\gamma(z)\xrightarrow[|R(t)|\to\infty]{}\tau \quad \text{a.s.};$$

$$\overline{\lim_{t\to\infty}}((C(t))^{-1}\sum_{z\in R(t)}\gamma(z)L_{z,t})\leq\tau \quad \text{a.s.}$$

We define:

$$(70)\qquad \varepsilon(t)=\tau^{-1}([C(t)]^{-1}\sum_{z\in R(t)}\gamma(z)L_{z,t})$$

Then we can decompose and estimate the exponent of (42) as follows:

$$(71)\qquad \frac{1}{t}\int_0^t\gamma(X_s)ds\leq\varepsilon(t)\tau\frac{C(t)}{t}+\lambda(1-\frac{C(t)}{t})$$

To proceed further note that by Jensen's inequality:

$$(72)\qquad E_\omega(\frac{1}{t}\log\hat{E}\exp(\int_0^t\gamma(X_s)ds))\leq\frac{1}{t}\log(E_\omega\hat{E}\exp(\int_0^t\gamma(X_s)ds))$$

Now use the fact that the $\{T_x(\cdot)\}_{x\in S}$ are averages of i.i.d. random variables to conclude that for $T_x(\cdot)\not\equiv\text{const}$ there exist $0<\tilde{\varepsilon}<\tilde{\tilde{\varepsilon}}$ such that:

$$(69')\qquad \text{Prob}((C(t))^{-1}\sum_{z\in R(t)}\gamma(z)L_{z,t}\geq\lambda-\tilde{\varepsilon})\leq c\cdot e^{-\tilde{\tilde{\varepsilon}}t}$$

Now it is straightforward to conclude with (68) (69') from (71) that the right handside of (72) is strictly smaller than λ in the limit. Now we obtain our assertion (66) from (72) with the dominated convergence theorem.

From Lemma 3 we obtain immediately theorem 1b). In order to obtain 1c) observe that $R(n)$ is "deterministic", it has the form $\{x,x+1,x+2,\ldots,x+n\}$ and therefore (69) and (71) allow to conclude that (16) holds.

To obtain theorem 1' observe first that Cramer's theorem on large deviations states that for a sum S_n of bounded random variables we have $\text{Prob}(S_n n^{-1}\leq\delta)\leq ce^{-\beta n}$ for some $\beta>0$ as long as δ is smaller than the drift. If $p(x,y)$ has finite range K then $K|R_n|$ is bigger than any component of the vector $(\sum_{i=1}^n X_i)$. This implies with Cramers theorem our assertion in the case of finite range. To obtain the general case note that the probability to have n jumps bigger than L has the form α^N with $\alpha<1$ for large L.

To show Proposition 2 observe first that $R(\bar{n})\supseteq[0,\max_{n\leq\bar{n}}(\sum_i^n X_i)]$. Then we apply (67)-(72) to obtain (19"). To see why (19') is true note that we can calculate as in (46)-(64), we only have to replace (62), now we obtain here a positive number which we can however calculate explicitly since we can calculate Π_A explicitly through solving a difference equation. We leave the straigthforward details to the reader.

To prove Proposition 4 observe that the inequality on the right of (23) follows from theorem 1 with Fatou's Lemma respectively with (72). The left inequality in(23) is implied by (42) and the following consequence of a theorem of Cramer on large deviations: There are $\tilde{\varepsilon},\tilde{\tilde{\varepsilon}}$ with $0<\tilde{\tilde{\varepsilon}}<\tilde{\varepsilon}$ such that $\mathrm{Prob}(n^{-1}\sum_{i=1}^{n} T_{x_i} \geq \tau+\tilde{\varepsilon}) \leq e^{-\tilde{\tilde{\varepsilon}}n}$ (The $T_x(\cdot)$ are averages of i.i.d. variables!, see (10)).

f) Proof of theorem 3:

In the symmetric situation the key to the whole theorem is the assertion that $\hat{E}(\eta_t^{\mu}(x))=o(e^{\lambda t})$. To prove this latter fact we use formula (42) and the set M_{ε} defined in (46) to write:

$$\hat{E}(\eta_t^{\mu}(x)) \leq \rho\, \hat{E}(\exp[-\varepsilon\int_0^t 1_{\{X_s^{\{x\}}\notin M_{\varepsilon}\}}ds])\, e^{\lambda t} \tag{73}$$

Now remember that $\{T_x(\omega)\}_{x\in S}$ is stationary, shiftergodic and the kernel $r(x,y)$ is irreducible. This allows to conclude from the fact that $T_x(\omega)\not\equiv \mathrm{const}$, that there exists an $\bar{\varepsilon}>0$ such that:

$$\int_0^t 1_{\{X_s^{\{x\}}\notin M_{\varepsilon}\}}ds \underset{t\to\infty}{\longrightarrow} +\infty \qquad \text{for all} \quad 0<\varepsilon\leq\bar{\varepsilon} \tag{75}$$

Putting (75) and (73) together yields (via the monotone convergence theorem) the assertion (36).

From (36) we obtain (33) via the following consideration: Suppose ν is an equilibrium for some ω and has the property that $\int\eta(x)d\nu\leq K$, $\forall x\in S$. Define the maps φ_n: $\eta\to\eta\wedge n$ and set $\nu_n=\varphi_n(\nu)$. By part c) we conclude with a comparison with the initial distribution $\delta_{\{\eta\equiv n\}}$ that (compare (76)) $\mathcal{L}(\eta_t^{\nu_n}) \underset{t\to\infty}{\overset{w}{\Longrightarrow}} \delta_{\{\eta\equiv 0\}}$ or ω belongs to a set with probability zero. Suppose ν is not in this exceptional set. It is now possible to define η_t^{ν}, $\eta_t^{\nu_n}$ on a common probability space such that (since $\lambda=0$!):

$$\hat{E}((\eta_t^{\nu}(x)-\eta_t^{\nu_n}(x))^+) \leq \hat{E}((\eta_0^{\nu}(x)-\eta_0^{\nu_n}(x))^+) \quad \forall t\in\mathbb{R}^+ \tag{76}$$

(to see this one couples as many transitions per site in both components as possible, for example in both components a particle moves from x to y at rate $(\eta^1(x)\wedge\eta^2(x))p(x,y)$ etc. For detailed expositions of these techniques see [6],[3]). Since the right side in (76) can be made arbitrarily small by making n large, we have that $\mathcal{L}(\eta_t^{\nu_n}) \underset{t\to\infty}{\longrightarrow} \delta_{\{\eta\equiv 0\}}$ implies that $\mathcal{L}(\eta_t^{\nu}) \underset{t\to\infty}{\longrightarrow} \delta_{\{\eta\equiv 0\}}$. This means that we must have that $\eta=\delta_{\{\eta\equiv 0\}}$ q.e.d.

It is clear that for $\lambda<0$, the process dies out locally a.s. since then $\hat{E}(\eta_t^{\mu}(x))\leq \rho e^{\lambda t}$, so that $\int_0^{\infty}1_{\{\eta_t^{\mu}(0)>0\}}dt<\infty$ a.s. For $\lambda=0$ we have only $\hat{E}(\eta_t^{\mu}(x)) \underset{t\to\infty}{\longrightarrow} 0$, according to (36).

It remains to show that for $\lambda>0$: $L(\eta_t^\mu) \underset{t\to\infty}{\longrightarrow} \delta_{\{\eta\equiv\infty\}}$. Observe first that there exist points $x \in S$, which have positive density such that $\underline{\lim}_{t\to\infty}(\eta_t^\mu(x))=+\infty$: just pick x in the center of a large enough cube with $T_y(\omega)\geq\delta'>0$ throughout and so that the expected number of descendents of a fixed particle at x at time 0, which remain also in that cube until time t, grows at least like $ce^{\delta t}$ with c,δ positiv. Now apply the martingale convergence theorem to obtain the divergence of that part of this family staying in the cube. Since the cube is finite, this implies immediately that $\eta_t^\mu(x)\to+\infty$ a.s.. Since $r(x,y)$ is irreducible we conclude that for all $x\in S$: $\underline{\lim}_{t\to\infty}\eta_t^\mu(x) = +\infty$.

5. Proof of theorem 2,2'

a) The case $\lambda>0$

The key to the proof in this case is the Lemma 4 below. We shall need the following quantities for our arguments: ((X_s) is introduced in 4a))

$$(77)\qquad h_t'(x,z) = \hat{E}\exp\left(\int_0^t \gamma(X_s^{\{x\}})ds\right)1_{\{X_t^{\{x\}}=z\}}$$

$$(78)\qquad \tilde{h}_t(x,y) = \sum_{z_1,z_2\in S} a(z_1,z_2)h_t'(x,z_1)h_t'(y,z_2)$$

Here $a(z_1,z_2)=K(q(z_1,z_2)+\tilde{q}(z_2,z_1)+\tilde{K}(\delta(z_1,z_2)+\sum_x\tilde{q}(z_1,x)\tilde{q}(z_2,x))$

Lemma 4: Suppose that $\lambda>0$, $b_2(x)+d_2(x)\equiv0$. Then we have:

$$(79)\qquad \hat{E}(\eta_t^\mu(x))^2 \leq f_t(x)+(f_t(x))^2+C\int_0^t(\tilde{h}_{t-s}(x,x))e^{\lambda s}ds$$

$$(80)\qquad \tilde{h}_t(x,y) = f_t(x)f_t(y)D_t(x,y) \quad \text{with } D_t(x,y) \underset{t\to\infty}{\longrightarrow} 0$$

(81) We can choose $\lambda'\in(\lambda/2,\lambda)$ such that there exists a $c>0$, $t_o,s_o<\infty$ such that:

$$f_t(x) \geq c\cdot f_{t-s}(x)e^{\lambda' s} \qquad \forall\, s\geq s_o\ ,\ t-s\geq t_o$$

From Lemma 4, (79) and Jensen's inequality we conclude:

$$(82)\qquad \rho^2 \leq \hat{E}(\xi_t^\mu(x))^2 \leq \rho^2+\tilde{c}\int_0^t D_{t-s}(x,x)e^{(\lambda-2\lambda')s}ds+o(1)$$

With part (80), (81) of the Lemma and the fact $\lambda-2\lambda'<0$ we obtain:

$$(83)\qquad \hat{E}(\xi_t^\mu(x))^2 \underset{t\to\infty}{\longrightarrow} (\hat{E}(\xi_t^\mu(x)))^2$$

This means:

(84) $\hat{Var}(\xi_t^\mu(x)) \xrightarrow[t\to\infty]{} 0$ q.e.d.

Proof of Lemma 4.

(85) Define $f_t(x,y) := \hat{E}(\eta_t^\mu(x)\eta_t^\mu(y)-\delta(x,y)\eta_t^\mu(x))$

With standard methods (compare [6]) one can show that:

(86) $\frac{d}{dt} f_t(x,y) = \hat{E}(G(\eta_t^\mu(x)\eta_t^\mu(y)-\delta(x,y)\eta_t^\mu(x)));\quad f_o(x,y)=\rho^2$

With an elementary calculation we evaluate the righthand side and get that for some positive b(x,y;z) (which could be given explicitly)

(87)
$$\begin{aligned}\frac{d}{dt} f_t(x,y) = {} & \varepsilon(x)(\sum_{z\in S} r(x,z)f_t(z,y)-f_t(x,y))+\\ & +\varepsilon(y)(\sum_{z\in S} r(y,z)f_t(x,z)-f_t(x,y))\\ & +\gamma(x)f_t(x,y)+\gamma(y)f_t(x,y)\\ & +\sum_z b(x,y;z)f_t(z)\end{aligned}$$

Now define a semigroup V_t on $L^\infty(S\times S)\otimes L^\infty(S)$ by the tupels of functions $\{f_t(x,y),f_t(z)\}_{x,y,z\in S}$ and a semigroup U_t on the same space by $\{\tilde{f}_t(x,y),f_t(z)\}_{x,y,z\in S}$. Here $\tilde{f}_t(x,y)$ are solutions of (87) but with the last line deleted. Denote by L_V, L_U the generators of V_t, U_t. A well-known formula says that:

(88) $V_t(f) = U_t(f)+\int_0^t U_{t-s}(L_V-L_U)V_s(f)ds$

By theorem 3c) we know that $f_t(x)\leq \rho e^{\lambda t}$. This leads with (88) and the evaluation of U_t via the Feynman-Kac formula (compare (44)!) to the assertion (79). To see (80) note that $a(z_1,z_2)$ is symmetric, positive $\sum_{z_2} a(z_1,z_2)\leq K<\infty$, so that with $D_t'(x,z):=(f_t(x))^{-1}h_t'(x,z)$ we can write

(89) $\tilde{D}_t(x,y) = \sum_{z_2} D_t'(x,z_2)(\sum_{z_1} a(z_1,z_2)D_t'(y,z_1))$

and it suffices therefore to show $(0\leq D_t'(x,z)\leq 1,\ \sum_z D_t'(x,z)=1!)$ that

(90) $D_t'(x,z) \xrightarrow[t\to\infty]{} 0 \quad \forall\ x,z$

Note that $h_t'(x,z)$ describes the quantity $\hat{E}(\eta_t^\alpha(x))$, $\alpha=\delta_{\{\eta\equiv\delta_z\}}$. With the fact that the average number of points with $\gamma(y)\leq\lambda-\varepsilon$ which a path from z to $\mathcal{C}(z+[-k,k]^n$ has to visit diverges for $k\to\infty$, we obtain (90).

To see (81) note that with (42) this relation amounts to showing (in view of 4c)!) that $\underline{\lim}_{t\to\infty}(\sum_{z\in M_\varepsilon} D_t(x,z))>0$ for all $\varepsilon\in[0,\bar{\varepsilon}]$ for some $\bar{\varepsilon}$. This can be done through coupling techniques and approximation through finite systems. We can't give the lengthy details here.

b) The case $T_x(\omega)\equiv$const.

Consider first the case $\lambda>0$. Then theorem 2' part (27') is obtained immediately following the calculations (77)-(84) and observing that $T_x(\omega)\equiv T$ implies (for arbitrary kernel) that:

$$(91)\qquad f_t(x) = e^{Tt} \quad , \quad h_t'(x,z) = e^{Tt}\text{Prob}(\{X_t^{\{x\}}=z\})$$

Due to this explicit formulas we can drop the assumption that μ is derived from a Poisson system. If in addition $\{X_t\}$ is transient, i.e. $\int_0^\infty \text{Prob}(\{X_t=z\})dt<\infty$, then (79) implies that:

$$(92)\qquad \int_0^\infty(\xi_t^\mu(0)-1)^2 dt < \infty \quad \text{a.s.}$$

Since we can construct a Poisson process N_h with parameter $(m+b)\eta_t^\mu(x)$ such that for $h\leq\bar{h}$, $c>0$ and with $\bar{N}_n=N_h\cdot(\hat{E}(\eta_t^\mu(x)))^{-1}$

$$(93)\qquad \xi_{t+h}^\mu(x) \geq (\xi_t^\mu(x)-\bar{N}_h)(\rho\, e^{\bar{h}T})^{-1}$$

the relation (92) implies that $\xi_t^\mu(0)$ converges to 1 a.s..

In the case $\lambda<0$ observe first that $\xi_t^\mu(0)$ can be represented in the form:

$$(94)\qquad \xi_t^\mu(0) = \sum_{i=0}^{N_t} Y_t^{(i)} \quad , \quad (Y_t^{(i)}) \text{ exchangeable Bernoulli-variables for given } N_t.\ N_t\to\infty\ .$$

(To see this imbedd the process in a growing system of Branching Random Walks where the deathrates are 0.) A wellknown theorem of de Finetti tells us now that the weak limit points of $\mathcal{L}(\xi^\mu_t(0))$ for $t\to\infty$ are mixed Poisson-distributions. By construction $\hat{E}(\xi_t^\mu(0))$ converges to ρ for $t\to\infty$ so that $\sup_t \hat{E}(\xi_t^\mu(0))^2<\infty$ implies that the weak limit points of $\mathcal{L}(\xi_t^\mu(0))$ are mixed Poisson with mean ρ. To obtain the result on the second moments we derive as in a) that:

$$(95)\qquad \hat{E}(\xi_t^{\mu}(0))^2 \leq (\hat{E}(\xi_t^{\mu}(0)))^2 + \hat{E}(\xi_t^{\mu}(0)) + C\sum_{z_1,z_2}\sum_{x,y\in I_t} a(z_1,z_2)\int_0^t e^{2\lambda(t-s)} e^{\lambda s} r_{t-s}(x,z_1) r_{t-s}(y,z_2)\,ds$$

Take for convenience $a(z_1,z_2)=\delta(z_1,z_2)$ then we have:

$$(96)\qquad \hat{E}(\xi_t^{\mu}(0))^2 \leq \rho^2+\rho+o(1)+C\sum_{x\in I_t} e^{-\lambda t}\int_0^t e^{2\lambda(t-s)} e^{\lambda s}\hat{r}_{t-s}(x,0)\,ds \leq K < \infty$$

A more careful analysis of the second moments shows that in fact $\lim_{t\to\infty} \hat{E}(\xi_t^{\mu}(0))^2 > \rho^2+\rho$ so that we obtain a mixed Poisson distribution in the limit. The case $\lambda<0$ is handled similar.

References

[1] Dawson, D.; Fleischmann, K.: On spatially homogeneous branching processes in random environment. Math.Nachr. 113 (1983), 249-257.

[2] Dawson, D.; Fleischmann, K.: A branching random walk in a random environment. Preprint (1984).

[3] Greven, A.: Critical phenomena for the Coupled Branching Process, Preprint (1985)

[4] Greven, A.: The Coupled Branching Process in Random Environment. To appear Ann.of Probability Vol.13

[5] Greven, A.: Phase transition for a class of Markov processes on $(\mathbb{N})^S$ To appear: Proceedings of the AMS-Summer Research Conference on: Mathematics of Phase Transition, (1984). Contemporary Mathematical Series.

[6] Liggett, T.; Spitzer, F.: Ergodic Theorems for Coupled Random Walks and other systems with locally interacting components. Z.Wahrscheinlichkeitstheorie verw. Gebiete 56 (1981), 443-468.

Andreas Greven
Institut für Angewandte Mathematik
Im Neuenheimer Feld 294
D-6900 Heidelberg 1
West-Germany

PERCOLATION PROCESSES AND DIMENSIONALITY

Geoffrey Grimmett
School of Mathematics
University of Bristol
England

Abstract

We discuss bond percolation on the cubic lattice $\mathbb{Z}^d$ in dimensions $d = 1,2,3$, paying particular attention to the ways in which such processes "evolve" as the dimension increases from $d = 1$ through $d = 2$ to $d = 3$.. There are many conjectures.

Contents

1. Introduction to bond percolation

There are three dimensions to the space which we inhabit, and consequently most spatial models of statistical physics are three-dimensional. On the other hand, the two-dimensional case is often more popular with mathematicians and the following trite table may indicate something of why this is so.

dimension	degree of difficulty	observation
1	none	no critical phenomena
2	considerable	critical phenomena
3	very great	critical phenomena

Thus physical processes in one dimension are generally soluble, often having exact solutions which exhibit no critical behaviour as the underlying parameters vary. At the other extreme, three-dimensional processes are often very difficult to study and much of our information about such processes is derived by approximation methods such as Monte Carlo simulation, series expansions and renormalization rather than by rigorous analytical techniques. In the intermediate case of two dimensions, however, such processes may sometimes be studied usefully, though with some difficulty, by rigorous mathematical arguments. Also, two is the least number of dimensions for which there is a truly spatial aspect, generally exhibiting critical phenomena. Two-dimensional processes can thus both be challenges and yield rewards, since they are often rather difficult but may sometimes reveal their secrets under pressure. It is interesting to study particular physical processes in detail as their dimensions increase from one, through two, to three, and it is the purpose of this paper to do this for the bond percolation process on the cubic lattice $\mathbb{Z}^d$.

The bond percolation process was introduced by Broadbent and Hammersley (1957) as a model for a porous stone. More recently it has become a fashionable subject for contributors to physics journals, and one reason for this is that it may be considered to be a model for ferromagnetism in which the interaction has been reduced to a minimum. Recent reviews include Kesten (1982), Smythe and Wierman (1978), Essam (1980) and Stauffer (1979).

Let $\mathbb{Z}^d$ be the d-dimensional cubic lattice with vertex set $\{(i_1,i_2,\ldots,i_d) : i_k = \ldots,-1,0,1,\ldots,\ k = 1,2,\ldots,d\}$ and edges joining vertices $\underline{i}$ and $\underline{j}$ whenever

$$\sum_{k=1}^{d} |i_k - j_k| = 1,$$

and let p be a number satisfying $0 < p < 1$. We declare each edge of $\mathbb{Z}^d$ to be <u>open</u> with probability p and <u>closed</u> otherwise, independently of all other edges. The set of vertices of $\mathbb{Z}^d$ together with the open edges (only) forms a subgraph of the cubic lattice, and the components of this subgraph are called <u>open clusters</u>. Percolation theory is concerned with properties of these open clusters and particularly with the ways in which such properties change as the density p of open edges increases from 0 to 1. Of great interest and importance is the so-called critical phenomenon which takes place for a particular intermediate value of p. Writing I for the event that there is an infinite open cluster and P_p for the probability function corresponding to a given value of p, it is not too difficult to show that there exists a number $\pi(d)$, depending on the dimension d, such that

$$P_p(I) = \begin{cases} 0 & \text{if } 0 \le p < \pi(d), \\ 1 & \text{if } \pi(d) < p \le 1, \end{cases}$$

and $\pi(d)$ is called the <u>critical probability</u> of the process. It is easy to see that $\pi(1) = 1$, so that there is no interesting critical phenomenon in one dimension. On the other hand, it may be shown that $0 < \pi(d) < 1$ if $d \ge 2$, so that all dimensions exceeding 1 exhibit critical behaviour. It seems to be exceedingly difficult to calculate $\pi(d)$ exactly, although the now celebrated Harris-Russo-Seymour-Welsh-Kesten theorem states that $\pi(\frac{1}{2}) = \frac{1}{2}$. It is thought that the sequence $\pi(1),\pi(2),\ldots$ is strictly decreasing and satisfies

$$d\pi(d) \to \tfrac{1}{2} \quad \text{as} \quad d \to \infty$$

(see Gaunt and Ruskin (1978)).

We place the emphasis of this paper upon the dimensionality of the bond percolation process, beginning with a discussion of the case $d = 1$. The usual percolation process is of little interest here, but we shall see that there is a more general type of process involving "long-range interactions" which enjoys a critical phenomenon; this new process is akin to a well-known one-dimensional model for ferromagnetism. We study the transition from $d = 1$ to $d = 2$ in some detail, paying particular attention to the problem of determining the "effective dimensions" of subsets of the square lattice $\mathbb{Z}^2$. Much less is known about the transition from $d = 2$ to $d = 3$ and there are several related entertaining conjectures.

2. One-dimensional percolation models and long-range effects

In this section we consider a more general one-dimensional percolation model than the usual bond percolation process. Let $\underline{p} = (p(1),p(2),\ldots)$ be a sequence of numbers satisfying $0 \le p(n) < 1$ for all n, and let $\mathbb{Z}$ be the set of all integers. We examine each distinct unordered pair (i,j) in turn, and either we join this pair by an edge with probability $p(|i-j|)$ or we leave the pair disconnected otherwise; this is done independently of all other unordered pairs. We denote by G the ensuing (random) graph. The usual bond percolation process on $\mathbb{Z}$ is retrieved by setting $p(n) = 0$ if $n \ge 2$, and thus the general model may be thought of as a percolation-type process with (possibly) long-range interaction.

Whereas the bond percolation process on $\mathbb{Z}$ is trivial to study and shows no critical behaviour, the model described above has a rich structure which depends largely upon the tail behaviour of the sequence $p(1),p(2),\ldots$. The following theorem describes the threshold between connectedness and disconnectedness of G.

Theorem 1. *The graph G is almost surely connected if*

$$\sum_{n=1}^{\infty} p(n) = \infty \tag{2.1}$$

and the greatest common divisor of $\{n : p(n) > 0\}$ equals 1; if either of these two conditions fails to hold then G is almost surely disconnected.

See Grimmett, Keane and Marstrand (1984) for the proof of this theorem and for more details of the threshold. Theorem 1 may be generalized in the obvious way to higher dimensions, replacing $\mathbb{Z}$ by $\mathbb{Z}^d$ where $d \ge 2$. It is believed that the corresponding result is valid for the subgraph of G on the smaller vertex set

$\{0,1,2,\ldots\}^d$, but no proof of this is known if $d \geq 2$. S. Kalikow has found an easier approach (unpublished) to the proof of Theorem 1; he extends the conclusion of the theorem to the subgraph of G on $\{0,1,2,\ldots\}$ but his argument is not easy to generalize to higher dimensions.

Even when $\Sigma_n p(n) < \infty$ it is possible that G contains an infinite component. Let $\theta(\underline{p})$ be the probability that the component of G containing the vertex 0 is infinite. It is not difficult to show that

$$P_{\underline{p}}(\text{G contains an infinite component}) = \begin{cases} 0 & \text{if } \theta(\underline{p}) = 0, \\ 1 & \text{if } \theta(\underline{p}) > 0, \end{cases}$$

where $P_{\underline{p}}$ is the probability function corresponding to the sequence $\underline{p}$ of edge-probabilities. This threshold between non-existence and existence of infinite components is akin to the usual bond percolation threshold described in the introduction. It turns out that there is a non-trivial threshold for the long-range process. First, there is a simple argument (see Schulman (1984)) showing that

$$\theta(\underline{p}) = 0 \quad \text{if} \quad \sum_{n=1}^{\infty} np(n) < \infty \tag{2.2}$$

and implying that all components are almost surely finite if $\Sigma_n np(n) < \infty$. The remaining cases are those for which the p(n)'s satisfy the conditions of neither (2.1) nor (2.2), and the natural case to study is when

$$p(n) \simeq \frac{c}{n^s} \quad \text{for large } n, \tag{2.3}$$

for constants c and s satisfying $1 < s \leq 2$. With a little thought we may see that, unlike in Theorem 1, both short-range and long-range interactions are important here; consequently we think of the sequence $\underline{p}$ as containing two "independent" quantities, being the number p(1) and the sequence (p(2),p(3),...) satisfying (2.3). The usual branching process argument (see Schulman (1984)) gives that all components are almost surely finite if p(1) and c are small enough:

$$\theta(\underline{p}) = 0 \quad \text{if} \quad \sum_{n=1}^{\infty} p(n) \leq \tfrac{1}{2}.$$

On the other hand, C. Newman and L. Schulman have shown (unpublished) that it is possible to have infinite components with positive probability.

Theorem 2. *Suppose that the sequence* $(p(2),p(3),\ldots)$ *satisfies* (2.3).

(i) *If* $1 < s < 2$ *and* $c > 0$, *then* $\theta(\underline{p}) > 0$ *for large values of* $p(1)$.

(ii) *If* $s = 2$ *and* $c > 2$, *then* $\theta(\underline{p}) > 0$ *for large values of* $p(1)$.

It seems that the case $s = 2$ provides the most interesting critical behaviour, especially in the light of the following result (unpublished) of M. Aizenman and C. Newman.

Theorem 3. *Under the hypothesis of Theorem 2,*

(i) *if* $s = 2$ *and* $c < 1$, *then* $\theta(\underline{p}) = 0$,

(ii) *if* $s = 2$, *then either* $\theta(\underline{p}) = 0$ *or* $\theta(\underline{p}) > c^{-\frac{1}{2}}$.

Thus, in the extreme case when $s = 2$, $\theta(\underline{p})$ is a discontinuous function of c at the critical point. It is currently an open problem to ascertain the critical value σ of c given by

$$\sigma = \inf\{c : \theta(\underline{p}) > 0 \text{ for large } p(1)\},$$

although we have from Theorems 2 and 3 that $1 \le \sigma \le 2$.

In summary, the usual bond percolation process in one dimension is of little or no interest, but a contrasting long-range model has a complicated and interesting theory. It is revealing to note the similarity between this model and a well-known one-dimensional model of statistical physics in which two vertices which are distance n apart enjoy a ferromagnetic interaction with strength $J(n) \approx Jn^{-s}$ for constants J and s. See Fröhlich and Spencer (1982) for recent results about the critical behaviour of this process.

3. Two-dimensional percolation and the transition from one to two dimensions

We return to the usual bond percolation process on $\mathbb{Z}^d$ and concentrate in this section on the planar case when $d = 2$. It is now well-known that the critical probability $\pi(2)$ in two dimensions is equal to $\frac{1}{2}$. If A is a subgraph of $\mathbb{Z}^2$, we may define the critical probability of the bond percolation process on A to be the number $\pi(A)$ given by

$$\pi(A) = \sup\{p : P_p(I(A)) = 0\}$$

where I(A) is the event that A contains an infinite open cluster. The critical probability of A is an indication of the "effective dimension" of the bond percolation process on A, two extreme cases being $\pi(\mathbb{Z}) = 1$ and $\pi(\mathbb{Z}^2) = \frac{1}{2}$. It turns out that $\pi(A)$ takes all values in the interval $[\frac{1}{2},1]$ as A varies over subsets of $\mathbb{Z}^2$.

Theorem 4. *Let* $0 \le c < \infty$ *and let* $f(x) = c \log(x+1)$ *for* $x \ge 0$. *The subgraph of* $\mathbb{Z}^2$ *on the set*

$$A(c) = \{(i,j) \in \mathbb{Z}^2 : 0 \le j \le f(i),\ i \ge 0\}$$

has critical probability $\pi(A(c)) = \nu(c)$ *where* ν *is a continuous and strictly decreasing function which maps* $[0,\infty)$ *onto* $(\frac{1}{2},1]$.

See Grimmett (1981, 1983) for a proof of this result. Fig. 1 contains a sketch of the function ν. Theorem 4 may be read as saying that the transition from one dimension to two dimensions is smooth, in the sense that the critical probability of $A(c)$ varies continuously between $\pi(\mathbb{Z}) = 1$ and $\pi(\mathbb{Z}^2) = \frac{1}{2}$ as c varies from 0 to ∞.

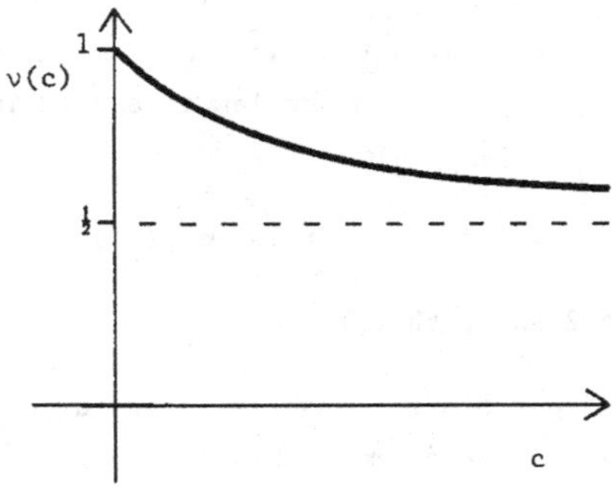

Fig. 1 A sketch of the function ν.

J. van den Berg has pointed out that the *strict* monotonicity of ν may be obtained by applying a general theorem of reliability theory (see Barlow and Proschan (1975)) as well as by the ad hoc argument of Grimmett (1983).

The conclusion of Theorem 4 may be contrasted with the results of Kesten (unpublished) and Hammersley and Whittington (1984) who have counted self-avoiding walks in wedges of $\mathbb{Z}^2$. Let w_n be the number of self-avoiding walks of length n in $\mathbb{Z}^2$ starting from the origin, and let $w_n(f)$ be the number of such walks which are confined to the subset

$$A(f) = \{(i,j) \in \mathbb{Z}^2 : 0 \le j \le f(i),\ i \ge 0\}$$

of $\mathbb{Z}^2$ where f is a non-negative function on $[0,\infty)$. It is well-known that the limit

$$\kappa = \lim_{n\to\infty} \frac{1}{n} \log w_n$$

exists; κ is called the *connective constant* of $\mathbb{Z}^2$. Hammersley and Whittington (1984) show the following theorem.

Theorem 5. If $f(x) \to \infty$ as $x \to \infty$, then

$$\frac{1}{n} \log w_n(f) \to \kappa \quad \text{as } n \to \infty.$$

That is to say, the connective constant of the wedge $A(f)$ is the same as the connective constant of the whole square lattice whenever the height of the wedge tends to infinity as one looks progressively to the right. In other words, "$f(x) \to \infty$ as $x \to \infty$" is a sufficient condition for $A(f)$ to be "effectively two-dimensional" with regard to the number of its self-avoiding walks.

4. Three-dimensional percolation and the transition from two to three dimensions

Bond percolation in three dimensions is a process which is perhaps as rich in conjectures and open problems as it is in known results. Pride of place amongst conjectures should go to the question of the uniqueness of the critical probability. Let $W(\underline{0})$ be the number of vertices of $\mathbb{Z}^3$ which are joined to the origin $\underline{0}$ by open paths (that is, by paths containing open edges only). The critical probability $\pi(3)$ may be defined by

$$\pi(3) = \sup\{p : P_p(W(\underline{0}) = \infty) = 0\}.$$

We may also define

$$\gamma(3) = \sup\{p : E_p(W(\underline{0})) < \infty\},$$

the supremum of all values of p for which $\underline{0}$ belongs to an open cluster with finite mean size. It is clear that $\gamma(3) \le \pi(3)$.

Conjecture 6. Using the above notation, it is the case that $\gamma(3) = \pi(3)$.

The transition from two to three dimensions is not well understood at present. There are (at least) two ways of formulating the problem, and the more important way is as follows. Let $\mathbb{Z}^2(k)$ denote the slice $\mathbb{Z}^2 \times \{1,2,\dots,k\}$ with thickness k cut from $\mathbb{Z}^3$, and let $\rho(k)$ be the critical probability of the bond percolation process on $\mathbb{Z}^2(k)$. Thus $\rho(1) = \pi(2) = \frac{1}{2}$ and $\rho(k+1) \le \rho(k)$ for $k = 1,2,\dots$. It is conjectured that the k-slice behaves more and more like the whole lattice $\mathbb{Z}^3$ as $k \to \infty$.

Conjecture 7. The limit $\rho = \lim_{k\to\infty} \rho(k)$ is given by $\rho = \pi(3)$.

It is clear that $\rho \ge \pi(3)$, but it does not seem to be at all easy to show equality here (see Kesten (1982) and Aizenman et al. (1983)).

An alternative approach to the transition from two to three dimensions resembles more the method of the previous section. Let f_2 and f_3 be non-negative, non-decreasing functions on $[0,\infty)$ and define the f-wedge of $\mathbb{Z}^3$ to be the subgraph of $\mathbb{Z}^3$ on the vertex set $\{(i,j,k) \in \mathbb{Z}^3 : 0 \le j \le f_2(i),\ 0 \le k \le f_3(i),\ i \ge 0\}$. We write $\pi(f)$ for the critical probability of the bond percolation process on this f-wedge. Very little is known about the behaviour of $\pi(f)$ in terms of the functions f_2 and f_3.

Theorem 8. *Let* $g(i) = (f_2(i) + 1)(f_3(i) + 1)$ *for* $i = 0,1,2,\ldots$.

(i) *If* $g(i)/\log i \to 0$ *then* $\pi(f) = 1$, *and*

(ii) *if* $g(i)/\log i \to \infty$ *then* $\pi(f) \le \frac{1}{2}$,

where the limits are taken as $i \to \infty$.

Hammersley and Whittington (1984) have proved part (i); part (ii) is a minor extension of another of their results. A slightly more sophisticated treatment provides some information about the behaviour of $\pi(f)$ in the critical case when $g(x) = a \log(x+1)$ and $a > 0$, giving that there exists $\nu(a) < 1$ such that

$$\max\{\pi(3),\ 1 - e^{-1/a}\} \le \pi(f) \le \nu(a)$$

in this case.

It is an open question to ascertain attractive conditions on f_2 and f_3 which imply that $\pi(f) = \pi(3)$. Perhaps it is enough to require that

$$f_2(i) \to \infty,\ f_3(i) \to \infty,\ f_2(i)f_3(i)/\log i \to \infty$$

as $i \to \infty$.

Another intriguing open question is whether or not there can be more than one infinite open cluster in $\mathbb{Z}^3$. Writing N for the number of such clusters, it is not difficult to show that almost surely exactly one of the following three possibilities must hold for any given value of p:

(i) $N = 0$, (ii) $N = 1$, (iii) $N = \infty$.

Clearly $N = 0$ almost surely if $p < \pi(3)$, and it is generally believed that $N = 1$ almost surely if $p > \pi(3)$.

Conjecture 9. $P_p(N = 0 \text{ or } N = 1) = 1$ *for all* p.

This conjecture is bound up with Conjecture 7, since H. Kesten can show (unpublished) that $P_p(N = 1) = 1$ if $p > \rho = \lim_{k\to\infty}\rho(k)$. See Newman and Schulman (1981 a,b) for some results and speculations about the possibility that there are infinitely many infinite open clusters for some value of p.

To every percolation process there corresponds a dual process. An observation of great importance in two dimensions is that the dual of a two-dimensional bond percolation process is also a bond percolation process, and this fact is extremely useful for planar models. Unfortunately, the dual of bond percolation on $\mathbb{Z}^3$ is a type of "random surface" process, and this is rather different from the original process. There are some quite difficult topological complications which have to be taken into account in studying this random surface model, and some of the first steps in doing so have been taken by Aizenman et al. (1983). We finish this survey with a simple related conjecture.

<u>Conjecture 10</u>. <u>The number</u> $W(\underline{0})$ <u>of vertices of</u> $\mathbb{Z}^3$ <u>which are joined to</u> $\underline{0}$ <u>by open paths satisfies</u>

$$E_p(W(\underline{0})1(W(\underline{0}) < \infty)) < \infty \tag{4.1}$$

<u>for</u> $p \neq \pi(3)$, <u>where</u> $1(A)$ <u>is the indicator function of the event</u> A.

That is to say, we conjecture that the mean size of the cluster containing $\underline{0}$ is finite whenever this cluster is finite itself, so long as $p \neq \pi(3)$. This is evidently true if $p < \gamma(3)$ but we know of no proof if $p > \pi(3)$, although it should not be too difficult to prove this if $p > \rho$. If this conjecture is true, then so are certain conjectured central limit theorems of Cox and Grimmett (1984). For instance, if $p > \pi(3)$ and (4.1) holds then the number of vertices of $\mathbb{Z}^3$ which are contained within a cube with side-length n and are joined by open paths to the boundary of this cube is asymptotically normally distributed as $n \to \infty$.

<u>References</u>

Aizenman, M., Chayes, J.T., Chayes, L., Fröhlich, J. and Russo, L. (1983), On a sharp transition from area law to perimeter law in a system of random surfaces, preprint.

Barlow, R.E. and Proschan, F. (1975), *Statistical theory of reliability and life testing*, Holt, Rinehart and Winston, New York.

Broadbent, S.R. and Hammersley, J.M. (1957), Percolation processes I. Crystals and mazes, *Proceedings of the Cambridge Philosophical Society* <u>53</u>, 629-641.

Cox, J.T. and Grimmett, G.R. (1984), Central limit theorems for associated random variables and the percolation model, *Annals of Probability* <u>12</u>, 514-528.

Essam, J.W. (1980), Percolation theory, *Reports on Progress in Physics* <u>43</u>, 833-912.

Fröhlich, J. and Spencer, T. (1982), The phase transition in the one-dimensional Ising model with $1/r^2$ interaction energy, *Communications in Mathematical Physics* <u>84</u>, 87-101.

Gaunt, D.S. and Ruskin, H. (1978), Bond percolation processes in d dimensions, *Journal of Physics A: Mathematical and General* <u>11</u>, 1369-1380.

Grimmett, G.R. (1981), Critical sponge dimensions in percolation theory, *Advances in Applied Probability* <u>13</u>, 314-324.

Grimmett, G.R. (1983), Bond percolation on subsets of the square lattice, and the threshold between one-dimensional and two-dimensional behaviour, *Journal of Physics A: Mathematical and General* 16, 599-604.

Grimmett, G.R., Keane, M. and Marstrand, J.M. (1984), On the connectedness of a random graph, *Mathematical Proceedings of the Cambridge Philosophical Society* 96, 151-166.

Hammersley, J.M. and Whittington, S.G. (1984), Self-avoiding walks in wedges, *Journal of Physics A: Mathematical and General*, to appear.

Kesten, H. (1981), Analyticity properties and power law estimates of functions in percolation theory, *Journal of Statistical Physics* 25, 717-756.

Kesten, H. (1982), *Percolation theory for mathematicians*, Birkhäuser, Boston.

Newman, C.M. and Schulman, L.S. (1981a), Infinite clusters in percolation models, *Journal of Statistical Physics* 26, 613-628.

Newman, C.M. and Schulman, L.S. (1981b), Number and density of percolating clusters, *Journal of Physics A: Mathematical and General* 14, 1735-1743.

Schulman, L.S. (1984), Long range percolation in one dimension, *Journal of Physics A: Mathematical and General* 16, L639-L641.

Smythe, R.T. and Wierman, J.C. (1978), *First-passage percolation on the square lattice*, Lecture notes in mathematics no. 671, Springer, Berlin.

Stauffer, D. (1979), Scaling theory of percolation clusters, *Physics Reports* 54, 1-74.

BIRTH AND DEATH PROCESSES WITH KILLING AND

APPLICATIONS TO PARASITIC INFECTIONS

K. P. Hadeler
Lehrstuhl für Biomathematik
Universität Tübingen
Auf der Morgenstelle 10
D - 7400 Tübingen

Summary: A birth and death process with killing and reestablishment of the total population is defined. For the process with constant rates the transient probability distribution can be obtained from a renewal equation. The ideas are applied to a model for parasitic infections.

Consider a situation where a host carries a finite number of parasites. Within the host the parasite population is governed by a birth and death process with immigration. Furthermore the host can die and later be replaced by another host which itself can carry parasites. Thus in addition to the states $i = 0,1,2,..$ of a host carrying i parasites there is a state ϕ which describes the absence of the host.

For the transition between these states we assume the following laws. If the system is in state $i \in \mathbb{N}_0$, then the probability of birth of a new individual is $\rho i \Delta t$, where $\rho \geq 0$. Similarly the probability of death of a parasite is $\sigma i \Delta t$, $\sigma > 0$, and the probability of immigration of a parasite is $\varphi(t) \Delta t$. Finally, the probability of death of the host is $(\alpha i + \mu) \Delta t$ with $\alpha > 0$, $\mu > 0$. If the system is in the state ϕ, then the probability of transition into the state i is $\delta_i \Delta t$, $\delta_i \geq 0$.

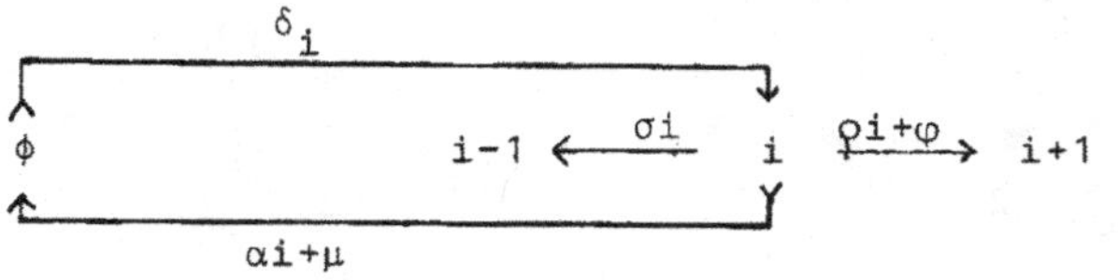

$$
\begin{array}{ccc}
& \delta_i & \\
\phi & \quad i-1 \xleftarrow{\ \sigma i\ } i \xrightarrow{\ \rho i + \varphi\ } i+1 & \quad (1) \\
& \alpha i + \mu &
\end{array}
$$

In the special case $\mu = 0$, $\delta_i = 0$ the process is a birth and death process with killing in the sense of Karlin and Tavaré [4], or Puri [5].

Let p_i, $i \in \mathbb{N}_0$, p_ϕ, be the probability that the system is in state i, or ϕ, respectively. The functions p_i satisfy the system of differential equations

$$\dot{p}_\phi = \sum_{i=0}^{\infty} (\alpha i+\mu)p_i - \gamma p_\phi , \qquad \gamma = \sum_{i=o}^{\infty} \delta_i , \tag{2}$$

$$\begin{aligned} \dot{p}_i = & -((\alpha+\sigma+\rho)i+\mu+\varphi(t))p_i \\ & +\rho(i-1)p_{i-1} + \varphi(t)p_{i-1} \\ & +\sigma(i+1)p_{i+1} + \delta_i p_\phi , \\ & \qquad i = 0,1,2,\ldots \end{aligned} \tag{3}$$

where formally $p_{-1} \equiv 0$.

For the generating function

$$u(t,z) = \sum_{i=0}^{\infty} p_i(t) z^i \tag{4}$$

and the function

$$U(t) = p_\phi(t) \tag{5}$$

one obtains a degenerate system of partial differential equations

$$\begin{aligned} u_t(t,z) = & -g(z)u_z(t,z) - \mu u(t,z) + \\ & + \varphi(t)(z-1)u(t,z) + h(z)U(t) \end{aligned} \tag{6}$$

$$\dot{U}(t) = \alpha u_z(t,1) + \mu u(t,1) - \gamma U(t) \tag{7}$$

where

$$g(z) = (\alpha+\sigma+\rho)z - \sigma - \rho z^2 , \tag{8}$$

$$h(z) = \sum_{i=0}^{\infty} \delta_i z^i , \quad \gamma = h(1) . \tag{9}$$

The immigration rate φ is allowed to depend on t. The equations satisfy the conservation law

$$\frac{\partial}{\partial t} u(t,z)\Big|_{z=1} + \dot{U}(t) = 0. \tag{10}$$

Together with the equations (6)(7) we consider the initial data

$$u(0,z) = u_o(z) \ , \quad U(0) = U_o \ . \tag{11}$$

Here it is assumed that $u_o(z)$ has a power series expansion with non-negative coefficients which converges in $z = 1$. Also let

$$u_o(1) + U_o = 1 \tag{12}$$

hold.

The characteristic differential equations corresponding to the equation (6) are

$$\frac{dt}{ds} = 1, \quad \frac{dz}{ds} = g(z) \ , \tag{13}$$

$$\frac{du}{ds} = \varphi(t)(z-1)u-\mu u+h(z)U \ . \tag{14}$$

The solution can be expressed in terms of the solution operator $G(s,z)$ of the Riccati equation $\dot{z} = g(z)$. In the case $\rho > 0$ let z_1, z_2 be the roots of the equation $g(z) = 0$,

$$z_{1,2} = \frac{1}{2\rho}\left[\alpha+\sigma+\rho \pm \sqrt{(\alpha+\sigma+\rho)^2 - 4\sigma\rho}\right] \tag{15}$$

$$z_2 \leq 1 \leq z_1 \ .$$

Define

$$\varkappa = \begin{cases} \rho(z_1 - z_2), & \rho > 0 \\ \alpha + \sigma \ , & \rho = 0 \end{cases} \tag{16}$$

If $\rho = 0$ then put $z_2 = \sigma/\varkappa$.
Then, for $\rho > 0$

$$G(t,z) = \frac{z_1(z-z_2) + z_2(z_1-z)e^{-\varkappa t}}{(z-z_2) + (z_1-z)e^{-\varkappa t}} \tag{17}$$

and for $\rho = 0$

$$G(t,z) = 1 - (1-z)e^{\varkappa t} - \frac{\alpha}{\varkappa}(1-e^{\varkappa t}) \tag{18}$$

The function G satisfies the equation

$$G_t(t,1) = \alpha G_z(t,1) \quad . \tag{19}$$

Define

$$\Gamma(t) = G(-t,1) - 1 \tag{20}$$

and

$$\begin{aligned} Q(t) &= -\int_0^t \Gamma(s)\,ds, \\ q(t) &= \int_0^t G_z(-s,1)\,ds \quad . \end{aligned} \tag{21}$$

For the characteristic curves one finds the equations

$$t = s \;, \quad z(s) = G(s,z_o) \;,$$

$$\frac{du}{ds} = \varphi(s)\,[G(s,z_o) - 1]u - \mu u + h(G(s,z_o))\,U(s) \;,$$

hence

$$\begin{aligned} u(t,z) &= u_o(G(-t,z))\,e^{\int_0^t [G(s-t,z)-1]\varphi(s)ds-\mu t} \\ &+ \int_0^t e^{\int_y^t [G(s-t,z)-1]\varphi(s)ds-\mu(t-y)}\, h(G(y-t,z))U(y)\,dy \quad . \end{aligned} \tag{22}$$

From this equation, for $z = 1$, and the conservation law

$$U(t) = 1 - u(t,1)$$

follows an integral equation for the function U,

$$\begin{aligned} U(t) &= 1 - u_o(G(-t,1))\,e^{\int_0^t \Gamma(t-s)\varphi(s)ds-\mu t} \\ &- \int_0^t e^{\int_y^t \Gamma(t-s)\,\varphi(s)\,ds - \mu(t-y)}\, h(G(y-t,1))U(y)\,dy \quad . \end{aligned} \tag{23}$$

If φ is constant, then the equation (23) is of convolution type. The kernel

$$k(t) = e^{\int_0^t \Gamma(s)ds\varphi - \mu t} h(G(-t,1)) \qquad (24)$$

is in principle known, though difficult to evaluate.

If t runs from 0 to ∞, then $h(G(-t,1))$ decreases from $h(1)$ to $h(z_2)$, and $h(1) > 0$ implies $h(z_2) > 0$. Hence k is integrable iff either $h \equiv 0$ or $\varphi + \mu > 0$. If k is not integrable then $U(t) \to 0$, otherwise from the integral equation

$$U(t) \to U_\infty = \frac{1}{1+K} \qquad (25)$$

where

$$K = \int_0^\infty k(t)dt. \qquad (26)$$

From the conditions for the stationary case,

$$\alpha u_z(1) + \mu u(1) - \gamma U = 0,$$

$$u(1) + U = 1 \ ,$$

it follows that

$$\alpha u_z(1) + \mu(1-U) - \gamma U = 0,$$

hence

$$u_z(1) = \frac{\gamma - \mu K}{\alpha(1+K)} \qquad (27)$$

(of course $K \leq \gamma/\mu$ holds). Thus the expected number of parasites per host is

$$\frac{u_z(1)}{u(1)} = \frac{\gamma - \mu K}{\alpha K} \ . \qquad (28)$$

It is of some interest to see how this expression depends on the parameters. It is easy to see that K is increasing in ρ, φ and decreasing in σ. If $h(z) \equiv \gamma$ then $u_z(1)/u(1)$ is decreasing in μ. The dependence on α is much more involved.
Higher moments can be easily formed. From

$$0 = -g'(z)u_z - g(z)u_{zz} - \mu u_z + \varphi u + \varphi(z-1)u_z + h'(z)U_\infty$$

it follows that

$$\alpha u_{zz}(1) = (\rho-\alpha-\sigma-\mu)u_z(1) + \varphi u(1) + h'(1)U_\infty .$$

The variance is

$$V = \frac{u_{zz}+u_z}{u} - \frac{u_z^2}{u^2}$$

$$= \frac{\rho-\sigma-\mu}{\alpha} \cdot \frac{\gamma-\mu K}{\alpha K} + \frac{\varphi}{\alpha} + \frac{h'(1)}{\alpha K} - \frac{(\gamma-\mu K)^2}{\alpha^2 K^2} \tag{29}$$

Application to the the theory of epidemics.

Consider a population of hosts which can carry parasites. The hosts are classified according to individual parasite load. Within a host the parasite population is governed by a birth and death process with immigration and killing of the host.

Again, let $\rho \geq 0$, $\sigma > 0$ be the birth and death rate of parasites within a host. Let $b > 0$ and $\mu > 0$ be the natural birth and death rate of a host in the absence of parasites. Let $\alpha > 0$ be the differential mortality of a host due to the presence of one parasite. Similarly, let $\omega \in [0,1]$ be the differential decrease in fertility due to the presence of one parasite. Let $\varphi(t)$ be the rate with which hosts acquire parasites or, equivalently, the immigration rate of parasites into hosts.

It is assumed that newborns do not carry parasites.

Let $n(t,i)$ be the number of hosts at time t carrying i parasites. The functions $n(t,i)$ satisfy the system of differential equations

$$\begin{aligned}\frac{dn(t,i)}{dt} = &- [\varphi(t) + (\alpha+\sigma+\rho)i + \mu]n(t,i) \\ &+ [\varphi(t) + \rho(i-1)]n(t,i-1) \\ &+ \sigma(i+1)n(t,i+1) \\ &+ \delta_{i0}U(t), \quad i \geq 0 ,\end{aligned} \tag{30}$$

where formally $n(t,-1) \equiv 0$. Here δ_{ij} is the Kronecker symbol.

The function U(t) is the time-dependent rate at which newborns enter the population. The function U can be specified in several ways.

Case 1: The birth rate is constant $U(t) \equiv U$.

Case 2: The birth rate U(t) depends on the state of the total population

$$U(t) = b \sum_{i=0}^{\infty} \omega^i n(t,i). \tag{31}$$

Case 3: The population lives in a habitat with given capacity $\bar{U}$. Some portion of the habitat is not occupied. This portion is increased if individuals die. On the other hand this vacant portion is filled up at a constant rate by newborns. In this case the function U satisfies an additional differential equation

$$\frac{dU(t)}{dt} = \sum_{i=0}^{\infty} (\mu+\alpha i)n(t,i) - U(t). \tag{32}$$

In this case the system (30)(32) admits a conservation law

$$\sum_{i=0}^{\infty} n(t,i) + U(t) \equiv \bar{U}. \tag{33}$$

The average parasite load of the population is given by

$$w(t) = \frac{\sum_{i=0}^{\infty} in(t,i)}{\sum_{i=0}^{\infty} n(t,i)}, \tag{34}$$

and the parasite acquisition rate is given by

$$\varphi(t) = \beta f(w(t)). \tag{35}$$

Here f is a given function, in general nonlinear, which describes the transmission by vectors, and β is the contact rate between hosts and vectors. In some respect these models are simplified versions of the models for parasitic infection in [1][2][3]. At present the age structure of the host population is neglected.

With the generating function

$$u(t,z) = \sum_{i=0}^{\infty} n(t,i)z^{i} \tag{36}$$

the equation (30) can be written

$$\frac{\partial u}{\partial t} + g(z)\frac{\partial u}{\partial z} - [\varphi(t)(z-1) - \mu]u - U = 0 \tag{37}$$

In case 1 the function U is a given constant, in case 2 U(t) is given by the expression

$$U(t) = bu(t,\omega) \tag{38}$$

and in case 3 the function U satisfies the differential equation

$$\frac{dU}{dt} = \mu u(t,1) + \alpha u_z(t,1) - U(t). \tag{39}$$

The average parasite load and the transmission law are given by

$$w(t) = \frac{u_z(t,1)}{u(t,1)} \tag{40}$$

and equation (35).

The initial condition reads

$$u(0,z) = u_o(z). \tag{41}$$

In case 3 also U(0) has to be prescribed,

$$U(0) = U_o. \tag{42}$$

We follow the approach of the first section. Suppose that the functions φ and U were known. Then u(t,z) can be represented as in (23). From (23) it follows that

$$u_z(t,z)$$

$$= u_o'(G(-t,z))e^{\int_o^t [G(s-t,z)-1]\varphi(s)ds-\mu t} G_z(-t,z)$$

$$+ u_o(G(-t,z))e^{\int_o^t [G(s-t,z)-1]\varphi(s)-\mu t} \int_o^t G_z(s-t,z)\varphi(s)ds$$

$$+ \int_0^t e^{\int_y^t [G(s-t,z)-1]\varphi(s)ds - \mu(t-y)} \int_y^t G_z(s-t,z)\,\varphi(s)\,ds\,U(y)\,dy \quad (43)$$

Define

$$\mathfrak{D} = u_0(G(-t,1))\,(L\varphi)(t,0) + \int_0^t (L\varphi)(t,y)\,U(y)\,dy \quad (44)$$

$$\begin{aligned}\mathcal{N} = {} & u_0'(G(-t,1))\,(L\varphi)(t,0)\cdot G_z(-t,1) \\ & + u_0(G(-t,1))\,(L\varphi)(t,0)\cdot \int_0^t G_z(s-t,1)\,\varphi(s)\,ds \\ & + \int_0^t (L\varphi)(t,y) \int_y^t G_z(s-t,1)\,\varphi(s)\,ds\; U(y)\,dy\end{aligned} \quad (45)$$

where

$$(L\varphi)(t,y) = e^{\int_y^t \Gamma(t-s)\varphi(s)ds - \mu(t-y)} \,. \quad (46$$

Then in case 1 the problem is reduced to the single integral equation for the function φ

$$\varphi(t) = \beta f(\mathcal{N}/\mathfrak{D}) \quad (47)$$

where $\mathcal{N}, \mathfrak{D}$ are given by (44)(45) with constant U.

In case 2 the problem leads to a system of two coupled integral equations for the functions φ and U.

$$\varphi = \beta f(\mathcal{N}/\mathfrak{D}) \quad (48a)$$

where $\mathcal{N}, \mathfrak{D}$ are given by (44) (45), now with variable U which satisfies

$$\begin{aligned}U(t) = {} & U_0(G(t,\omega))\,e^{\int_0^t [G(s-t,\omega)-1]\varphi(s)ds - \mu t} \\ & + b \int_0^t e^{\int_y^t [G(s-t,\omega)-1]\varphi(s)ds - \mu(t-y)} \,.\, U(y)\,dy\end{aligned} \quad (48b)$$

Also in case 3 the problem reduces to two coupled equations,

$$\varphi = \beta f(\mathcal{N}/\mathcal{D}) \tag{49a}$$

where $\mathcal{N}$, $\mathcal{D}$ are given by (44), (45), now with variable U satisfying

$$U(t) = \bar{U} - u_o(G(-t,1))(L\varphi)(t,0) - \int_o^t (I\varphi)(t,y)U(y)dy \tag{49b}$$

Following similar lines as in the existence proof of [1] or [3], one can show

Theorem 1: Let $f : [0,\infty) \to [0,\infty)$ be a continuously differentiable function such that $f(0) = 0$, $f(w) > 0$ for $w > 0$, $f(w) \leq f_o w$, where f_o is a constant. Let $u_o = u_o(z)$ be a generating function such that $u_{oz}(1)$ exists. Then each of the integral equations (47), (48), (49) has a unique nonnegative solution, which exists for all times $t \geq 0$.

Each of the problems (48), (49), (50) has stationary or persistent solutions which can be characterized in terms of the constant parasite acquisition rate φ.

Case 1: For all $\beta > 0$ there is the solution $\varphi = 0$ which describes an uninfected population. Also there is a branch of non-trivial solutions parametrized by $\varphi > 0$,

$$\beta = \frac{\varphi}{f(\mathcal{N}(\varphi)/\mathcal{D}(\varphi))} \tag{50}$$

where (cf. (20) (21) (43))

$$\mathcal{D}(\varphi) = \int_o^\infty e^{-Q(t)\varphi - \mu t} dt \tag{51}$$

$$\mathcal{N}(\varphi) = \int_o^\infty e^{-Q(t)\varphi - \mu t} q(t)\, dt\varphi \tag{52}$$

Case 2: In this case one does not expect stationary solutions but rather persistent solutions. For all $\beta > 0$ there is the solution $\varphi = 0$ describing a noninfected population. The population grows exponentially with exponent $\lambda = b-\mu$.

There is also a branch of non-trivial solutions parametrized by $\varphi > 0$. For given $\varphi > 0$ one first obtains from equation (48b) the corresponding exponent $\lambda(\varphi)$ of population growth. $\lambda(\varphi)$ is the unique positive solution of the equation

$$b \int_0^\infty e^{\int_0^y [G(-\tau,\omega)-1]\,d\tau \; - \mu y - \lambda y}\, dy = 1. \tag{53}$$

Then one introduces this φ into equation (48a) and finds the corresponding β from

$$\varphi = \beta f \left(\frac{\int_0^\infty e^{-Q(t)\varphi - \mu t + \lambda(\varphi) t} q(t)\, dt\, \varphi}{\int_0^\infty e^{-Q(t)\varphi - \mu t + \lambda(\varphi) t}\, dt} \right). \tag{54}$$

<u>Case 3</u>: In this case there are stationary solutions. For all $\beta > 0$ there is the solution $\varphi = 0$ describing a noninfected population. Also, there is a branch of non-trivial solutions parametrized by $\varphi > 0$. Along this branch the quantities β and U are given by

$$\beta = \frac{\varphi}{f(\mathcal{N}(\varphi)/\mathfrak{D}(\varphi))} \tag{55}$$

and

$$U = \frac{\overline{U}}{1+\mathfrak{D}(\varphi)}, \tag{56}$$

whereby $\mathcal{N}(\varphi)$, $\mathfrak{D}(\varphi)$ are defined in (51) (52).

Thus, in all three cases, there is a trivial branch of noninfected populations and a nontrivial branch parametrized by $\varphi > 0$. In general this branch is not monotone, i.e. for some value of β there may be several nontrivial solutions.

References

[1] Hadeler, K.P., Dietz, K.: Nonlinear hyperbolic partial differential equations for the dynamics of parasite populations. Comp. Math. Appl. 9, (Nr.3) 415-430 (1983).

[2] Hadeler, K.P.: Integral equations for infections with discrete parasites: Hosts with Lotka birth law. In: Levin, S., Hallam, T. (eds.) Mathematical Ecology, Trieste. Lecture Notes in Biomath. 54, Springer-Verlag, 1984, p.356-365.

[3] Hadeler, K.P., Dietz, K.: Population dynamics of killing parasites which reproduce in the host. J. Math. Biol. 21 (1984) 45-65.

[4] Karlin, S., Tavaré, S.: Linear birth and death processes with killing. J. Appl. Prob. 19, 477-487 (1982).

[5] Puri, P.S.: A method for studying the integral functionals of stochastic processes with applications III. Proc. Sixth Berkeley Symp. Math. Stat. Prob. Vol. III, 481-500, UCLA Press (1972).

LIMIT THEOREMS FOR MULTITYPE BRANCHING RANDOM WALKS*

Gail Ivanoff
Department of Mathematics
University of Ottawa
Ottawa, Canada

§0. Introduction

The multitype branching random walk (MBRW) may be described as a discrete-time multitype branching process in which the offspring resulting from a branch move instantaneously to new positions. Temporal limit theorems for the MBRW have been studied by Prehn and Roder (1977), Fleischmann and Prehn (1978) and Ivanoff (1983) for the case in which the mean matrix of the branching process is irreducible and aperiodic, and by Ivanoff (1982) for the case in which the mean matrix is periodic or reducible. In this paper, we will extend some of the limit theorems of Ivanoff (1982) and (1983) for MBRW's beginning with a superposition of Poisson random fields to include the non-critical and reducible cases not previously considered.

The paper is organized in the following manner: section 1 provides background on the model and certain definitions and notation. In section 2, it is proven that if the initial random fields are suitably renormalized, the MBRW converges in distribution to a (possibly trivial) limit. Another limit theorem is given in section 3 for the critical reducible case in which the limit is trivial. Sketches of the proofs of the results in sections 2 and 3 are given in section 4.

§1. The Multitype Branching Random Walk

The model described in this section is as in Ivanoff (1982) and (1983). Let $D = R^d$ be d-dimensional Euclidean space, let $\mathcal{B}(D)$ be the Borel sets of D.

The MBRW begins at the 0th generation with an initial distribution (possibly random) of particles on D. At each successive generation,

*This research was supported by a grant from the Natural Sciences and Engineering Research Council of Canada.

all of the particles currently in the system branch according to some law which does not vary with time or location, and then the offspring instantaneously move to new positions, again according to some law which is homogeneous in time and space. We make the following assumptions and definitions:

(1) Each particle migrates and branches independently.

(2) At each generation, the probability that a type i particle produces j_1 type 1 particles,..., j_q type q particles is $p^i(j_1,\ldots,j_q)$. Let $E^i_k(\ell_1,\ldots,\ell_k)$ denote the kth order joint factorial moment of the numbers of type $\ell_1,\ldots,\ell_k$ particles ($\ell_i \in \{1,\ldots,q\}$, $i = 1,\ldots,k$) produced from the branch of a type i particle. Let $m_{ij} = E^i_1(j)$, and let $M = (m_{ij})$ be the mean matrix of the branching. Denote by m^n_{ij} the ijth entry of M^n. Let ρ be the largest positive eigenvalue of M. As usual, the process will be called supercritical, critical, or subcritical if ρ is > 1, = 1, or < 1, respectively. Finally, let $\underset{\sim}{\mu}$ and $\underset{\sim}{\chi}$ be right and left eigenvectors of M with eigenvalue ρ such that $\underset{\sim}{\mu}\cdot\underset{\sim}{\chi} = \underset{\sim}{\mu}\cdot\underset{\sim}{1} = 1$.

(3) The distribution of the migration process will be assumed to be absolutely continuous with respect to Lebesgue measure, with density function $g(x,y)$ (i.e., the probability that a particle at $x \in D$ moves to a set $A \in \mathcal{B}(D)$ is $\int_A g(x,y)dy$). Assume that $g(x,y) = g(y,x) = g(0,y-x) = g(x-y,0)$. Let $g_n(x,y)$ denote the nth convolution of $g(\cdot,\cdot)$. Assume that the transition density has mean 0 and covariance matrix Γ.

For $A \in \mathcal{B}(D)$, let the random (row) vector $\underset{\sim}{N}_n(A|a,i) = (N_n(A,1|a,i),\ldots,N_n(A,q|a,i))$, where $N_n(A,j|a,i)$ denotes the number of type j particles in A at generation n, given initially a single type i particle at $a \in D$.

The MBRW beginning with Poisson random fields (PRFs) obeys the laws defined above, with an initial state consisting of a superposition of independent PRFs of each type of particle. Let $\lambda_i dx$ be the intensity measure of the ith random field. (Lebesgue measure is denoted by "dx".) Let $\underset{\sim}{\lambda} = (\lambda_1,\ldots,\lambda_q)$. Denote the corresponding point process at the nth generation by $\underset{\sim}{N}_n(\cdot)$.

We consider temporal limit theorems for $\underset{\sim}{N}_n$ in three situations: for M irreducible and aperiodic, for M periodic with period K, and for M reducible. In the reducible case, for clarity it will be assumed that the mean matrix is of the form

$$M = \begin{bmatrix} M(1) & 0 \\ M(2,1) & M(2) \end{bmatrix} \tag{1.1}$$

where M(1) and M(2) are irreducible and aperiodic, and $M(2,1) \neq 0$.

For further details and results about the MBRW and general multitype point processes, see Ivanoff (1981), (1982) and (1983).

§2. The Limit Distributions

Let $\underset{\sim}{N}_n^*$ denote the point process generated by the nth generation of a MBRW, beginning with a superposition of PRF's whose intensities are specified by the vector $\underset{\sim}{\lambda}(n) = (\lambda_1(n),\ldots,\lambda_q(n))$. In general, $\underset{\sim}{\lambda}(n) = c_n\underset{\sim}{\lambda}$, for some sequence of constants (c_n), and $\underset{\sim}{\lambda} = (\lambda_1,\ldots,\lambda_q)$ is an arbitrary fixed vector.

We make the following assumption:

(A) For all i,j,k such that $1 \leq i,j,k, \leq q$, $E_2^i(j,k) < \infty$.

Under assumption (A), we will prove the following theorems:

Theorem 2.1: Assume that (A) holds and that M is irreducible and aperiodic. Then if $\underset{\sim}{\lambda}(n) = \rho^{-n}\underset{\sim}{\lambda}$, $\underset{\sim}{N}_n^*$ converges in distribution to a limiting point process as $n \to \infty$. The limit is non-trivial if $\rho = 1$ and $\sum_{k=1}^{\infty} g_{2k}(x_1,x_2) < \infty$ or if $\rho < 1$. The limit is trivial if $\rho = 1$ and $\sum_{k=1}^{\infty} g_{2k}(x_1,x_2) = \infty$ or if $\rho > 1$. (A point process $\underset{\sim}{N}$ is trivial if $P(N(A,i) > 0) = 0$ for all bounded sets $A \in \mathcal{B}(D)$, and all $i = 1,\ldots,q$)

Comment 2.2: It seems likely that a limit exists if (A) does not hold. It is certainly true in the critical case (cf. Prehn and Roder (1977)). However, (A) allows us to give a very simple direct proof of Theorem 2.1 involving probability generating functionals and allows us to show immediately the non-triviality of the limits.

Comment 2.3: For a subcritical case, Fleischmann and Prehn (1978) prove a somewhat different limit theorem. Rather than renormalizing the initial random field, they rescale D itself to maintain a constant intensity measure. The effect of this scaling is to slow down the migration process to the extent that the limiting point process is equivalent (in distribution) to a Poisson cluster process, whose clusters are groups of particles all at the location of the cluster centre. By renormalizing the initial random field only, we allow the migration process to continue to play a role. In particular, the limiting random field (if non-trivial) will not be completely random, as is the case with the limit found by Fleischmann and Prehn (1978).

Corollary 2.4: Assume that (A) holds and that M is irreducible and periodic with period K. Then if $\underset{\sim}{\lambda}(n) = \rho^{-n}\underset{\sim}{\lambda}$, $\underset{\sim}{N}_{s+nK}^*$ converges in distribution as $n \to \infty$ to a limiting point process $\underset{\sim}{N}_s^{\infty}$, $s = 1,\ldots,K$. The limits are each non-trivial if $\rho = 1$ and $\sum_{k=1}^{\infty} g_{2k}(x_1,x_2) < \infty$ or if

$\rho < 1$. The limits are trivial if $\rho = 1$ and $\sum_{k=1}^{\infty} g_{2k}(x_1,x_2) = \infty$ or if $\rho > 1$.

In the reducible case, if M is of the form (1.1), let ρ_i denote the maximal eigenvalue of $M(i)$, $i = 1,2$. Note that $\rho = \max(\rho_1,\rho_2)$.

Corollary 2.5: If M is reducible and of the form (1.1) then:

a) Theorem 2.1 is valid if $\rho = \rho_1 > \rho_2$ or $\rho = \rho_2 > \rho_1$.

b) If i) $\rho = \rho_1 = \rho_2 < 1$ or if ii) $\rho = \rho_1 = \rho_2 = 1$ and $\sum_1^n k^2 g_{2k}(x_1,x_2) = o(n)$, then for $\lambda(n) = \lambda n^{-1}\rho^{-n}$, N_n^* converges in distribution as $n \to \infty$ to a nontrivial point process consisting only of particles with types in the class with mean matrix $M(1)$.

c) If i) $\rho = \rho_1 = \rho_2 > 1$ or if ii) $\rho = \rho_1 = \rho_2 = 1$, $d = 1$ or 2, then for $\lambda(n) = \lambda n^{-1}\rho^{-n}$, N_n^* converges in distribution to a trivial point process.

Comment 2.6: Corollary 2.5 is less general than the preceding results. In b), it seems that if $\rho = \rho_1 = \rho_2 = 1$, $\sum_1^n k^2 g_{2k}(x_1,x_2) = O(n)$ should be sufficient, but the stronger condition is required by our method of proof. This leaves open the following question: What happens if, for example, $g(x_1,x_2) = (2\pi)^{-d/2} e^{-(x_1-x_2)^2/2}$, $\rho = \rho_1 = \rho_2 = 1$ and $d = 3$ or 4? We will partially answer this question in the next section.

§3. The Critical Reducible Case: Other Limit Theorems

We consider a simple example of the critical MBRW with a reducible mean matrix and $\rho_1 = \rho_2 = 1$. We assume 2 types of particles. At each generation, type 1 particles remain unchanged. Upon branching, a type 2 particle produces either 2 type 2 offspring or 1 type 1 offspring, each with probability 1/2. All particles of both types migrate according to a Gaussian law: $g(x,y) = (2\pi)^{-d/2} e^{-(x-y)^2/2}$. Without loss of generality, we assume that initially there is a PRF of type 2 particles only, with intensity λdx. We shall consider only the limiting behaviour of $N_n(A,1)$, since $N_n(A,2)$ is simply a critical (single-type) BRW.

Theorem 3.1: Assume that $d = 4$. Then for the model defined above and $\lambda(n) = \lambda n^{-1}$, N_n^* converges in distribution as $n \to \infty$ to a non-trivial poin process.

Theorem 3.2: Assume that $d = 1,2$ or 3, and that A is a bounded set in

$\mathcal{B}(D)$. Then for the model defined above and $\lambda(n) =_{\infty} \lambda n^{-1}$, there exists a non-increasing function $f(\cdot)$ on $(0,\infty)$ such that $\int_0^{\infty} f(x)dx = 1/2$ and at all continuity points x of f, $x > 0$,

$$\lim_{n\to\infty} n^{c/2}\lambda^{-1}F_n'(x) = f(x),$$

where $F_n'(x) = P(n^{-c/2}|A|^{-1}N_n^*(A,1) > x)$, $c = d-4$.

Corollary 3.3: Under the conditions of Theorem 3.2, for any bounded set $A \in \mathcal{B}(D)$,

$$\lim_{n\to\infty} P(N_n^*(A,i) > 0) = 0, \quad i = 1,2.$$

(In other words, the limit is trivial.)

Comment 3.4: The model used here is extremely simple in order to keep the calculations in the proofs reasonably straightforward. It should not be difficult to extend Theorems 3.1 and 3.2 to include more general branching mechanisms. In fact, we conjecture that if $\rho = 1$ and has multiplicity $r > 1$ and if the migration process is Gaussian, then letting $\lambda(n) = \lambda n^{-(r-1)}$, N_n^* converges to a non-trivial limit only if $d \geq 2r$.

§4. Sketch of Proofs

Let $G_n(\cdot|y,i)$ denote the probability generating functional (PGF) of $N_n(\cdot|y,i)$. Using (2.1)-(2.3) of Ivanoff (1982) and an iterative argument, the following lemma may be proven:

Lemma 4.1: For arbitrary $n, m \in Z^+$, and $\phi = 1 - (1-s)\chi_A$, $0 \leq s \leq 1$, $A \in \mathcal{B}(D)$, A bounded, (χ_A is the indicator function of A),

$$\sum_{i=1}^{q}\sum_{r=1}^{q} \lambda_i m_{ir}^n \int_D 1-G_m(\phi|y,r)dy$$

$$- \frac{1}{2}\sum_{\ell=m+1}^{m+n}\sum_{i=1}^{q}\sum_{r=1}^{q}\sum_{s=1}^{q}\sum_{t=1}^{q}\sum_{h=1}^{q}\sum_{k=1}^{q} \lambda_i m_{ir}^{n+m-\ell}$$

$$\times E_2^r(s,t) m_{sh}^{\ell-1} m_{tk}^{\ell-1} \int_A\int_A g_{2\ell}(u_1,u_2)du_1du_2$$

$$\leq \sum_{i=1}^{q} \lambda_i \int_D 1-G_{n+m}(\phi|y,i)dy$$

$$\leq \sum_{i=1}^{q}\sum_{r=1}^{q} \lambda_i m_{ir}^n \int_D 1-G_m(\phi|y,r)dy. \tag{4.1}$$

Proof of Theorem 2.1: Using (2.3)-(2.5) of Ivanoff (1983) and (2.5) of Ivanoff (1982), it is readily shown that the sequence of first cumulant densities of $(\underset{\sim}{N}_n^*)$ converges to a non-zero limit and if $\rho = 1$, $\sum_{k=1}^{\infty} g_{2k}(x_1,x_2) < \infty$ or if $\rho < 1$, that the second order cumulant densities remain uniformly bounded. In this case, the sequence $(\underset{\sim}{N}_n^*)$ is tight and any limit is non-trivial. Then Lemma 4.1 may be used to show that the sequence of PGF's of $(\underset{\sim}{N}_n^*)$ has a unique limit. (The PGF of $\underset{\sim}{N}_n^*$ is defined by (2.3) of Ivanoff (1982).)

To complete the proof, the triviality of the limit if $\rho = 1$ and $\sum_k g_{2k}(x_1,x_2) = \infty$ is a result of the instability of the "cluster distribution" in the terminology of Prehn and Roder (1977). If $\rho > 1$, without loss of generality assume that A is a sphere of radius r in D with centre at the origin. Let $A_n = nA$. Let $R_n^1(A)$ denote the number of particles in A at generation n whose original ancestors were in A_n, and let $R_n^2(A)$ denote the remaining number of particles in A at n. Simple probabilistic arguments involving the initial distribution and migration process show that $R_n^i(A)$ converges to 0 in probability, $i = 1,2$. □

The proofs to Corollaries 2.4 and 2.5 follow roughly the same line as that of Theorem 2.1. The condition $\sum_1^n k^2 g_{2k}(x_1,x_2) = o(n)$ in the critical case in b) of Corollary 2.5 is used with Lemma 4.1 to show that the sequence of PGF's of $(\underset{\sim}{N}_n^*)$ has a unique limit. For Corollary 2.5 c) in the case $\rho = 1$, $d = 1$ or 2, we make use of a result due to Foster and Ney (1976) on the rate of convergence to zero of the probability that any one particle alive at generation 0 has offspring alive at generation n.

Proofs of Theorems 3.1 and 3.2: For the proofs of Theorems 3.1 and 3.2, we define $h_k(a,n) = |A|^{-k} n^{-c(k-1)/2 - 1} E((N_n(A,1|a,2))^k)$, where $c = d-4$. Recursive formulae for the h_k's are readily obtained for the model described, and replacing the discrete time parameter with a continuous one, the following integral approximations are found: for all $t > 0$,

$g_t(a,y) = (2\pi t)^{-d/2} e^{-(a-y)^2/2t}$,

$$H_1(a,t) = \begin{cases} (1/2)|A|^{-1} \int_A g_t(a,y)dy, & t \geq 1 \\ 0, & t < 1 \end{cases}$$

$$H_k(a,t) = \begin{cases} (1/2)t^{-c(k-1)/2}|A|^{-k} \int_A g_t(a,y)dy \\ \quad +(1/2)\sum_{i=1}^{k-1} \binom{k}{i} t^{-c(k-1)/2 - 1} \int_0^{t-1} s^{c(k-2)/2 + 2} \\ \quad \times \int_D g_{t-s-1}(a,y) \int_D H_i(x_1,s)g(y,x_1)dx_1 \\ \quad \times \int_D H_{k-i}(x_2,s)g(y,x_2)dx_2dyds, & , t \geq 1 \\ 0 & , t < 1. \end{cases}$$

The general approach to be used is similar to that used by Durrett (1979) to prove his equation (8.1), although the technical details of the proofs are different.

It may be shown inductively that the H_k's satisfy the following lemma:

<u>Lemma 4.2</u>:

a) For all $t > 0$, $\int_D H_k(x,t)g(a,x)dx \leq L(k) \int_A g_{t+1}(a,x)dx$, where if $d = 1,2$ or 3, $L(k) = \begin{Bmatrix} |A|^{-k}k! \text{ if } |A| < 1 \\ |A|^{k-1}k! \text{ if } |A| \geq 1 \end{Bmatrix}$, and if $d = 4$,

$L(k) = \begin{Bmatrix} |A|^{-k}(2k-3)!! \text{ if } |A| < 1 \\ |A|^{k-1}(2k-3)!! \text{ if } |A| \geq 1 \end{Bmatrix}$, and $(2k-3)!!$ is interpreted as 1 for $k = 1$.

b) Let $a_t = at^{1/2}$. For all $a \in D$, $k \geq 1$,

$$t^{d/2} \int_D H_k(y,t)g(a_t,y)dy \to h_k(a) \text{ as } t \to \infty, \text{ where}$$

$$h_1(a) = (1/2)g(0,a), \text{ and}$$

$$h_k(a) = 1/2\Big[\delta(4-d)|A|^{-(k-1)}g(0,a) + \sum_{i=1}^{k-1} \binom{k}{i} \int_0^1 \int_D g_{1-u}(a,w_u)h_i(w)h_{k-i}(w)dwdu\Big]$$

for $k \geq 2$ and where $\delta(x) = \begin{Bmatrix} 1, & x = 0 \\ 0, & x \neq 0 \end{Bmatrix}$, $w_u = wu^{1/2}$.

c) i) $\int_D H_k(a,t)da \to h_k$ as $t \to \infty$, where

$$h_1 = 1/2, \text{ and}$$

$$h_k = (c(k-1)+2)^{-1}\Big[\delta(4-d)|A|^{-(k-1)} + \sum_{i=1}^{k-1} \binom{k}{i} \int_D h_i(w)h_{k-i}(w)dw\Big], \quad k \geq 2.$$

ii) $\sum_{k=1}^{\infty} h_k s^k/k!$ has a positive radius of convergence.

d) $\lim_{n\to\infty} \int_D h_k(a,n)da = \lim_{t\to\infty} \int_D H_k(a,t)da.$

The proofs of Theorems 3.1 and 3.2 follow from Lemma 4.2, using techniques similar to those used in the proofs of Theorems 7.1 and 7.2, respectively, of Dawson and Ivanoff (1978). □

That Corollary 3.3 is a consequence of Theorem 3.2 may be shown by the same sort of argument as used by Durrett (1979), pg. 376.

References

Dawson, D.A. and Ivanoff, B.G. (1978). Branching diffusions and random measures. In Advances in Probability (A. Joffe and P. Ney, Eds.), Vol. 5, 61-104, Dekker, New York.

Durrett, R. (1979). An infinite particle system with additive interactions. Adv. Appl. Prob., 11, 355-383.

Ivanoff, B.G. (1981). The multitype branching diffusion. J. Multiv. Anal., 11, 289-318.

Ivanoff, B.G. (1982). The multitype branching random walk, II. J. Multiv. Anal., 12, 526-548.

Ivanoff, B.G. (1983). The multitype branching random walk, I. Canadian J. Statist., 11, 245-257.

Fleischman, J. (1978). Limiting distributions for branching random fields. Trans. Amer. Math. Soc., 239, 353-389.

Fleischmann, K. and Prehn, U. (1978). Limit theorems for spatially homogeneous branching processes with a finite set of types, II. Math. Nachr., 82, 277-296.

Foster, J. and Ney, P. (1976). Decomposable critical multi-type branching processes. Sankhya A, 38, 28-37.

Prehn, U. and Roder, B. (1977). Limit theorems for spatially homogeneous branching processes with a finite set of types, I (in Russian). Math. Nachr., 80, 37-86.

ON THE REPRODUCTION RATE OF THE SPATIAL GENERAL EPIDEMIC

Kari Kuulasmaa
Department of Epidemiology
National Public Health Institute
Mannerheimintie 166
SF-00280 Helsinki
Finland

1. Introduction

We consider a spatial *general epidemic* model $GE(Z^d,\alpha,\mu,F)$ defined by Mollison [7] as follows: At time zero there is an infected individual at the origin of the d-dimensional integer lattice Z^d. All other *sites*, i.e. elements of Z^d, are occupied by healthy individuals. The infected individual emits germs in a Poisson process of rate α, $\alpha \geq 0$, until she is removed after a random infectious period determined by F, a probability distribution defined on the positive real axis. Each germ emitted goes independently of the other germs to a site whose location with respect to the parent site is chosen according to the *contact distribution* μ, a probability density defined on $Z^d \setminus \{0\}$. If a healthy individual gets a germ she becomes infected and starts to emit germs until she is removed. After an individual is removed her site remains empty thereafter. If an infected individual or an empty site receives a germ nothing happens. The germ emission processes and lifetimes of different individuals are mutually independent.

As general epidemics usually, also this model (if $d \geq 2$) has threshold behaviour: for some parameter there is a critical value such that below the value extinction of the disease is almost certain whereas above the critical value the infection spreads for ever with positive probability. The *infection rate* α is a natural choice for the critical parameter according to the definition of the model above. The threshold behaviour of the model with respect to α has been investigated in [4], [5], [6] and [7]. Here we consider an alternative critical parameter which, though mathematically more complicated because it does not appear in the definition of the model, has an easier interpretation than the infection rate.

2. The reproduction rate

The alternative parameter we consider is the reproduction rate which is defined as the expected number of different sites that receive a germ from any particular infective. In non-spatial branching or birth and death processes, which in a sense correspond to the spatial general epidemic, the reproduction rate is the same as the expected number of germs emitted by an infective. In the spatial general epidemic only the germs landing on healthy sites are effective. The reproduction rate has previously been considered the critical parameter in spatial epidemic models in [8] and [9].

In the general epidemic $GE(Z^d,\alpha,\mu,F)$ the probability that site $v \in Z^d$ receives a germ from the origin is

$$\int_0^\infty (1 - e^{-\alpha t \mu(v)})dF(t).$$

Hence the reproduction rate is

$$RR = RR(\alpha) = \sum_{v \in Z^d} \int_0^\infty (1 - e^{-\alpha t \mu(v)})dF(t) = \sum_{v \in Z^d} [1 - \phi(\alpha\mu(v))], \quad (2.1)$$

where ϕ is the Laplace transform of F:

$$\phi(x) = \int_0^\infty e^{-xt} dF(t), \qquad x \geq 0.$$

The following lemma characterizes the relationship between the reproduction rate $RR(\alpha)$ and the infection rate α.

Lemma. If $RR(\alpha)$ is finite for some $\alpha > 0$, then $RR(\alpha)$ is a continuous strictly increasing function from $[0,\infty)$ to $[0,\infty)$.

Proof. If $RR(\alpha_0)$ is finite then $RR(\alpha)$ is clearly strictly increasing in $[0,\alpha_0]$. Furthermore, $RR(2\alpha_0)$ is finite: Jensen's inequality and the fact that $1 - x^2 \leq 2(1 - x)$ for every $x \in \mathbf{R}$ imply

$$RR(2\alpha_o) = \sum_v [1 - \int_0^\infty e^{-2\alpha_o t\mu(v)} dF(t)]$$

$$\leq \sum_v [1 - \{\int_0^\infty e^{-\alpha_o t\mu(v)} dF(t)\}^2]$$

$$\leq \sum_v 2[1 - \int_0^\infty e^{-\alpha_o t\mu(v)} dF(t)] = 2RR(\alpha_o).$$

This implies that if $RR(\alpha) < \infty$ for some $\alpha > 0$ then $RR(\alpha) < \infty$ for every $\alpha \geq 0$, and $RR(\alpha)$ is strictly increasing in $[0,\infty)$. The Laplace transform ϕ is continuous. If $RR(\alpha) < \infty$ for every α, then the sum of the right hand side of equation (2.1) converges uniformly on every interval $[0,\alpha]$ and thus $RR(\alpha)$ is continuous on every interval $[0,\alpha]$. Hence $RR(\alpha)$ is continuous on $[0,\infty)$.

From the lemma it follows that if $RR(\alpha)$ is finite for some α, and hence for every α, we can use $RR(\alpha)$ as a parameter of the model instead of α. In practical applications the reproduction rate is usually finite. In [4] it was shown that the finiteness of $RR(\alpha)$ for some α is a sufficient condition for the critical infection rate, and thus for the critical reproduction rate, to be greater than zero.

In [4] it was also shown that when considering the threshold behaviour of the general epidemic one can forget the time structure and reduce the process to a locally dependent random graph, which is a generalization of a bond percolation process. The main advantage of $RR(\alpha)$ compared to α is that it is determined by the random graph. α instead depends strongly on the time structure: it is inversely proportional to the time unit used.

3. Examples from the two-dimensional space

In a similar non-spatial model, essentially in a branching process, the critical reproduction rate is one. In a spatial model it may be very different. Indeed, if d=1 extinction of the disease is almost certain for any α and hence for any $RR(\alpha)$ ([2]). The simplest examples of general epidemics with threshold we get from the two-dimensional space. In [4] the two-dimensional nearest neighbour general epidemic where μ is 1/4 at each of the four nearest neighbours of the origin was investigated. The constant lifetime case corresponds to the bond percolation on the square lattice, for which the critical probability for a bond to be open is 1/2([3]). Hence the critical reproduction rate of this general epidemic is 2. If F is exponential with mean one the probability for an infective to send a germ to any fixed neighbour is $\alpha/(\alpha+4)$. Hence $RR(\alpha) = 4\alpha/(\alpha+4)$. The critical infection rate is known to be at least 4 and estimated to be between 4.5 and 5 ([4]). Hence the critical reproduction rate is at least 2 and estimated to be between 2.1 and 2.2. Note that if we change the mean of the lifetime distribution F the critical infection rate changes as well but the critical reproduction rate remains unchanged. Typically, among general epidemics with the same contact distribution which is uniform in a finite set of sites the constant lifetime epidemic has smallest critical reproduction rate (This is a direct consequence of theorem 4.1 of [4]).

The dependence of the critical reproduction rate on the contact distribution is another comparison of interest. The other general epidemics for which the critical reproduction rates are known exactly are the constant lifetime epidemics corresponding to the bond percolations on triangular and hexagonal lattices([10]). The critical reproduction rates of these are $12\sin(\pi/18) \doteqdot 2.08$ and $3-6\sin(\pi/18) \doteqdot 1.96$ respectively. These figures are all about twice the critical value of a non-spatial model. An interesting open question is whether the critical reproduction rate of a two-dimensional spatial general epidemic can be arbitrarily close to one. Note that in the examples above the critical value is largest on the triangular lattice, when the number of neighbours of a site was largest.

Ball [1] has considered the generalization of the general epidemic where each site has n individuals. A germ landing on a site chooses a particular individual of the site with probability 1/n. It is shown that if n tends to infinity the critical reproduction rate converges to one. This, however, does not seem to indicate what can happen if the number of individuals on each site is kept constant and only the contact distribution is varied.

References

[1] Ball, F. (1983) The threshold behaviour of epidemic models. J. Appl. Prob. **20**, 227-241.
[2] Kelly, F.P. (1977) In discussion of Mollison (1977), 318-319.
[3] Kesten, H. (1980) The critical probability of bond percolation on the square lattice equals 1/2. Commun. Math. Phys. **74**, 41-59.
[4] Kuulasmaa, K. (1982) The spatial general epidemic and locally dependent random graphs. J. Appl. Prob. **19**, 745-758.
[5] Kuulasmaa, K. (1983) Locally dependent random graphs and their use in the study of epidemic models. Preprint.
[6] Kuulasmaa, K. and Zachary, S. (1984) On spatial general epidemics and bond percolation processes. J. Appl. Prob. **21**, to appear.
[7] Mollison, D. (1977) Spatial contact models for ecological and epidemic spread. J. R. Statist. Soc. **B 39**, 283-326.
[8] Mollison, D. (1981) The importance of demographic stochasticity in population dynamics. In The Mathematical Theory of the Dynamics of Biological Populations II, ed. R.W. Hiorns and D.L. Cooke. Academic Press, London, 99-107.
[9] Mollison, D. and Kuulasmaa, K. (1985) Spatial epidemic models: theory and simulations. To appear in The Population Dynamics of Wildlife Rabies, ed. P.J. Bacon. Academic Press, London.
[10] Wierman, J.C. (1981) Bond percolation on honeycomb and triangular lattices. Adv. Appl. Prob. **13**, 298-313.

NEAREST PARTICLE SYSTEMS: RESULTS AND OPEN PROBLEMS

Thomas M. Liggett
U.C.L.A.

1. Introduction. Nearest particle systems are certain continuous time Markov processes on $X=\{0,1\}^Z$ (where Z is the set of integers). Flips occur from zero to one and from one to zero at each site. The characteristic property of these systems is that the rate at which a coordinate flips depends on the rest of the configuration only through the distances to the nearest sites to the right and left which have the value one. These processes were introduced by Spitzer (1977), and have been the subject of a number of papers since then. A substantial theory of these processes has been developed, but a number of open problems still remain. This paper presents an overview of the theory as it stands today, and a description of some of the more important open problems which should be addressed. No detailed proofs will be provided here, but in many cases the main ideas of the proof will be given. Complete proofs can be found in the original papers, or in Chapter VII of Liggett (1985).

Configurations $\eta \in X$ will be given an occupancy interpretation: $\eta(x)=1$ means that $x \in Z$ is occupied, while $\eta(x)=0$ means that x is vacant. For $x \in Z$ and $\eta \in X$, let $\ell_x(\eta)$ and $r_x(\eta)$ be the distances from x to the nearest occupied site to the left and right respectively:

$$\ell_x(\eta) = x-\max\{y<x:\eta(y)=1\},$$

and

$$r_x(\eta) = \min\{y>x:\eta(y)=1\}-x.$$

These are allowed to take the value ∞ if there is no occupied site to the left or right. Let $\{\beta(\ell,r):1\le\ell,r\le\infty\}$ be a collection of nonnegative numbers which satisfy

$$\sup_{\ell,r} \beta(\ell,r)<\infty, \qquad \beta(\ell,r) = \beta(r,\ell),$$

$$\beta(1,\infty) = \beta(\infty,1)>0, \quad \text{and} \quad \beta(\infty,\infty) = 0.$$

These play the role of birth rates for the system. The death rates will be taken to be identically one for simplicity. For $x \in Z$ and $\eta \in X$, define $\eta_x \in X$ by $\eta_x(x) = 1-\eta(x)$ and $\eta_x(y) = \eta(y)$ for all $y\neq x$. The transition rates for the process are then given by

$$\eta\to\eta_x \text{ at rate } \begin{cases} \beta(\ell_x(\eta),r_x(\eta)) & \text{if} \quad \eta(x) = 0 \\ 1 & \text{if} \quad \eta(x) = 1. \end{cases}$$

We will discuss both infinite and finite versions of this process. The infinite system is the process η_t with state space

$$X' = \{\eta \in X: \sum_{x<0} \eta(x) = \sum_{x>0} \eta(x) = \infty\}.$$

This Markov process was constructed by Gray (1978). The values of $\beta(\ell,r)$ for $\ell=\infty$ or $r=\infty$ are clearly irrelevant for its definition. The process remains on X' for all time because the death rates are bounded. The semigroup of this process will be denoted by $S(t)$. The finite system is a Markov chain A_t on

$$Y = \{\eta \in X: \sum_x \eta(x)<\infty\},$$

which is identified with the finite subsets A of Z via $A=\{x:\eta(x)=1\}$. It is for its definition that the values of $\beta(\ell,r)$ for $\ell=\infty$ or $r=\infty$ are used. In order that A_t remain on Y for all time, we will assume that

(1.1) $$\sum_{\ell=1}^{\infty} \beta(\ell,\infty)<\infty$$

whenever we are discussing the finite system.

We will find that the class of nearest particle systems contains both reversible and nonreversible examples. It should not be surprising that a lot more can be said about reversible systems than about general ones. Results about reversible systems then lead to conjectures about nonreversible ones. Our discussion will be organized into four parts, according to whether the system is reversible or general, and whether it is finite or infinite.

This is a good time to mention some examples to which we will return from time to time:

<u>Example 1</u>: <u>The contact process</u>. This process was introduced by Harris (1974). A survey of this topic was given by Griffeath (1981), and a fairly complete treatment of it appears in Chapter VI of Liggett (1985). The contact process is special in that its flip rates have only nearest neighbor dependence. The birth rates are given by

$$\beta(\ell,r) = \begin{cases} b & \text{if} \quad \ell=r=1, \\ \frac{b}{2} & \text{if} \quad \ell=1 \quad \text{or} \quad r=1 \quad \text{and} \quad \ell+r>2, \quad \text{and} \\ 0 & \text{if} \quad \ell\geq 2 \quad \text{or} \quad r\geq 2. \end{cases}$$

<u>Example 2</u>: <u>First reversible process</u>. The birth rates are given by

$$\beta(\ell,r) = \frac{\beta(\ell)\beta(r)}{\beta(\ell+r)} = c\left(\frac{1}{\ell} + \frac{1}{r}\right)^p,$$

where $\beta(n) = cn^{-p}$. Also $\beta(\ell,\infty) = c\ell^{-p}$ is defined by continuity. Here c and p are positive parameters. Note that we must assume $p>1$ in order that the finite process be defined.

Example 3: Second reversible process. The birth rates are given by

$$\beta(\ell,r) = \frac{\beta(\ell)\beta(r)}{\beta(\ell+r)} \quad \text{if} \quad \ell,r<\infty, \quad \text{and}$$

$$\beta(\ell,\infty) = \beta(\ell) \quad \text{if} \quad \ell<\infty,$$

where

$$\beta(n) = b\,\frac{(2n-2)!}{(n-1)!n!}\,\frac{1}{4^n}.$$

This seemingly unnatural choice is of interest because it is the only reversible process which satisfies

$$(1.2) \qquad \sum_{\ell+r=n+1} \beta(\ell,r) = b \quad \text{for} \quad 1\leq n<\infty,$$

and

$$\sum_{\ell=1}^{\infty} \beta(\ell,\infty) = \frac{b}{2}.$$

Note that $\beta(n) \sim \frac{b}{4\sqrt{\pi}}\, n^{-3/2}$ as $n\uparrow\infty$, so that both the finite and infinite systems are defined in this case.

Example 4: Uniform birth process. Here

$$\beta(\ell,r) = \frac{b}{\ell+r-1} \quad \text{for} \quad \ell,r<\infty.$$

Only the infinite process will be defined in this case. The interpretation is that each maximal interval of zeros in η has a total birth rate b, which is spread out uniformly over all the sites in that interval.

Example 5: Centered birth process. In this case,

$$\beta(\ell,r) = \begin{cases} b & \text{if} \quad \ell=r \\ \frac{b}{2} & \text{if} \quad |\ell-r|=1 \\ 0 & \text{if} \quad |\ell-r|\geq 2. \end{cases}$$

Again only the infinite version of the process will be considered. The interpretation is similar to that in the previous example, except that the births all occur at the midpoint of the interval of zeros.

All the above examples except the second satisfy (1.2). We will be particularly interested in processes with this property. A nearest particle system is said to be attractive if $\beta(\ell,r)$ is a decreasing function of ℓ and r. Attractive systems are very amenable to monotonicity arguments - see Chapter III of Liggett (1985), for example. All the above examples except the last are attractive.

The main problem of interest for nearest particle systems is to determine which survive and which die out. The finite system is said to survive if

$$\lim_{t\to\infty} P^A(A_t\neq\emptyset)>0$$

for all $A\neq\emptyset$, and to die out otherwise. The infinite system is said to survive if it has an invariant measure concentrating on X'. In the attractive case, this is equivalent to

$$\lim_{t\to\infty} P^1(\eta_t(0)=1)>0,$$

where 1 is the configuration $\eta\equiv 1$. This problem of survival will be resolved completely in the reversible case, and partially in the general case. In an effort to stimulate further work, various special cases and refinements of this problem and related ones will be formulated at appropriate places in this paper.

2. Reversible finite systems. Let $q(A,B)$ be the Q matrix for the finite nearest particle system A_t with birth rates $\beta(\ell,r)$, which are assumed to satisfy (1.1).

Definition 2.1. A strictly positive function π on Y is said to be a reversible measure for A_t provided that

$$\pi(A)q(A,B) = \pi(B)q(B,A)$$

for all $A,B\neq\emptyset$.

The following characterization of reversible finite nearest particle systems is elementary, and can be found in Chapter VII of Liggett (1985).

Theorem 2.2. A_t has a reversible measure π if and only if there is a strictly positive function $\beta(\ell)$ on $\{1,2,...\}$ such that

$$\sum_{\ell=1}^{\infty} \beta(\ell)<\infty,$$

$$(2.3)\qquad \beta(\ell,r) = \frac{\beta(\ell)\beta(r)}{\beta(\ell+r)} \quad \text{for } 1\le \ell, r<\infty, \text{ and}$$

$$(2.4)\qquad \beta(\ell,\infty) = \beta(\ell) \quad \text{for } 1\le \ell<\infty.$$

In this case, π is given (up to constant multiples) by $\pi(\{x\}) = 1$ and

$$(2.5)\qquad \pi(A) = \prod_{i=1}^{n-1} \beta(x_{i+1} - x_i),$$

where $A = \{x_1,\ldots,x_n\}$ and $x_1<x_2<\cdots<x_n$ for $n\ge 2$.

For the remainder of this section, we will assume that $\beta(\ell,r)$ is given by (2.3) and (2.4). The next result gives a necessary and sufficient condition for survival in this case. It was proved by Griffeath and Liggett (1982). The condition is given in terms of the parameter

$$(2.6)\qquad \lambda = \sum_{\ell=1}^{\infty} \beta(\ell).$$

Let τ be the hitting time of $\emptyset$, and let $*$ denote a generic singleton.

Theorem 2.7. (a) A_t survives if and only if $\lambda>1$.

(b) If $\lambda>1$, then

$$(2.8)\qquad \frac{\lambda-1}{\lambda} \le P^*(\tau=\infty) \le \left|\lambda\log\frac{\lambda-1}{\lambda}\right|^{-1}.$$

Idea of the proof. The main tool is a comparison theorem for reversible Markov chains. Roughly speaking, this theorem says that increasing the transition rates of a reversible Markov chain in such a way that the reversibility is maintained, has the effect of making the chain more transient (e.g. in this case, the effect is to increase the value of $P^*(\tau=\infty)$). For the first comparison, set all interior birth and death rates (i.e, those corresponding to transitions which do not change the diameter of A) equal to zero. The modified chain has the property that the cardinality $|A_t|$ is a Markov chain on $\{0,1,2,\ldots\}$ with transition rates

$n\to n+1$ at rate 2λ, and

$n\to n-1$ at rate 2.

This chain is transient if and only if $\lambda>1$, and its probability of never returning to 0 when started at 1 is $1-\lambda^{-1}$. This gives half of statement (a), and the first inequality in (2.8). For the second comparison it is important to identify elements of Y which are translates of one another. Now add new "infinite" transition rates between all pairs of (equivalence classes of) points in Y which have the same cardinality. The effect of this is to guarantee that every time the cardinality of A_t changes, the new distribution of A_t is "immediately" the one given by the $\pi(\cdot)$ in (2.5) restricted to sets of that cardinality. The cardinality of this modified chain is then again a Markov chain on $\{0,1,\ldots\}$ with transition rates $n\to n+1$ at rate

$$\frac{\sum_{|A|=n} \pi(A)[\text{total birth rate in } A]}{\sum_{|A|=n} \pi(A)}$$

$$= \frac{\sum_{d_1,\ldots,d_{n-1}\geq 1} \left[\prod_{i=1}^{n-1} \beta(d_i)\right]\left[2\lambda + \sum_{i=1}^{n-1} \sum_{\ell+r=d_i} \beta(\ell,r)\right]}{\sum_{|A|=n} \pi(A)}$$

$$= 2\lambda + \frac{\sum_{i=1}^{n-1} \sum_{d_1,\ldots,d_n\geq 1} \prod_{i=1}^{n} \beta(d_i)}{\lambda^{n-1}} = (n+1)\lambda, \quad \text{and}$$

$n\to n-1$ at rate

$$\frac{\sum_{|A|=n} \pi(A)[\text{total death rate in } A]}{\sum_{|A|=n} \pi(A)} = n.$$

This chain is also transient if and only if $\lambda>1$, and its probability of never returning to 0 when started at 1 is $\left|\lambda\log \frac{\lambda-1}{\lambda}\right|^{-1}$. This gives the other half of (a) and the second inequality in (2.8).

The inequalities in (2.8) appear to be the best estimates which can be obtained without imposing further assumptions. It would be nice to know the exact asymptotic behavior of $P^*(\tau=\infty)$ as $\lambda\downarrow 1$ for various one parameter families of birth rates. Liggett (1983b) proved the following result along these lines.

<u>Theorem 2.9</u>. Consider the one parameter family of birth rates $\beta_\lambda(\ell,r)$ given by (2.3) and (2.4) in terms of $\beta_\lambda(\ell)=\lambda f(\ell)$, where $f(\cdot)$ is strictly positive and satisfies

$$\sum_{n=1}^{\infty} f(n)=1, \tag{2.10}$$

$$\frac{f(n)}{f(n+1)} \downarrow 1 \quad \text{as} \quad n\uparrow\infty, \quad \text{and} \tag{2.11}$$

$$\sum_{n=1}^{\infty} n^2 f(n)<\infty. \tag{2.12}$$

Then there is a constant C so that

$$P^*(\tau=\infty) \le C(\lambda-1) \quad \text{for} \quad \lambda\ge 1. \tag{2.13}$$

<u>Idea of the proof</u>. The proof uses a comparison similar to that used in the proof of Theorem 2.7. This time we set the rates of all the interior transitions equal to "infinity." For this comparison chain, every time the diameter of A_t changes, the interior of A_t is immediately placed in its stationary distribution. Thus the diameter of the comparison chain is a Markov chain on $\{0,1,\ldots\}$, which turns out to have the following transition rates:

$$\begin{cases} 1\to 0 & \text{at rate } 1 \\ k\to\ell & \text{at rate} \quad \lambda f(\ell-k) \quad \text{if} \quad \ell>k\ge 1 \\ k\to\ell & \text{at rate} \quad \dfrac{\lambda f(k-\ell)\rho(\ell)}{\rho(k)} \quad \text{if} \quad k>\ell\ge 1, \end{cases} \tag{2.14}$$

where $\rho(k)$ can be computed recursively via $\rho(1)=1$ and

$$\rho(k)=\lambda\sum_{\ell=1}^{k-1}\rho(\ell)\, f(k-\ell).$$

In the context of Theorem 2.7, the comparison technique led to chains which only jumped to nearest neighbors on $\{0,1,\ldots\}$. This is not true in the present case, so the analysis of the comparison chain is less elementary. By the renewal theorem, $\rho(k)$ is asymptotic to a constant multiple of s^{-k}, where s is the unique solution of

$$1 = \lambda \sum_{\ell=1}^{\infty} f(\ell)s^{\ell}.$$

Therefore the chain with rates (2.14) is very close to being a random walk when it is near $+\infty$. Under the second moment assumption (2.12), a comparison with that random walk can be carried out. Once this is done, (2.13) is obtained as an application of known potential theoretic results for the random walk.

Problem 1. Under the assumptions of Theorem 2.9, show that

$$\lim_{\lambda \downarrow 1} \frac{P^*(\tau=\infty)}{\lambda-1}$$

exists. Combining (2.8) and (2.13), this limit must be positive and finite.

Problem 2. Determine the asymptotic behavior of $P^*(\tau=\infty)$ as $\lambda \downarrow 1$ when (2.12) fails. It would already be of interest to find an example for which

$$\lim_{\lambda \downarrow 1} \frac{P^*(\tau=\infty)}{\lambda-1} = \infty.$$

Theorems 2.7 and 2.9 give some quantitative results concerning the system just above the critical value 1. A rather simple computation due to Griffeath and Liggett (1982) provides the following information in the subcritical case.

Theorem 2.15. If $\lambda<1$, then

(a) $E^*(\tau) = (1-\lambda)^{-1}$, and

(b) $E^* \int_0^{\tau} |A_t|dt = (1-\lambda)^{-2}$.

Problem 3. How do the higher moments of τ behave as $\lambda \uparrow 1$?

3. General finite systems. Consider now a general finite nearest

particle system A_t, and let

$$b(n) = \sum_{\ell+r=n+1} \beta(\ell,r) \quad \text{if} \quad 1 \le n < \infty, \quad \text{and}$$

$$b(\infty) = \sum_{\ell=1}^{\infty} \beta(\ell,\infty) + \sum_{r=1}^{\infty} \beta(\infty,r).$$

The idea is that the size of these quantities should indicate the propensity for survival of the system. The following result was proved by Liggett (1984).

Theorem 3.1. (a) If $b(n) \le 1$ for all $1 \le n \le \infty$, then A_t dies out. (b) For every number $b>2$, there is a finite nearest particle system which survives and satisfies $b(n)=b$ for all $1 \le n \le \infty$. (c) If $b(n) \ge 4$ for all $1 \le n \le \infty$, then A_t survives.

Idea of the proof. Parts (a) and (b) are easy. For part (a), it is enough to note that the cardinality $|A_t|$ is a nonnegative supermartingale, which must converge to zero. For part (b), apply Theorem 2.7 to Example 3. The proof of part (c) is harder. Let μ be the stationary renewal process on X corresponding to the probability density $f(n) = F(n) - F(n+1)$, where

$$F(n) = \frac{(2n-2)!}{(n-1)!n!} 4^{-n+1}, \quad n \ge 1.$$

Define h on Y by

$$(3.2) \qquad h(A) = \mu\{\eta \in X : \eta(x)=1 \text{ for some } x \in A\}.$$

Then $h(\emptyset)=0$ and $\lim_{|A| \to \infty} h(A) = 1$. In order to prove the survival of A_t, it is enough to show that

$$(3.3) \qquad \frac{d}{dt} E^A h(A_t) \Big|_{t=0} \ge 0$$

for all $A \in Y$, since then $h(A_t)$ is a submartingale. The proof of (3.3) is based on a comparison with the contact process. Holley and

Liggett (1978) showed that

$$(3.4) \qquad \frac{d}{dt} E^A h(B_t)\Big|_{t=0} \geq 0$$

for all $A \in Y$, where B_t is the contact process from Example 1 with $b=4$. To deduce (3.3) from (3.4), it suffices to show that the difference is nonnegative. Letting $\tilde{\beta}(\ell,r)$ be the birth rates for B_t, write

$$\frac{d}{dt} E^A h(A_t)\Big|_{t=0} - \frac{d}{dt} E^A h(B_t)\Big|_{t=0}$$

$$(3.5) \qquad = \sum_{x \notin A} [\beta(\ell_x(A), r_x(A)) - \tilde{\beta}(\ell_x(A), r_x(A))][h(A \cup x) - h(A)],$$

where $\ell_x(A)$ and $r_x(A)$ are the distances from x to the nearest points in A to the left and right respectively. Now write the complement of A as the union of maximal intervals I_k, and consider the sum of the terms in (3.5) over those x's in a single I_k. This sum will be nonnegative since $b(n) \geq 4$ for all n provided that $h(A \cup x)$ is a concave function of x on that I_k. This concavity can be checked using the explicit knowledge we have about μ.

Problem 4. Prove that $h(A \cup x)$ is a concave function of x on connected components of A^c if h is defined as in (3.2) with μ replaced by the upper invariant measure for the contact process (whenever it survives). For this choice, (3.4) is automatically true, with equality. The advantage of this is that it would yield part (c) of Theorem 3.1 with 4 replaced by the smallest b for which Example 1 survives. This is known to be between 3 and 4, and is thought to be about 3.3 (See Chapter VI of Liggett (1985) for more on this point.)

Problem 5. What is the smallest b such that there is a finite nearest particle system which survives and satisfies $b(n)=b$ for all $1 \leq n \leq \infty$? The best that can be done with reversible systems is to find examples which survive with $b>2$. It would seem that one should be able to do better with nonreversible systems. Can one get arbitrarily close to $b=1$?

Problem 6. Does the finite critical system always die out? This is true in the reversible case by Theorem 2.7. Even for the contact process, this is a hard open problem.

4. Reversible infinite systems. We now begin the discussion of infinite systems - those with state space X'. First is the characterization of reversible infinite systems, which is due to Spitzer (1977).

Definition 4.1. A probability measure μ on X' is said to be reversible for η_t if

$$\int fS(t)g\,d\mu = \int gS(t)f\,d\mu$$

for all f,g which depend on finitely many coordinates, and all $t\geq 0$.

Theorem 4.2. Suppose $\beta(\ell,r)>0$ for all $1\leq \ell,r<\infty$. Then η_t has a reversible measure if and only if there is a strictly positive probability density function $\beta(n)$ on $\{1,2,...\}$ with finite mean such that

$$\beta(\ell,r) = \frac{\beta(\ell)\beta(r)}{\beta(\ell+r)} \quad \text{for } 1\leq \ell,\ r<\infty. \tag{4.3}$$

In this case, the unique reversible measure is the stationary renewal measure μ_β corresponding to that density.

Note that (4.3) is not changed if $\beta(n)$ is replaced by $\beta(n)s^n$ for some $s>0$. Thus if $\beta(\ell,r)$ has the form (4.3) for some positive function $\beta(n)$, the condition should be read as asserting the existence of an $s>0$ so that

$$\sum_{n=1}^{\infty} \beta(n)s^n = 1 \quad \text{and} \quad \sum_{n=1}^{\infty} n\beta(n)s^n<\infty. \tag{4.4}$$

A simple argument based on Theorem 4.2 gives the following result (see Liggett (1983a) or Liggett (1985)):

Corollary 4.5. Suppose the process is attractive and $\beta(\ell,r)$ has the form (4.3) for some positive function $\beta(\ell)$. If there is no $s>0$ so that (4.4) holds, then the system dies out.

Combining Theorem 4.2 and Corollary 4.5, we obtain the following analogue of Theorem 2.7 for infinite systems.

Corollary 4.6. Suppose $\beta(\ell,r)$ has the form (4.3) for some strictly positive $\beta(n)$ which satisfies

$$\frac{\beta(n)}{\beta(n+1)} \downarrow 1 \quad \text{as} \quad n\uparrow\infty.$$

Define λ as in (2.6). Then the infinite system survives if either $\lambda>1$, or if $\lambda=1$ and

$$\sum_{n=1}^{\infty} n\beta(n)<\infty. \tag{4.7}$$

Otherwise it dies out.

Applying Corollary 4.6 to Example 2 yields the following conlusions for the infinite process: If $p\leq 1$, the process survives for all $c>0$. If $1<p\leq 2$, the process survives if and only if $c\sum_{n=1}^{\infty} n^{-p}>1$. If $p>2$, the process survives if and only if $c\sum_{n=1}^{\infty} n^{-p}\geq 1$. When applied to Example 3, it implies that the process survives if and only if $b>2$.

Problem 7. Comparing Theorem 2.7 with Corollary 4.6, we see that in the reversible case, the finite and infinite systems have the same critical value, although their behavior at that critcal value may differ. Are the critical values of the finite and infinite systems the same if the processes are not reversible? This is true for the contact process by duality.

When the infinite system has a reversible measure, it is of interest to know whether it can have other invariant measures (concentrating on X'). The following partial answer was obtained by Liggett (1983a) using free energy (relative entropy) techniques.

Theorem 4.8. Suppose $\beta(\ell,r)$ is of the form (4.3), where $\beta(n)$ is a strictly positive probability density with finite mean. Suppose also that the process is attractive, so that

$$\gamma = \lim_{n\to\infty} \frac{\beta(n)}{\beta(n+1)} \geq 1$$

exists. If $\gamma>1$, or if $\gamma=1$ and

$$\sum_{n=1}^{\infty} \frac{\beta^2(n)}{\beta(2n)} < \infty,$$

then μ_β is the only invariant measure which is translation invariant and concentrates on X'.

Using Theorem 4.8, it is not hard to show that certain reversible systems are ergodic (i.e. converge to the unique invariant measure for any initial distribution as $t\uparrow\infty$). This is true in Example 2 for $0<p<1$.

Since the upper invariant measure for an infinite reversible system is an explicitly known renewal measure, a lot can be said about the behavior just above the critical value. For example, consider the context of Corollary 4.6, where $\beta(n)$ is a constant multiple of n^{-p} for some $p>1$. Let $\rho(\lambda)$ be the density of the upper invariant measure. Then $\rho(\lambda)$ has the following asymptotic behavior as $\lambda\downarrow 1$:

$$\rho(\lambda) \sim c(\lambda-1)^{\frac{2-p}{p-1}} \qquad \text{if} \quad 1<p<2,$$

$$\rho(\lambda) \sim c\,|\log(\lambda-1)|^{-1} \qquad \text{if} \quad p=2,$$

$$\rho(\lambda) - \rho(1) \sim c(\lambda-1)^{p-2} \qquad \text{if} \quad 2<p<3,$$

$$\rho(\lambda) - \rho(1) \sim c(\lambda-1)|\log(\lambda-1)| \quad \text{if} \quad p=3, \quad \text{and}$$

$$\rho(\lambda) - \rho(1) \sim c(\lambda-1) \qquad \text{if} \quad p>3.$$

This was worked out in Chapter VII of Liggett (1985).

Problem 8. In the context of Corollary 4.6, how rapidly does the process converge to δ_0 as $t\uparrow\infty$ if $\lambda<1$. Presumably the convergence is exponentially fast. If so, how rapidly does

$$\int_0^\infty P^1[\eta_t(0)=1]\,dt$$

tend to ∞ as $\lambda\uparrow 1$?

Problem 9. Prove convergence theorems for these processes for general initial distributions. Most interesting in this context is the critical nonergodic case: the case treated in Corollary 4.6 when $\lambda=1$ and (4.7) is satisfied. In this case, the process survives starting from

$\eta\equiv 1$, but dies out starting from finite configurations. What happens, for example, if the initial distribution is a product measure with small density? Rather complete convergence results were proved by Durrett (1980) for the contact process.

5. <u>General infinite systems</u>. Here we discuss the analogues of the results in Section 3 for infinite systems. Define $b(n)$ as in that section. Part (c) of the next theorem is a generalizaiton of results due to Bramson and Gray (1981). Its proof appears in Chapter VII of Liggett (1985).

<u>Theorem 5.1</u>. (a) If $b(n)\leq 1$ for all $1\leq n<\infty$, then

$$\lim_{t\to\infty} \mu S(t) = \delta_0$$

for all translation invariant μ on X'. If in addition the process is attractive, then it dies out. (b) For every number $b>2$, there is an infinite nearest particle system which survives and satisfies $b(n)=b$ for all $1\leq n<\infty$. (c) Suppose that

$$\varliminf_{n\to\infty} \frac{1}{n} \sum_{\ell+r=n} \beta(\ell,r)[n\log n - \ell\log\ell - r\log r] > 2\log 2.$$

Then

$$\inf_{t>0} \frac{1}{t} \int_0^t P^1[\eta_s(0) = 1]ds > 0.$$

If in addition the process is attractive, then η_t survives.

<u>Idea of the proof</u>. Part (a) is based on the following simple computation:

$$\frac{d}{dt} \mu S(t)\{\eta:\eta(0)=1\} = -\mu S(t)\{\eta:\eta(0)=1\}$$

$$+ \sum_{n=1}^{\infty} b(n)\, \mu S(t)\{\eta:\eta(0)=1,\eta(1)=0,\ldots,\eta(n)=0,\eta(n+1)=1\}$$

$$\leq -\mu S(t)\{\eta:\eta(0)=\eta(1)=1\}.$$

For part (b), apply Corollary 4.6 to Example 3. Part (c) is more interesting, and is based on a differential inequality satisfied by the following moment of the distribution of the interparticle distances:

$$h(\mu) = \sum_{n=1}^{\infty} (n\log n)\ \mu\{\eta:\eta(0)=1,\eta(1)=0,\dots,\eta(n)=0,\eta(n+1)=1\}.$$

Applying the criterion of part (c) of the above theorem to two of our examples, one finds that it is satisfied for $b>4\log 2(\approx 2.77)$ in Example 4, and for $b>2$ in Example 5. These are the examples discussed by Bramson and Gray in their paper.

Problem 10. Note that the assumptions in parts (c) of Theorems 3.1 and 5.1 are similar, but that neither implies the other. Is there a single condition of this type which is sufficient in both contexts? In particular, is $b_n \geq 4$ for all $1 \leq n < \infty$ sufficient for the survival of the infinite system? One way of proving this would be to show that survival of the finite system implies survival of the infinite system.

Problem 11. Show for a class of examples including Example 5 that survival of the infinite system for a given b implies its survival for all larger b. This is immediate for attractive systems, but Example 5 is not attractive.

Problem 12. Show that moving birth rates more toward the center of an interval of zeros makes it more likely that the process will survive. For example, let b_1, b_3, b_4 and b_5 be the critical values for the infinite versions of Examples 1,3,4 and 5 respectively. (These exist in the first three cases by attractiveness, but is not known to exist in the fourth - see Problem 11.) Show that

$$b_1 \geq b_3 \geq b_4 \geq b_5. \tag{5.2}$$

Known bounds for these critical values are

$$3 \leq b_1 \leq 4, \qquad b_3 = 2$$

$$1 \leq b_4 \leq 2.8, \qquad 1 \leq b_5 \leq 2.$$

Therefore, some of the inequalities in (5.2) are true. However, the intention here is that the solution to this problem provide a general monotonicity statement, not that it be based on improved bounds for critical values.

Problem 13. The same as Problem 5 for infinite systems.

REFERENCES

Bramson, M. and Gray, L. (1981). A note on the survival of the long-range contact process. Ann. Probab. 9, 885-890.

Durrett, R. (1980). On the growth of one dimensional contact processes. Ann. Probab. 8, 890-907.

Gray, L. (1978). Controlled spin-flip systems. Ann. Probab. 6, 953-974.

Griffeath, D. (1981). The basic contact process. Stochastic Process. Appl. 11, 151-186.

Griffeath, D. and Liggett, T.M. (1982). Critical phenomena for Spitzer's reversible nearest particle systems. Ann. Probab. 10, 881-895.

Harris, T.E. (1974). Contact interactions on a lattice. Ann. Probab. 2, 969-988.

Holley, R. and Liggett, T.M. (1978). The survival of contact processes. Ann. Probab. 6, 198-206.

Liggett, T.M. (1983a). Attractive nearest particle systems. Ann. Probab. 11, 16-33.

Liggett, T.M. (1983b). Two critical exponents for finite reversible nearest particle systems. Ann. Probab. 11, 714-725.

Liggett, T.M. (1984). Finite nearest particle systems. Z. Wahrsch. Verw. Gebiete, 68,65-73

Liggett, T.M. (1985). Interacting Particle Systems, Springer Verlag, New York.

Spitzer, F. (1977). Stochastic time evolution of one dimensional infinite particle systems. Bull. Amer. Math. Soc. 83, 880-890.

The preparation of this paper was supported in part by NSF Grant MCS83-00836

NEUTRAL MODELS OF GEOGRAPHICAL VARIATION

Thomas Nagylaki
Department of Molecular Genetics and Cell Biology
The University of Chicago
920 East 58th Street
Chicago, Illinois 60637, USA

ABSTRACT

The amount and pattern of genetic variability in a geographically structured population under the joint action of migration, mutation, and random genetic drift is studied. The monoecious, diploid population is subdivided into panmictic colonies that exchange gametes. In each deme, the rate of self-fertilization is equal to the reciprocal of the number of individuals in that deme. Generations are discrete and nonoverlapping; the analysis is restricted to a single locus in the absence of selection; every allele mutates to new alleles at the same rate. It is shown that if the population is at equilibrium, the number of demes is finite, and migration does not alter the deme sizes, then population subdivision produces interdeme differentiation and the mean homozygosity and the effective number of alleles exceed their panmictic values. The equilibrium and transient states of the island, circular stepping-stone, and infinite linear stepping-stone models are investigated in detail.

1. INTRODUCTION

Since natural populations are frequently distributed in space, the analysis of the evolutionary consequences of population subdivision is an important problem in population genetics. Here, we study only neutral models of geographical variation; see Nagylaki (1984) for recent references to the interesting literature on models that involve selection or genotype-dependent migration. The reader unfamiliar with population genetics should consult Nagylaki (1984) for background, orientation, and references to monographs.

We assume that a monoecious, diploid population is subdivided into a finite or infinite number of panmictic colonies that exchange gametes in a fixed pattern; colony j contains N_j adults. Generations are discrete and nonoverlapping; the analysis is restricted to a single locus in the absence of selection; every allele mutates to new alleles at the same rate u ($0 \leq u \leq 1$). We measure time, t (= 0, 1, 2, ...), in generations. Random genetic drift operates through population regulation.

To begin the life cycle, every one of the N_j adults in deme j produces the same very large number of gametes, which then disperse. Complete random union of gametes follows. Therefore, a proportion $1/N_j$ of the zygotes whose gametes originate in deme j are produced by self-fertilization. Mutation is next, and finally population regulation returns the number of individuals in deme j to N_j. Let $f_{jk}(t)$ denote the probability that two distinct genes chosen at random from adults just before gametogenesis in generation t, one from colony j and one from colony k, are the same allele. The probability of allelic identity is (in principle) a measurable functional of our rather complicated Markov chain. Its complement, $h_{jk} = 1 - f_{jk}$, is an index of the amount and pattern of genetic diversity in the population. In particular, f_{jj} and h_{jj} represent the expectations of the homozygosity and heterozygosity in deme j, respectively. We summarize our model in the life cycle below.

Adults	$\xrightarrow{\text{gametogenesis}}$	Gametes	$\xrightarrow{\text{dispersion}}$	Gametes	$\xrightarrow{\text{fertilization}}$
N_j, f_{jk}		$\infty, -$		$\infty, -$	

Zygotes	$\xrightarrow{\text{mutation}}$	Zygotes	$\xrightarrow{\text{regulation}}$	Adults
$\infty, -$		$\infty, -$		N_j, f'_{jk}

Let m_{jk} designate the probability that a gamete in deme j after dispersion was produced in deme k. Since the gametes disperse independently, after one generation we have (Malécot, 1951; Sawyer, 1976; Nagylaki, 1980)

$$f'_{jk} = v\left[\sum_{\ell n} m_{j\ell} m_{kn} f_{\ell n} + \sum_{\ell} m_{j\ell} m_{k\ell} (2N_\ell)^{-1}(1 - f_{\ell\ell})\right], \qquad (1)$$

where $v = (1 - u)^2$ and the prime signifies the next generation. We placed mutation after migration only for definiteness; actually, (1) holds if mutation occurs at any time between gametogenesis and regulation. Population regulation during this period would have no effect if it were sufficiently weak to leave very large numbers of gametes and zygotes. Clearly, (1) also applies to a model with $2N_j$ haploid individuals in deme j, as noted by Sawyer (1976).

It is easy to see that if $0 \leq f_{jk}(0) \leq 1$ for every j and k, then $0 \leq f_{jk}(t) \leq 1$ for every j, k, and t.

If $u > 0$, then as $t \to \infty$, $f_{jk}(t) \to \hat{f}_{jk}$, the unique equilibrium of (1), at least as fast as v^t (Nagylaki, 1980). Furthermore, some genetic variability is preserved: from (1) we infer that $\hat{f}_{jk} < 1$ for some j and k.

Now suppose that $u = 0$. If the number of colonies is finite and the backward migration matrix M is ergodic, some allele will be fixed with probability one in a finite time, which implies that $f_{jk}(t) \to 1$ as $t \to \infty$ for every j and k. The last conclusion also holds if the population occupies an infinite lattice in one or two (but not three or more) dimensions, the number of adults in each deme is the same ($N_j = N$), and the migration pattern is homogeneous ($m_{jk} = m_{j-k}$) and strongly aperiodic and has a finite covariance matrix (Weiss and Kimura, 1965; Nagylaki, 1976; Sawyer, 1976).

Our aim here is to complement Nagylaki (1984), where general properties of (1) were treated. In Section 2, we shall establish some new general results concerning the expected homozygosity and the effective number of alleles. Since detailed biological insight comes mainly from particular models, in Sections 3, 4, and 5 we shall examine three special cases: the island, circular stepping-stone, and infinite linear stepping-stone models, respectively. Consult Nagylaki (1984) for references to investigations of other migration patterns.

2. THE HOMOZYGOSITY AND THE EFFECTIVE NUMBER OF ALLELES

We assume here that the population is at equilibrium, the number of demes is finite, and migration is conservative, i.e., it does not change the deme sizes. The models in Sections 3 and 4 exemplify conservative migration. In order to have a nontrivial equilibrium, we posit that $0 < u < 1$. We demonstrate that population subdivision produces interdeme differentiation and increases the mean homozygosity and the effective number of alleles.

At equilibrium, (1) reads

$$\hat{f}_{jk} = v\left[\sum_{\ell n} m_{j\ell} m_{kn} \hat{f}_{\ell n} + \sum_{\ell} s_{\ell} m_{j\ell} m_{k\ell}\right], \tag{2}$$

where

$$s_{\ell} = (1 - \hat{f}_{\ell\ell})/(2N_{\ell}). \tag{3}$$

Defining the matrix A and the vector $\underset{\sim}{b}$ by

$$A = vM\otimes M, \qquad b_{jk} = v \sum_{\ell} s_{\ell} m_{j\ell} m_{k\ell}, \tag{4}$$

where the notation signifies that A is a Kronecker product, we rewrite (2) as the vector equation

$$\underset{\sim}{\hat{f}} = A\underset{\sim}{\hat{f}} + \underset{\sim}{b}. \tag{5}$$

Treating b as known, we can solve (5) immediately:

$$\underset{\sim}{\hat{f}} = (I - A)^{-1}\underset{\sim}{b} = \sum_{p=0}^{\infty} A^p \underset{\sim}{b}, \tag{6}$$

in which $\underline{I}$ denotes the identity matrix. Since M is stochastic, its spectral radius is one. Hence, the spectral radius of A is $v < 1$, which implies convergence of the sum in (6). Substituting (4) into (6) leads to

$$\hat{f}_{jk} = \sum_q s_q \sum_{p=0}^{\infty} v^{p+1} m_{jq}^{(p+1)} m_{kq}^{(p+1)}, \tag{7}$$

where $m_{jk}^{(p+1)} = (M^{p+1})_{jk}$. Malécot (1948, Sect. 3.3) has obtained essentially the same result. From (7) we infer at once, for <u>any</u> migration pattern,

$$\hat{f}_{jk} \le \sum_q s_q \sum_{p=0}^{\infty} v^{p+1}\tfrac{1}{2}\{[m_{jq}^{(p+1)}]^2 + [m_{kq}^{(p+1)}]^2\}$$

$$= \tfrac{1}{2}(\hat{f}_{jj} + \hat{f}_{kk}); \tag{8}$$

equality holds if and only if $m_{jq}^{(p+1)} = m_{kq}^{(p+1)}$ for every p and q.

We specialize now to <u>conservative</u> migration. Define the proportion of adults in deme j, the total population number, the mean probability of identity, and the mean homozygosity:

$$\kappa_j = N_j/N_T, \qquad N_T = \sum_j N_j, \tag{9}$$

$$\hat{\bar{f}} = \sum_{jk} \kappa_j \kappa_k \hat{f}_{jk}, \qquad \hat{\bar{f}}_0 = \sum_j \kappa_j \hat{f}_{jj}. \tag{10}$$

Averaging (8) shows that the mean probability of identity cannot exceed the mean homozygosity:

$$\hat{\bar{f}} \le \frac{1}{2} \sum_{jk} \kappa_j \kappa_k (\hat{f}_{jj} + \hat{f}_{kk}) = \hat{\bar{f}}_0. \tag{11}$$

Thus, population subdivision produces interdeme differentiation. Equality occurs in (11) if and only if

$$m_{jk}^{(p+1)} = c_k^{(p+1)} \tag{12}$$

for every j, k, and p, for some $c_k^{(p+1)}$. For conservative migration (Nagylaki, 1980),

$$\kappa_k = \sum_j \kappa_j m_{jk}. \tag{13}$$

On the one hand, if (12) holds, we take p = 0 and substitute $m_{jk} = c_k$ into (13) to conclude that $c_k = \kappa_k$ for every k, which means that the population is panmictic. On the other hand, if we posit panmixia, then $c_k = \kappa_k$ for every k. But if $m_{jk}^{(p)} = \kappa_k$ for every j and k, then

$$m_{jk}^{(p+1)} = \sum_\ell m_{j\ell}^{(p)} m_{\ell k} = \kappa_k, \tag{14}$$

so (12) holds. Therefore, equality occurs in (11) if and only if the entire population mates at random.

For conservative migration (Crow and Maruyama, 1971; Nagylaki, 1982),

$$\hat{\bar{f}} = \frac{v(1 - \hat{\bar{f}}_0)}{2N_T(1 - v)} . \tag{15}$$

Combining (11) and (15) yields

$$\hat{\bar{f}}_0 \geq \frac{v}{v + 2N_T(1 - v)} = f_r, \tag{16}$$

where f_r signifies the expected homozygosity for a panmictic population (Malécot, 1946, 1948; Kimura and Crow, 1964). Thus, the expected homozygosity is at least as great as for panmixia; equality holds in (16) if and only if the entire population mates at random.

Now, (16) gives

$$1 - \hat{\bar{f}}_0 \leq \frac{2N_T(1 - v)}{v + 2N_T(1 - v)} . \tag{17}$$

Inserting (17) into (15), we get

$$\hat{\bar{f}} \leq f_r, \tag{18}$$

i.e., the mean probability of identity is decreased by population subdivision. For the effective number of alleles (Kimura and Crow, 1964; Maruyama, 1970), we obtain

$$n_e = 1/\hat{\bar{f}} \geq 1 + 2N_T(v^{-1} - 1) = n_e^r, \tag{19}$$

where n_e^r designates the panmictic value; equality holds in (19) if and only if the population mates at random. In an infinite panmictic population with K alleles, it is trivial to prove that $n_e \leq K$, with equality if and only if all alleles are equally frequent (Nagylaki,

1977, p. 35). Equation (19) shows that population subdivision raises this index of genetic diversity.

The inequalities (18) and (19) can fail for nonconservative migration. In the strong-migration limit, an effective population number $N_e < N_T$ appears in the theory, and we fix M and let $u \to 0$ and $N_e \to \infty$ so that $N_e u$ remains fixed. One finds (Nagylaki, 1980)

$$\hat{f}_{jk} \to 1/(1 + 4N_e u) > 1/(1 + 4N_T u). \tag{20}$$

Since the right side of (20) is the limit of f_r as $u \to 0$ and $N_T \to \infty$ with $N_T u$ fixed, we conclude that (18) and (19) must be reversed in this case.

3. THE ISLAND MODEL

We assume that n ($\geq$ 2) demes, each of which comprises N individuals, exchange gametes with no spatial effect on dispersion, i.e., if the migration rate is m ($0 < m < 1$), every colony receives a proportion $m/(n - 1)$ of its immigrants from each of the other colonies (Moran, 1959; Maruyama, 1970; Maynard Smith, 1970):

$$m_{jk} = \begin{cases} 1 - m, & j = k, \\ m/(n - 1), & j \neq k. \end{cases} \tag{21}$$

Guided by (21), we suppose

$$f_{jk} = \begin{cases} f_0, & j = k, \\ f_1, & j \neq k. \end{cases} \tag{22}$$

From (1) we can verify that if (22) holds initially, it remains valid. Therefore, by the uniqueness of the equilibrium for $u > 0$, $\hat{f}_{jk}$ satisfies (22). Substituting (21) and (22) into (1) leads to the much simpler system (Maruyama, 1970; Latter, 1973; Nei, 1975, pp. 121-122; Nagylaki, 1983)

$$f_0' = v[ac + a(1 - c)f_0 + (1 - a)f_1], \tag{23a}$$

$$f_1' = v[bc + b(1 - c)f_0 + (1 - b)f_1], \tag{23b}$$

in which

$$a = (1 - m)^2 + \frac{m^2}{n - 1}, \qquad b = \frac{1 - a}{n - 1}, \qquad c = \frac{1}{2N}. \tag{24}$$

We treat separately the equilibrium and convergence of (23).

3.1. Equilibrium

The exact equilibrium of (23) is easily derived (Maruyama, 1970; Latter, 1973; Nei, 1975, pp. 121-122; Nagylaki, 1983), but rather complicated. To obtain detailed biological insight, we make the reasonable approximations that migration is weak and mutation is weak relative to the stronger one of migration and random drift:

$$m \ll 1, \qquad u \ll \max(m, 1/N). \tag{25}$$

Then

$$\hat{f}_0 \approx \frac{m + u(n-1)}{m + u(4mN_T + n - 1)}, \tag{26a}$$

$$\hat{f}_1 \approx \frac{m}{m + u(4mN_T + n - 1)}, \tag{26b}$$

where $N_T = nN$ represents the total population number. From (10), (19), and (26) we deduce the effective number of alleles

$$n_e = \left[\frac{1}{n}\hat{f}_0 + \left(1 - \frac{1}{n}\right)\hat{f}_1\right]^{-1} \approx \frac{n[m + u(4mN_T + n - 1)]}{nm + (n-1)u}. \tag{27}$$

Now we discuss the expected heterozygosity, genetic variation, and local differentiation in terms of unidirectional and bidirectional implications. The probabilities of nonidentity in state are $\hat{h}_j = 1 - \hat{f}_j$, $j = 1,2$. All the results follow from (26) and (27).

The heterozygosity is high or low in accordance with the conditions

$$\text{high} \Longleftrightarrow \hat{f}^0 \ll 1 \Longleftrightarrow 4muN_T \gg m + nu, \tag{28a}$$

$$\text{low} \Longleftrightarrow \hat{h}_0 \ll 1 \Longleftrightarrow 4muN_T \ll m + nu. \tag{28b}$$

For the genetic diversity, we have

$$\text{high} \Longleftrightarrow n_e \gg 1 \Longleftrightarrow u(4mN_T + n) \gg \max(m, u) \Longleftrightarrow \hat{f}_1 \ll 1, \tag{29a}$$

$$\text{low} \Longleftrightarrow n_e \approx 1 \Longleftrightarrow u(4mN_T + n) \ll m \Longleftrightarrow \hat{h}_1 \ll 1. \tag{29b}$$

From (28) and (29) we obtain the biologically reasonable relations

$$\text{high heterozygosity} \Longrightarrow \text{high diversity}, \tag{30a}$$

$$\text{low diversity} \Longrightarrow \text{low heterozygosity}; \tag{30b}$$

the converse implications are false. To examine the differentiation of gene frequencies among demes, we separate the cases of high and low diversity and employ the criteria of Kimura and Maruyama (1971). If the genetic diversity is high, the conditions for strong and weak interdeme differentiation read

$$\text{strong} \Longleftrightarrow \hat{f}_0/\hat{f}_1 \gg 1 \Longleftrightarrow nu \gg m \Longleftarrow \hat{h}_0 \ll 1, \tag{31a}$$

$$\text{weak} \Longleftrightarrow \hat{f}_0/\hat{f}_1 \approx 1 \Longleftrightarrow nu \ll m \Longrightarrow \hat{f}_0 \ll 1. \tag{31b}$$

For low diversity, (29b) yields $\hat{h}_0 \leq \hat{h}_1 \ll 1$, so $\hat{h}_1/\hat{h}_0$ is a much more sensitive measure of differentiation than is $\hat{f}_0/\hat{f}_1$. Therefore, we get

$$\text{strong} \Longleftrightarrow \hat{h}_1/\hat{h}_0 \gg 1 \Longleftrightarrow 4mN \ll 1, \tag{32a}$$

$$\text{weak} \Longleftrightarrow \hat{h}_1/\hat{h}_0 \approx 1 \Longleftrightarrow 4mN \gg 1. \tag{32b}$$

Equations (25) to (32) were derived in Nagylaki (1983). The differentiation criteria (31) and (32) may be combined in the form

$$4mN \ll \max(1, 4uN_T) \Longleftrightarrow \text{strong}, \tag{33a}$$

$$4mN \gg \max(1, 4uN_T) \Longleftrightarrow \text{weak}. \tag{33b}$$

Returning to (27), we assume $u \ll m$ and derive (Nagylaki, 1983)

$$n_e \approx 1 + u[4N_T + (n/m)]. \tag{34}$$

Thus, (34) holds if $u \ll m \ll 1$. Maruyama (1970) presented this formula with the constraints $n \gg 1$ and $u \ll m$; he also discussed its biological implications.

3.2. Convergence

Li (1976) has obtained the complete solution of (23). We offer only one observation before making approximations. Write the eigenvalues that control the convergence of (23) as $v\lambda_\pm$; λ_+ and λ_- are real and $1 > \lambda_+ > \lambda_- \geq 0$. One can demonstrate from the explicit formula for λ_+ that

$$\lambda_+ \geq 1 - (2N_T)^{-1}, \tag{35}$$

with equality if and only if the population is panmictic, i.e., $m = (n - 1)/n$. Since the right side of (35) is precisely the rate of convergence for panmixia (Wright, 1931; Malécot, 1946, 1948; Kimura, 1963), we infer that population subdivision retards convergence here. If there is no mutation, this means that the rate of loss of genetic variability is slower than for random mating.

Despite its intuitive appeal, the conclusion (35) is not completely general: In the strong-migration limit, for nonconservative migration, the reduced dominant eigenvalue corresponding to λ_+ is approximately (Nagylaki, 1980)

$$1 - (2N_e)^{-1} < 1 - (2N_T)^{-1}. \tag{36}$$

The eigenvalues are fairly simple if migration is weak or strong:

Assuming

$$m \ll \max(1/\sqrt{N},\ \sqrt{n}/N), \tag{37}$$

one finds

$$\lambda_+ \approx 1 - \frac{2m}{(n-1)(1+4mN)}, \tag{38a}$$

$$\lambda_- \approx \left(1 - \frac{1}{2N}\right)(1 - 2m) \approx 1 - \frac{1}{2N} - 2m, \tag{38b}$$

where the second approximation in (38b) holds if $m \ll 1$ and $N \gg 1$.

Assuming

$$m \gg n/N, \tag{39}$$

we have

$$\lambda_+ \approx 1 - \frac{1}{2N_T}, \tag{40a}$$

$$\lambda_- \approx (1 - nb)\left(1 - \frac{n-1}{2N_T}\right) \approx 1 - \frac{2nm}{n-1} - \frac{n-1}{2N_T}, \tag{40b}$$

where the last approximation holds if $m \ll 1$ and $N \gg 1$. General results for the strong-migration limit imply not only (40a), but also that the asymptotic transient pattern is approximately uniform (Nagylaki, 1980)

$$\lim_{N\to\infty} \lim_{t\to\infty} [f_0(t) - \hat{f}_0]/[f_1(t) - \hat{f}_1] = 1. \tag{41}$$

4. THE CIRCULAR STEPPING-STONE MODEL

We suppose that n colonies, each of which comprises N individuals, form a closed loop (Malécot, 1951). This arrangement might be a mathematical idealization of an atoll; demes around a mountain, lake, or shore of an island; or colonies of amphibious or shallow-water organisms in a large, deep lake or around an island. Starting at an arbitrary colony, we number the colonies 0, 1, 2,... counter-clockwise and 0, -1, -2,... clockwise. Dispersion is homogeneous: $m_{jk} = m_{j-k}$; thus, m_j signifies the probability of displacement by j demes. The probability that the separation between two gametes changes by j reads (Maruyama, 1971; Nagylaki, 1974a)

$$r_j = \sum_{k=-\infty}^{\infty} m_k m_{j+k}\ . \tag{42}$$

Observe that r is even (Nagylaki, 1974a, 1978a): $r_{-j} = r_j$. The homogeneity of dispersion leads us to posit homogeneity of the probabilities of allelic identity: $f_{jk} = \tilde{f}_{j-k}$. We can verify that (1) preserves initial homogeneity. Therefore, by the uniqueness of the equilibrium for $u > 0$, $\hat{f}_{jk} = \hat{\tilde{f}}_{j-k}$. Homogeneity reduces (1) to

$$\tilde{f}_j' = v \sum_{k=-\infty}^{\infty} [r_k \tilde{f}_{j-k} + (2N)^{-1}(1 - \tilde{f}_0) r_{j-kn}]. \tag{43}$$

By evenness and periodicity, we have

$$\tilde{f}_{-j} = \tilde{f}_j, \qquad \tilde{f}_{j+kn} = \tilde{f}_j \tag{44}$$

$(k = 0, \pm 1, \pm 2, \ldots)$.

Discrete Fourier analysis enables us to express the equilibrium solution of (43) as an explicit finite sum (Malécot, 1951, 1965, 1975; Nagylaki, 1983), but we proceed at once to the diffusion approximation.

4.1. The Diffusion Approximation

We scale space and time according to

$$x = j\varepsilon, \qquad T = \lambda t, \qquad f(x,T) = \tilde{f}_j(t). \tag{45}$$

In the new units, ε represents the distance between adjacent colonies and λ corresponds to one generation; the circumference of the habitat is $L = n\varepsilon$. Put $u_0 = u/\lambda$ and $\rho_0 = N\lambda/\varepsilon$, where u_0 and ρ_0 signify the mutation rate in the new time units and a scaled population density, respectively. The population density in the new length units is $\rho = N/\varepsilon = \rho_0/\lambda$. Let $\varepsilon \to 0$, $\lambda \to 0$, $n \to \infty$, $u \to 0$, and $N \to \infty$ so that λ/ε^2, L, u_0, and ρ_0 remain fixed. For migration, we require the diffusion hypotheses

$$\lim_{\lambda \to 0} \frac{\varepsilon^2}{\lambda} \sum_{j:\ |j|<\theta/\varepsilon} j^2 r_j = \sigma_0^2, \tag{46a}$$

$$\lim_{\lambda \to 0} \frac{1}{\lambda} \sum_{j:\ |j|\geq\theta/\varepsilon} r_j = 0 \tag{46b}$$

for every fixed $\theta > 0$. (Recall that r_j is even, and hence has mean zero.) Clearly, σ_0^2 is the variance of the change in gametic separation per new time unit in the new length units; the corresponding variance in generations is $\sigma^2 = \lambda\sigma_0^2$. From (42) it is easy to prove that the variance of r_j is twice that of m_j. Thus, mutation ($u \propto \varepsilon^2$), migration ($\sigma \propto \varepsilon$), and random drift ($1/N \propto \varepsilon$) must be weak, and the number of demes must be large ($n \propto 1/\varepsilon$).

For any function $\chi(x)$ with a continuous second derivative, (46) leads to (cf. Feller, 1971, pp. 333-335)

$$\sum_k r_k \chi[(j-k)\varepsilon] = \chi(j\varepsilon) + \tfrac{1}{2}\lambda\sigma_0^2 \frac{d^2\chi}{dx^2}(j\varepsilon) + o(\lambda) \tag{47}$$

as $\lambda \to 0$. Substituting (47) into (43) produces

$$f(x,T+\lambda) = (1 - 2\lambda u_0 + \lambda^2 u_0^2)\Big\{f(x,T) + \tfrac{1}{2}\lambda\sigma_0^2 \frac{\partial^2 f}{\partial x^2} + o(\lambda)$$
$$+ \left[\frac{1 - f(0,T)}{2N}\right] \sum_{k=-\infty}^{\infty} \delta_{j,kn} + O(\lambda/N)\Big\} \tag{48}$$

as $\lambda \to 0$, where the summand designates the Kronecker delta. Replacing $\delta_{j,kn}$ by $\varepsilon\delta(\varepsilon j - \varepsilon kn) = \varepsilon\delta(x-kL)$, where $\delta(x)$ denotes the Dirac delta function, rearranging (48), and letting $\lambda \to 0$, we find

$$f_T = -2u_0 f + \tfrac{1}{2}\sigma_0^2 f_{xx} + (2\rho_0)^{-1}[1 - f(0,T)] \sum_{k=-\infty}^{\infty} \delta(x-kL); \tag{49}$$

here and below, subscripts signify partial differentiation. Now, (44) translates to

$$f(-x,T) = f(x,T), \qquad f(x+kL,T) = f(x,T) \tag{50}$$

($k = 0, \pm 1, \pm 2, \ldots$). From (49) and (50) we easily deduce the boundary-value problem

$$f_T = -2u_0 f + \tfrac{1}{2}\sigma_0^2 f_{xx}, \qquad 0 < x < \tfrac{1}{2}L, \tag{51a}$$

$$-2\rho_0\sigma_0^2 f_x(0+,T) = 1 - f(0,T), \tag{51b}$$

$$f_x(\tfrac{1}{2}L,T) = 0 \tag{51c}$$

for $T > 0$. We take $f(x,0)$ as given.

See Malécot (1967), Maruyama (1971), and Nagylaki (1974a,b) for heuristic derivations of partial differential equations for this problem and for similar ones. Note that there are serious difficulties with the formulation in continuous space and discrete time of models similar to the stepping-stone model (Felsenstein, 1975; Kingman, 1977; Sudbury, 1977; Sawyer and Felsenstein, 1981) and that the diffusion approximation fails in more than one spatial dimension (Fleming and Su, 1974; Nagylaki, 1974b, 1978b).

Before solving (51), we establish if $0 \leq f(x,0) \leq 1$ for every x, then $0 \leq f(x,T) \leq 1$ for every x and T, i.e., a biologically sensible initial condition yields a biologically sensible solution. We consider (51) on $0 \leq T \leq T^* < \infty$ and show first that $f(x,T) \leq 1$. By the maximum principle for parabolic partial differential equations (Protter and Weinberger, 1967, pp. 173-174), if f has a positive maximum, it must occur on $T = 0$, $x = 0$ or $x = \tfrac{1}{2}L$. By assumption, $f(x,0) \leq 1$. If the maximum occurs at $(0,T_1)$ for some T_1, the maximum principle implies that $f_x(0+, T_1) < 0$, so (51b) informs us that

$f(0,T_1) \le 1$. The maximum cannot occur at $(\frac{1}{2}L, T_2)$ for any T_2, because $f_x(\frac{1}{2}L, T_2) > 0$ would contradict (51c). Essentially the same argument proves that $-f(x,T)$ cannot have a positive maximum.

We solve separately for the equilibrium and transient components of f.

4.2. Equilibrium

From (51) we easily obtain the equilibrium

$$\hat{f}(x) = \frac{\cosh[\alpha(1 - \frac{2x}{L})]}{\cosh\alpha + \beta\sinh\alpha}, \qquad 0 \le x \le \tfrac{1}{2}L, \tag{52}$$

where

$$\alpha = \sqrt{u_0}L/\sigma_0 = \sqrt{u}L/\sigma, \qquad \beta = 4\rho_0\sigma_0\sqrt{u_0} = 4\rho\sigma\sqrt{u} \tag{53}$$

represent dimensionless parameters. In particular, (52) gives the homozygosity

$$\hat{f}(0) = 1/(1 + \beta\tanh\alpha) \tag{54}$$

and the mean probability of identity

$$\hat{\hat{f}} = \frac{1}{L^2}\int_{-L/2}^{L/2}\int_{-L/2}^{L/2} \hat{f}(x-y)dx\,dy = \frac{1}{\alpha(\beta + \coth\alpha)} . \tag{55}$$

Consequently, the effective number of alleles reads

$$n_e = 1/\hat{\hat{f}} = \alpha(\beta + \coth\alpha). \tag{56}$$

Equations (52) to (56) are derived in Nagylaki (1974). Set $\hat{h}(x) = 1 - \hat{f}(x)$. For high and low genetic diversity, we employ $\hat{f}(0)/\hat{\hat{f}}$ and $\hat{\hat{h}}/\hat{h}(0)$, respectively, as indices of local differentiation (Kimura and Maruyama, 1971). These formulae enable us parallel the biological discussion in Section 3.1.

The heterozygosity is high or low in accordance with the conditions

$$\text{high} \Longleftrightarrow \beta\tanh\alpha \gg 1 \Longleftrightarrow \beta \gg 1 \text{ and } \alpha\beta \gg 1, \tag{57a}$$

$$\text{low} \Longleftrightarrow \beta\tanh\alpha \ll 1 \Longleftrightarrow \beta \ll 1 \text{ or } \alpha\beta \ll 1. \tag{57b}$$

For the genetic diversity, we have

$$\text{high} \Longleftrightarrow \alpha(\beta + \coth\alpha) \gg 1 \Longleftrightarrow \alpha \gg 1 \text{ or } \alpha\beta \gg 1 \Longleftarrow \hat{f}(0) \ll 1, \tag{58a}$$

$$\text{low} \Longleftrightarrow \alpha(\beta + \coth\alpha) \approx 1 \Longleftrightarrow \alpha \ll 1 \text{ and } \alpha\beta \ll 1 \Longrightarrow \hat{h}(0) \ll 1. \tag{58b}$$

Thus, the intuitive relations (30) hold again. We can summarize the criteria for interdeme differentiation of gene frequencies as

$$\beta \ll \alpha \max(1, \alpha\beta) \Longleftrightarrow \text{strong}, \tag{59a}$$

$$\beta \gg \alpha \max(1, \alpha\beta) \Longleftrightarrow \text{weak}. \tag{59b}$$

It is very instructive to compare (57)-(59) with the corresponding results (28), (29), and (33) for the island model. To effect the comparison, we rewrite (57)-(59) for symmetric nearest-neighbor migration at rate m per generation. Since $\sigma^2 = 2m\varepsilon^2$, (53) becomes

$$\alpha = n\sqrt{\frac{u}{2m}}, \qquad \beta = 4N\sqrt{2mu}, \qquad \alpha\beta = 4uN_T, \tag{60}$$

where $N_T = nN$. Therefore, the criteria (57) for the heterozygosity now read

$$\text{high} \Longleftrightarrow 4N\sqrt{2mu} \gg 1 \quad \text{and} \quad 4uN_T \gg 1, \tag{61a}$$

$$\text{low} \Longleftrightarrow 4N\sqrt{2mu} \ll 1 \quad \text{or} \quad 4uN_T \ll 1. \tag{61b}$$

The conditions (58) for the genetic diversity have the form

$$\text{high} \Longleftrightarrow n\sqrt{u} \gg \sqrt{2m} \quad \text{or} \quad 4uN_T \gg 1, \tag{62a}$$

$$\text{low} \Longleftrightarrow n\sqrt{u} \ll \sqrt{2m} \quad \text{and} \quad 4uN_T \ll 1. \tag{62b}$$

For genetic differentiation, from (59) we obtain

$$8mN \ll n \max(1, 4uN_T) \Longleftrightarrow \text{strong}, \tag{63a}$$

$$8mN \gg n \max(1, 4uN_T) \Longleftrightarrow \text{weak}. \tag{63b}$$

Since we are interested in the influence of population subdivision, we assume that $u \leq m$. Then we find that high heterozygosity, low diversity, and weak differentiation in the circular stepping-stone model imply the same in the island model, whereas low heterozygosity, high diversity, and strong differentiation in the island model imply the same in the circular stepping-stone model. We conclude, as expected intuitively, that the heterozygosity is lower, whereas the genetic diversity and interdeme differentiation are higher, in the circular stepping-stone model than in the island model (Nagylaki, 1983).

4.3. Convergence

We return to (51), subtract $\hat{f}(x)$, and rescale to eliminate dimensions:

$$f(x,T) = \hat{f}(x) - e^{-2u_0\tau}\phi(\xi,\tau), \tag{64a}$$

$$x = a\xi, \qquad T = T_0\tau, \tag{64b}$$

$$a = 2\rho_0\sigma_0^2, \qquad T_0 = (2\rho_0\sigma_0)^2, \qquad \gamma = L/(2a). \tag{64c}$$

Inserting (64) into (51), we deduce the boundary-value problem

$$\phi_\tau = \frac{1}{2}\phi_{\xi\xi}, \qquad 0 < \xi < \gamma, \quad \tau > 0, \tag{65a}$$

$$\phi_\xi(0+,\tau) = \phi(0,\tau), \qquad \tau > 0, \tag{65b}$$

$$\phi_\xi(\gamma,\tau) = 0, \qquad \tau > 0, \tag{65c}$$

$$\phi(\xi,0) = \phi_0(\xi), \qquad 0 \le \xi \le \gamma. \tag{65d}$$

To solve (65), we introduce the Green function $g(\xi,\eta,\tau)$, which satisfies (65a)-(65c) with the initial condition

$$g(\xi,\eta,0) = \delta(\xi-\eta). \tag{66}$$

Then

$$\phi(\xi,\tau) = \int_0^\gamma g(\xi,\eta,\tau)\phi_0(\eta)d\eta. \tag{67}$$

Regarding the Laplace transform

$$G(\xi,\eta,s) = \int_0^\infty e^{-s\tau}g(\xi,\eta,\tau)d\tau \tag{68}$$

as a function of ξ, we obtain from (65) and (66)

$$sG - \delta(\xi-\eta) = \frac{1}{2}\frac{d^2G}{d\xi^2}, \qquad 0 < \xi < \gamma, \tag{69a}$$

$$\frac{dG}{d\xi}(0+,\eta,s) = G(0,\eta,s), \tag{69b}$$

$$\frac{dG}{d\xi}(\gamma,\eta,s) = 0. \tag{69c}$$

The problem (69) has the unique solution

$$G(\xi,\eta,s) = [\cosh(\sqrt{2s}\,\gamma) + \sqrt{2s}\,\sinh(\sqrt{2s}\,\gamma)]^{-1} \cdot$$
$$\{\cosh[\sqrt{2s}(\gamma-|\xi-\eta|)] + \cosh[\sqrt{2s}(\gamma-\xi-\eta)]$$
$$+ (1/\sqrt{2s})[\sinh[\sqrt{2s}(\gamma-|\xi-\eta|)] - \sinh[\sqrt{2s}(\gamma-\xi-\eta)]]\}. \tag{70}$$

To invert the Laplace transform (68), we examine G in the complex s-plane. Since changing the sign of $\sqrt{2s}$ does not affect (70), G has no cuts. The poles of (70) occur where the denominator vanishes:

$$\sqrt{2s_j} = ip_j, \qquad s_j = -\tfrac{1}{2}p_j^2, \tag{71a}$$

$$\cot\gamma p_j = p_j, \tag{71b}$$

$j = 0, 1, 2, \ldots$; we define $p_j > 0$. From (71b) we infer for $j = 0, 1, 2, \ldots$ the bounds

$$j\pi/\gamma < p_j < (j + \tfrac{1}{2})\pi/\gamma. \tag{71c}$$

Finally, on a circle centered at the origin and bounded away from the poles,

$$|G(\xi,\eta,s)| \leq (\text{const.})/\sqrt{|s|}. \tag{72}$$

Therefore, the inversion contour along the imaginary axis can be completed by a semi-circle in the left half-plane (Apostol, 1974, p. 468), and we find

$$g(\xi,\eta,\tau) = \sum_{j=0}^{\infty} e^{-p_j^2\tau/2}\psi_j(\xi)\psi_j(\eta), \tag{73a}$$

where

$$\psi_j(\xi) = \left[\frac{2(1+p_j^2)}{1+\gamma(1+p_j^2)}\right]^{1/2} \cos[p_j(\gamma-\xi)], \tag{73b}$$

$0 \leq \xi,\eta \leq \gamma$.

Setting $\tau = 0$ in (73a) and recalling (66) yields the completeness relation

$$\delta(\xi-\eta) = \sum_{j=0}^{\infty} \psi_j(\xi)\psi_j(\eta), \tag{74}$$

the normalization

$$\int_0^{\gamma} \psi_j(\xi)\psi_k(\xi)d\xi = \delta_{jk}, \tag{75}$$

which can be verified directly from (71b) and (73b), follows. Substituting (73a) into (67), we conclude

$$\phi(\xi,\tau) = \sum_{j=0}^{\infty} c_j e^{-p_j^2\tau/2}\psi_j(\xi), \tag{76a}$$

where

$$c_j = \int_0^{\gamma} \phi_0(\eta)\psi_j(\eta)d\eta. \tag{76b}$$

As $\tau \to \infty$, (76a) gives

$$\phi(\xi,\tau) \sim c_0 e^{-p_0^2\tau/2}\psi_0(\xi). \tag{77}$$

In view of (71c), $0 < \gamma p_0 < \pi/2$. Therefore, (71b) reveals that $p_0 < 1/(\gamma p_0)$, whence $p_0 < 1/\sqrt{\gamma}$. Since returning to our original variables shows that $\frac{1}{2}\gamma\tau = t/(2N_T)$, we infer that population subdivision retards convergence in this model. The remarks below (35) are relevant here.

For identity by descent, all genes are initially distinguished: $f(x,0) = 0$; hence (64a), (65d), (52), (53), (64b), and (64c) yield

$$\phi_0(\xi) = \hat{f}(x) = \frac{\cosh[\beta(\gamma - \xi)]}{\cosh\alpha + \beta\sinh\alpha} . \tag{78}$$

Substituting (73b) and (78) into (76b) and using (53), (64c), and (71b), we obtain

$$c_j = \frac{p_j \sin\gamma p_j}{\beta^2 + p_j^2}\left[\frac{2(1 + p_j^2)}{1 + \gamma(1 + p_j^2)}\right]^{1/2} . \tag{79}$$

Since (78) depends on the mutation rate, so does (79); in the absence of mutation, $\beta = 0$ in (79).

Very small and very large values of $\gamma = L/(4\rho_0\sigma_0^2)$ are of particular interest:

If $\gamma >> 1$, corresponding to a large habitat, weak migration, or low population density, then (71b) and (73b) reduce to ($j = 0, 1, 2, ...$)

$$p_j \approx (2j + 1)\frac{\pi}{2\gamma} , \tag{80a}$$

$$\psi_j(\xi) \approx (-1)^j\sqrt{\frac{2}{\gamma}}\sin\left[(2j + 1)\frac{\pi\xi}{2\gamma}\right] . \tag{80b}$$

From (80a) we obtain the decay rates in (76a):

$$\tfrac{1}{2}p_j^2\tau \approx \tfrac{1}{2}[(2j + 1)\pi\sigma/L]^2 t. \tag{81}$$

To convert (80b) to our original variables, it suffices to observe that

$$\gamma = \frac{L}{4\rho\sigma^2} , \qquad \frac{\xi}{2\gamma} = \frac{x}{L} . \tag{82}$$

If $\gamma << 1$, corresponding to a small habitat, strong migration, or high population density, then (71b) and (73b) simplify to ($j = 1, 2, ...$)

$$p_0 \approx 1/\sqrt{\gamma} , \qquad p_j \approx j\pi/\gamma , \tag{83a}$$

$$\psi_0(\xi) \approx 1/\sqrt{\gamma}, \qquad \psi_j(\xi) \approx (-1)^j\sqrt{2/\gamma}\cos(j\pi/\gamma). \tag{83b}$$

Therefore, the decay rates in (76a) read

$$\tfrac{1}{2}p_0^2\tau \approx t/(2N_T), \qquad \tfrac{1}{2}p_j^2\tau \approx 2(j\pi\sigma/L)^2 t. \tag{84}$$

The expressions for p_0 and $\psi_0(\xi)$ follow from general results for the strong-migration limit (Nagylaki, 1980).

Maruyama (1971) derived heuristically the leading eigenvalue and the corresponding eigenvector for $\gamma << 1$ and $\gamma >> 1$. All the eigenvalues and eigenvectors were deduced in Nagylaki (1974a) from a less rigorous formulation than the one presented here; the complete solution (76) was not constructed there.

5. THE INFINITE LINEAR STEPPING-STONE MODEL

The stepping-stone model on an infinite lattice has been extensively investigated by techniques applicable in discrete space and time (see Nagylaki, 1984 for references), and its unidimensional case has been analyzed in the diffusion approximation (Nagylaki, 1974b, 1978a). Here, we derive and extend the latter results by letting the circumference tend to infinity in the circular stepping-stone model.

We assume that colonies of N individuals are located at 0, ±1, ±2, ... Such a long linear habitat might represent organisms along or in a river, close to a seashore, or along a mountain range.

5.1. Equilibrium

As $L \to \infty$, (52) converges to (Nagylaki, 1974a, 1978a)

$$\hat{f}(x) = \frac{e^{-2\sqrt{u}x/\sigma}}{1+\beta} = \frac{e^{-\beta\xi}}{1+\beta}, \qquad x \geq 0. \tag{85}$$

This classical result has been established by asymptotic methods without the diffusion approximation (Malécot, 1950, 1965; Weiss and Kimura, 1965; Nagylaki, 1976; Sawyer, 1977).

5.2. Convergence

As $L \to \infty$, (70) converges to $(\xi, \eta \geq 0)$

$$G(\xi,\eta,s) = \frac{e^{-\sqrt{2s}|\xi-\eta|}}{\sqrt{2s}} + \left(\frac{2}{1+\sqrt{2s}} - \frac{1}{\sqrt{2s}}\right) e^{-\sqrt{2s}(\xi+\eta)}, \tag{86}$$

which can be confirmed by solving the pertinent boundary-value problem on the half-line. The inversion of this Laplace transform (Erdélyi, 1954, p. 246) yields

$$g(\xi,\eta,\tau) = \frac{1}{\sqrt{2\pi\tau}}\left[e^{-(\xi+\eta)^2/(2\tau)} + e^{-(\xi-\eta)^2/(2\tau)}\right]$$

$$- e^{\xi+\eta+\frac{\tau}{2}}\operatorname{erfc}\left(\frac{\xi+\eta}{\sqrt{2\tau}} + \sqrt{\frac{\tau}{2}}\right), \tag{87}$$

where

$$\operatorname{erfc} y = \frac{2}{\sqrt{\pi}} \int_y^\infty e^{-z^2} dz. \tag{88}$$

Equation (67) now reads ($\xi \geq 0$)

$$\phi(\xi,\tau) = \int_0^\infty g(\xi,\eta,\tau)\phi_0(\eta)d\eta. \tag{89}$$

See Nagylaki (1974b, 1978a) for a direct derivation of this solution. Recall from (64a) and (65d) that

$$\phi_0(\xi) = \hat{f}(x) - f_0(\xi), \tag{90}$$

in which $f_0(\xi) = f(x,0)$ and $\hat{f}(x)$ is given by (85).

We turn to the most interesting aspect of our solution, its behavior for long times ($\tau \to \infty$). It is easy to prove that

$$g(\xi,\eta,\tau) \sim \sqrt{\frac{2}{\pi\tau^3}}(1 + \xi)(1 + \eta) \tag{91}$$

as $\tau \to \infty$ with ξ and η fixed. We shall see that the cases with and without mutation behave very differently.

No Mutation ($u_0 = 0$)

Here $\hat{f}(x) = 1$, so $\phi(\xi,\tau)$ is the expected heterozygosity. We use (89) and (90) to write

$$\phi(\xi,\tau) = \phi_1(\xi,\tau) - \phi_2(\xi,\tau), \tag{92}$$

where

$$\phi_1(\xi,\tau) = \int_0^\infty g(\xi,\eta,\tau)d\eta, \tag{93a}$$

$$\phi_2(\xi,\tau) = \int_0^\infty g(\xi,\eta,\tau)f_0(\eta)d\eta. \tag{93b}$$

Since $f_0(\xi) = 0$ for identity by descent, $\phi_1(\xi,\tau)$ is the exact solution in this special case. Invoking (87) and integrating by parts in (93a), we obtain (Nagylaki, 1978a)

$$\phi_1(\xi,\tau) = \operatorname{erf}\left(\frac{\xi}{\sqrt{2\tau}}\right) + e^{\xi+\frac{\tau}{2}}\operatorname{erfc}\left(\frac{\xi}{\sqrt{2\tau}} + \sqrt{\frac{\tau}{2}}\right), \tag{94}$$

where $\operatorname{erf} y = 1 - \operatorname{erfc} y$. As $\tau \to \infty$ with ξ fixed (Nagylaki, 1978)

$$\phi_1(\xi,\tau) \sim \sqrt{\frac{2}{\pi\tau}}\,(1 + \xi). \tag{95}$$

Next, we demonstrate that under the very weak and biologically reasonable assumption that $f_0(\xi) \to 0$ as $\xi \to \infty$, (95) generalizes to

$$\phi(\xi,\tau) \sim \sqrt{\frac{2}{\pi\tau}}(1 + \xi) \tag{96}$$

as $\tau \to \infty$ with ξ fixed. We must prove that

$$\phi_2(\xi,\tau) = o(1/\sqrt{\tau}) \tag{97}$$

as $\tau \to \infty$ with ξ fixed. Put

$$\phi_2(\xi,\tau) = \phi_{21}(\xi,\tau) + \phi_{22}(\xi,\tau), \tag{98}$$

where

$$\phi_{21}(\xi,\tau) = \int_0^X g(\xi,\eta,\tau)f_0(\eta)d\eta, \tag{99a}$$

$$\phi_{22}(\xi,\tau) = \int_X^\infty g(\xi,\eta,\tau)f_0(\eta)d\eta, \tag{99b}$$

and X is fixed. Given $\theta > 0$, by our assumption there exists $X > 0$ such that $f_0(\eta) < \theta$ for $\eta \geq X$. From (91) and (99a) we have

$$\phi_{21}(\xi,\tau) = O(\tau^{-3/2}) \tag{100}$$

as $\tau \to \infty$ with ξ fixed. Using either the maximum principle for parabolic partial differential equations (Protter and Weinberger, 1967, pp. 173-174) or the inequality (Feller, 1968, p. 175)

$$\operatorname{erfc} x < e^{-x^2}/(\sqrt{\pi}\, x), \tag{101}$$

we can show that $g(\xi,\eta,\tau) \geq 0$. Therefore, (99b) and (93a) yield

$$0 \leq \phi_{22}(\xi,\tau) < \theta \int_X^\infty g(\xi,\eta,\tau)d\eta < \theta\phi_1(\xi,\tau); \tag{102}$$

(95) and (102) imply

$$\phi_{22}(\xi,\tau) = o(1/\sqrt{\tau}) \tag{103}$$

as $\tau \to \infty$ with ξ fixed. Equations (98), (100), and (103) establish (97).

Nonzero Mutation ($u_0 > 0$)

Suppose $f_0(\xi) = O(\xi^{-2-\kappa})$ as $\xi \to \infty$ for some $\kappa > 0$. Then (85) and (90) inform us that $\phi_0(\xi) = O(\xi^{-2-\kappa})$ as $\xi \to \infty$, and hence (91) shows that (89) converges uniformly for $\tau \geq \tau_0 > 0$. Consequently, from (89) and (91) we deduce (Nagylaki, 1974b, 1978a)

$$\phi(\xi,\tau) \sim \sqrt{\frac{2}{\pi\tau^3}}(1 + \xi)\int_0^\infty (1 + \eta)\phi_0(\eta)d\eta \tag{104}$$

as $\tau \to \infty$ with ξ fixed.

If $f_0(\xi)$ decreases sufficiently slowly as $\xi \to \infty$, $\phi(\xi,\tau)$ can decay more slowly than $\tau^{-3/2}$ as $\tau \to \infty$: By (85) and (91), the contribution of $\hat{f}(x)$ to (89) is $O(\tau^{-3/2})$. If the population is initially homogeneous, $f_0(\xi) = 1$, then (89), (90), (93a), and (95) immediately reveal

$$\phi(\xi,\tau) \sim -\sqrt{\frac{2}{\pi\tau}}(1 + \xi) \tag{105}$$

as $\tau \to \infty$ with ξ fixed.

For identity by descent, $f_0(\xi) = 0$, and (85), (87), (89), and (90) lead to

$$\phi(\xi,\tau) = \frac{e^{\xi+\frac{\tau}{2}}}{1 - \beta^2}\operatorname{erfc}\left(\sqrt{\frac{\tau}{2}} + \frac{\xi}{\sqrt{2\tau}}\right)$$

$$+ \frac{1}{2}e^{\beta^2\tau/2}\left[\frac{e^{-\beta\xi}}{1+\beta}\operatorname{erfc}\left(\beta\sqrt{\frac{\tau}{2}} - \frac{\xi}{\sqrt{2\tau}}\right) - \frac{e^{\beta\xi}}{1-\beta}\operatorname{erfc}\left(\beta\sqrt{\frac{\tau}{2}} + \frac{\xi}{\sqrt{2\tau}}\right)\right] \tag{106a}$$

for $\beta \neq 1$ and

$$\phi(\xi,\tau) = \frac{1}{4}e^{\tau/2}\left[e^{-\xi}\operatorname{erfc}\left(\sqrt{\frac{\tau}{2}} - \frac{\xi}{\sqrt{2\tau}}\right) + (1 + 2\xi + 2\tau)e^{\xi}\operatorname{erfc}\left(\sqrt{\frac{\tau}{2}} + \frac{\xi}{\sqrt{2\tau}}\right)\right]$$

$$- \sqrt{\frac{\tau}{2\pi}}\, e^{-\xi^2/(2\tau)} \tag{106b}$$

for $\beta = 1$. Either (104) or (106) yields (Nagylaki, 1978a)

$$\phi(\xi,\tau) \sim \sqrt{\frac{2}{\pi\tau^3}}\,\frac{1 + \xi}{\beta^2} \tag{107}$$

as $\tau \to \infty$ with ξ fixed. Notice that (107) is inversely proportional to the mutation rate.

The results (95), (96), and (107) hold for $\xi \gg 1$ in the discrete problem. Consult Malécot (1975), Nagylaki (1976), and Sawyer (1976) for proofs of (95); Nagylaki (1976) and Sawyer (1976) both derive (96) and (107); refer to Nagylaki (1976) for the discrete analogue of (104).

REFERENCES

Apostol, T. M. 1974. Mathematical Analysis, 2nd edition. Addison-Wesley, Reading, Mass.

Crow, J. F., and Maruyama, T. 1971. The number of neutral alleles maintained in a finite, geographically structured population. Theor. Pop. Biol. 2, 437-453.

Erdélyi, A. 1954. Tables of Integral Transforms, Vol. I. McGraw-Hill, New York.

Feller, W. 1968. An Introduction to Probability Theory and Its Applications, Vol. I, 3rd edition. Wiley, New York.

Feller, W. 1971. An Introduction to Probability Theory and Its Applications, Vol. II, 2nd edition. Wiley, New York.

Felsenstein, J. 1975. A pain in the torus: some difficulties with models of isolation by distance. Am. Nat. 109, 359-368.

Fleming, W. H., and Su, C.-H. 1974. Some one-dimensional migration models in population genetics theory. Theor. Pop. Biol. 5, 431-449.

Kimura, M. 1963. A probability method for treating inbreeding systems, especially with linked genes. Biometrics 19, 1-17.

Kimura, M., and Crow, J. F. 1964. The number of alleles that can be maintained in a finite population. Genetics 49, 725-738.

Kimura, M., and Maruyama, T. 1971. Pattern of neutral polymorphism in a geographically structured population. Genet. Res. 18, 125-131.

Kingman, J. F. C. 1977. Remarks on the spatial distribution of a reproducing population. J. Appl. Prob. 14, 577-583.

Latter, B. D. H. 1973. The island model of population differentiation: A general solution. Genetics 73, 147-157.

Li, W.-H. 1976. Effect of migration on genetic distance. Am. Nat. 110, 841-847.

Malécot, G. 1946. La consanguinité dans une population limitée. Compt. Rend. Acad. Sci. 222, 841-843.

Malécot, G. 1948. Les mathématiques de l'hérédité. Masson, Paris. (Extended translation: The Mathematics of Heredity. Freeman, San Francisco, 1969.)

Malécot, G. 1950. Quelques schémas probabilistes sur la variabilité des populations naturelles. Ann. Univ. Lyon, Sci., Sect. A, 13, 37-60.

Malécot, G. 1951. Un traitement stochastiques des problèmes linéaires (mutation, linkage, migration) en Génétique de Population. Ann. Univ. Lyon, Sci., Sect. A, 14, 79-117.

Malécot, G. 1965. Évolution continue des fréquences d'un gène mendélien (dans le cas de migration homogène entre groupes d'effectif fini constant). Ann. Inst. H. Poincaré, Sect. B, 2, 137-150.

Malécot, G. 1967. Identical loci and relationship. Proc. Fifth Berk. Symp. Math. Stat. Prob. 4, 317-332.

Malécot, G. 1975. Heterozygosity and relationship in regularly subdivided populations. Theor. Pop. Biol. 8, 212-241.

Maruyama, T. 1970. Effective number of alleles in a subdivided population. Theor. Pop. Biol. 1, 273-306.

Maruyama, T. 1971. The rate of decrease of heterozygosity in a population occupying a circular or linear habitat. Genetics 67, 437-454.

Maynard Smith, J. 1970. Population size, polymorphism, and the rate of non-Darwinian evolution. Am. Nat. 104, 231-237.

Moran, P. A. P. 1959. The theory of some genetical effects of population subdivision. Aust. J. Biol. Sci. 12, 109-116.

Nagylaki, T. 1974a. Genetic structure of a population occupying a circular habitat. Genetics 78, 777-790.

Nagylaki, T. 1974b. The decay of genetic variability in geographically structured populations. Proc. Natl. Acad. Sci. USA 71, 2932-2936.

Nagylaki, T. 1976. The decay of genetic variability in geographically structured populations. II. Theor. Pop. Biol. 10, 70-82.

Nagylaki, T. 1977. Selection in One- and Two-Locus Systems. Springer, Berlin.

Nagylaki, T. 1978a. The geographical structure of populations. In Studies in Mathematics. Vol. 16: Studies in Mathematical Biology. Part II (S. A. Levin, ed.). Pp. 588-624. The Mathematical Association of America, Washington.

Nagylaki, T. 1978b. A diffusion model for geographically structured populations. J. Math. Biol. 6, 375-382.

Nagylaki, T. 1980. The strong-migration limit in geographically structured populations. J. Math. Biol. 9, 101-114.

Nagylaki, T. 1982. Geographical invariance in population genetics. J. Theor. Biol. 99, 159-172.

Nagylaki, T. 1983. The robustness of neutral models of geographical variation. Theor. Pop. Biol. 24, 268-294.

Nagylaki, T. 1984. Some mathematical problems in population genetics. In Proc. Symp. Appl. Math. Vol. 30: Population Biology (S. A. Levin, ed.). Pp. 19-36. American Mathematical Society, Providence, R. I.

Nei, M. 1975. Molecular Population Genetics and Evolution. North-Holland, Amsterdam.

Protter, M. H., and Weinberger, H. F. 1967. Maximum Principles in Differential Equations. Prentice-Hall, Englewood Cliffs, N. J.

Sawyer, S. 1976. Results for the stepping-stone model for migration in population genetics. Ann. Prob. 4, 699-728.

Sawyer, S. 1977. Asymptotic properties of the equilibrium probability of identity in a geographically structured population. Adv. Appl. Prob. 9, 268-282.

Sawyer, S., and Felsenstein, J. 1981. A continuous migration model with stable demography. J. Math. Biol. 11, 193-205.

Sudbury, A. 1977. Clumping effects in models of isolation by distance. J. Appl. Prob. 14, 391-395.

Weiss, G. H., and Kimura, M. 1965. A mathematical analysis of the stepping-stone model of genetic correlation. J. Appl. Prob. 2, 129-149.

Wright, S. 1931. Evolution in Mendelian populations. Genetics 16, 97-159.

Stochastic Measure Diffusions as Models of Growth and Spread.

Gerd Rosenkranz
Sonderforschungsbereich 123
Universität Heidelberg
Im Neuenheimer Feld 293
6900 Heidelberg

1. Introduction.

The primary purpose of this paper is to introduce measure valued diffusion processes as approximations for certain spacetime population models which can roughly be characterized as follows: The individuals (particles) are distributed spatially at discrete points on the line or at the nodes of some higher dimensional lattice and are subject to birth, death, and migration to neighbouring points. Instead of considering such processes directly we show that a sequence of these processes speeded up and in coarse spatial scale converge weakly to measure valued diffusion processes, i.e. to Markov processes with continuous paths in some measure space.

The systematic application of diffusion processes to population processes dates back to a Berkeley lecture of Feller (1951). He indicated heuristically that a sequence of properly scaled birth and death processes converges to a onedimensional diffusion process. It was hoped that some of the asymptotic properties of the diffusion process (extinction, explosion) correspond to the behaviour of the process originally under consideration. This approximation scheme has been applied to less tractable branching processes for example in Jagers (1971), Keiding (1975, 1976), Kurtz (1979), and Wofsy (1980), among others.

As to spacetime processes, a limiting measure valued diffusion process was obtained for the first time in Dawson (1975). He considered a system of Brownian particles in R^d reproducing independently according to a simple birth and death process. The qualitative behaviour of the limiting diffusion (called the branching measure diffusion) was studied in Dawson (1977). The main result was that a critical branching measure diffusion has an invariant distribution only if $d \geq 3$. It has been shown in Dawson and Hochberg (1979) that under the same condition the support of the process has Lebesgue measure zero. In a subsequent paper (Dawson, 1980), the qualitative behaviour of a spacetime analogue of the logistic model has been investigated.

Diffusion processes also play an important role in population genetics theory. (See for example Crow and Kimura (1970), Ethier (1976), Karlin and McGregor (1964), Sato (1976) and the references therein.) Fleming and Viot (1979) considered an infinite allele model which was defined in terms of a probability measure valued diffusion process. Results on the qualitative behaviour of this model were obtained in Dawson and Hochberg (1982, 1983).

In the present paper we show that the limiting diffusions of some lattice processes with interaction exist. In a forthcoming paper some of the qualitative properties of the resulting diffusion limits will be studied.

The particle processes we have in mind can be described as follows:

(A.1) Spatial structure.

(A.1.1) The particles are located at the nodes (sites) of a finite subset Z of the d-dimensional lattice $\mathbf{Z}^d$. A point $y \in \mathbf{Z}^d$ is called a neighbour of $x \in Z$, if $y \in N_x := \{y: |x - y| = 1\}$, where $|.|$ denotes Euclidian norm. N_x is called the neighbourhood of x.

(A.1.2) There can be more than one particle at each point $x \in Z$.

(A.2) Growth (branching) mechanism.

(A.2.1) Each particle has a random lifespan. At the end of its life it can either split into two new particles or can die without progeny.

(A.2.2) The split and the death rate may depend on the location of a particle as well as on the whole distribution of the other particles.

(A.3) Dispersion mechanism.

(A.3.1) The particles born at x jump to some point y of $\{x\} \cup N_x$.

(A.3.2) The choice of y may depend on both x and y.

(A.3.3) Particles are reflected at the boundary of Z.

We mainly consider two variants of the general model described above. In the first case, called Model I in what follows, we assume

(A.3.1') One of the particles born at x remains at the place of its birth.

(A.3.2') The other particle chooses a position $y \in N_x$ with equal probability.

This model may serve to describe the growth of cells in a tissue. Hereby the stability of the arrangement of such a cell system is mathematically represented by a finite connected subset of $\mathbf{Z}^d$. Spread of the whole tissue is only possible if new cells are born. By (A.2.2) interactions between the particles are possible.

We mention that we do not forbid that one site of the lattice is occupied by more than one particle as it is the case in similar models (Williams and Bjerkness (1972), Schürger and Tautu (1976) or Bramson and Griffeath (1980)). It should be possible perhaps after some further elaboration to study cellular systems producing columns like for example epidermis (see Potten, 1980). Model I is more related to the birth, death, and migration process of Bailey (1968).

With the second model (Model II) we want to describe growth and spread in some abstract type space: each lattice point corresponds to some characteristic of the particle. We assume:

(A.3.1") Both particles born at x choose the same site $y \in \{x\} \cup N_x$.

(A.3.2") The probability of a jump to y depends on both x and y.

The jump of the particles to a neighbouring site is interpreted as a change of type after birth. This type change need not to be restricted to genetic events only. Multistage carcinogenesis for example supposes a linear sequence of reversible cell metamorphoses specifying the stages. In addition, there is some evidence for a so called differentiating network, where the common source for normal and malignant metamorphoses consists of some stem cells (see Tautu, 1980). Such an assumption would require a multidimensional type as in the genetic context.

2. Notation and the main result.

If X is an arbitrary topological space then $\mathbb{C}(X)$ denotes the linear space of continuous functions and $\mathbb{C}_b(X)$ the Banach space of continuous bounded functions on X equipped with the usual sup-norm $\|.\|$. $\mathbb{C}_c(X)$ denotes the subset of $\mathbb{C}_b(X)$ whose elements have compact support. If $X \subset \mathbb{R}^d$ we also define $\mathbb{C}_b^k$, the space of k times continuously differentiable functions with bounded partial derivatives. The nonnegative members of each of these subsets will be denoted by the subscript "+".

Let M(X) denote the space of Borel measures on X. We assume M(X) to be equipped with the vague topology, i.e. the mapping $\xi \to \langle f,\xi\rangle := \int f d\xi$ is continuous for all $f \in \mathbb{C}_c(X)$. If $X \subset \mathbb{R}^d$, M(X) is a Polish space. If X is compact, then M(X) is locally compact and its topology has a countable basis. Let

$$\|\xi\| := \sup_{f\in\mathbb{C}_c(X),\|f\|\le 1} \langle f,\xi\rangle$$

denote the total variation norm of $\xi \in M(X)$. Then the set $K_R := \{\xi: \|\xi\| \le R\}$ is (vaguely) compact. By δ_x, $x \in X$, we denote Dirac measure, i.e. $\langle f,\delta_x\rangle = f(x)$.

Let $S := [a_1, b_1] \times \ldots\ldots [a_d, b_d]$, $a_j, b_j \in \mathbf{Z}$, $a_j < b_j$ for every $j = 1,\ldots,n$, and $\Omega := \mathbf{D}_{M(S)}[0,\infty)$ the space of cadlag functions on M(S) equipped with the Skorohod topology (see Kurtz, 1975). A Borel measurable mapping $\Xi : [0,\infty)\times\Omega \to M(S)$ defined by $\Xi(t,\omega) := \omega(t)$, $\omega \in \Omega$, is called the canonical measure valued process on Ω. Its distribution is determined by a probability measure P on F, the Borelalgebra on Ω. The space of all probability measures on F is assumed to be equipped with the weak topology (see Billingsley, 1968).

In what follows we consider sequences of processes Ξ_N, $N \in \mathbb{N}$, with state space $M(S_N)$ such that $Z = S_1$ and Ξ_1 is the process corresponding to Model I or Model II. S_N is defined in the following way. Let

$$k_{N,j} := \sup\left\{n \in \mathbb{N}:\ a_j + nN^{-1/2} \le b_j\right\}$$

$$\nu_{N,j} := (b_j - a_j)/k_{N,j}$$

and

$$S_{N,j} := \left\{a_j + n\nu_{N,j}:\ n = 0,\dots,k_{N,j}\right\}.$$

Then $S_N := S_{N,1} \times \dots\dots \times S_{N,d}$. Note that

$$N^{-1/2} \le \nu_{N,j} \le N^{-1/2} + o(N^{-1/2}).$$

For further reference we define vectors $\eta_{N,j} \in \mathbb{R}^d$ by $\eta_{N,j,l} = \delta_{jl}\nu_{N,j}$. ($\delta_{jl}$ denotes Kronecker delta.) Since $M(S_N)$ is a subspace of $M(S)$, all these processes can be defined on the same probability space Ω. We assume that the particle mass of the N-th process is N^{-1}; hence we regard those measures $\xi_N \in M(S_N)$ which have the representation $\xi_N = \frac{1}{N}\sum m_{x_N}\delta_{x_N}$ with $m_{x_N} = \xi_N(\{x_N\})$. To get a nondegenerate limit if $N \to \infty$, we have to speed up time by a factor N^2.

Let λ_N, μ_N : $S_N \times M(S_N) \to \mathbb{R}_+$ be the birth and death rates of the N-th process and σ_N : $S_N \times S_N \to \mathbb{R}_+$ the probability of a location or type change. According to (A.3.1) $\sigma_N(x_N,y_N) = 0$ if $|x_N - y_N| > N^{-1/2}$. Let $\xi_N := \frac{1}{N}\sum m_{x_N}\delta_{x_N}$. In both models a jump from ξ_N to $\xi_N - \frac{1}{N}\delta_{x_N}$ (a death) occurs with intensity $m_{x_N}\mu_N(x_N,\xi_N)$. Now define

$$l_N(x_N,x_N+\eta_{N,j}) := (2d)^{-1}\delta_{x_N+\eta_{N,j}}(S_N)\left[2 - \delta_{x_N-\eta_{N,j}}(S_N)\right]$$

and

$$l_N(x_N,x_N-\eta_{N,j}) := (2d)^{-1}\delta_{x_N-\eta_{N,j}}(S_N)\left[2 - \delta_{x_N+\eta_{N,j}}(S_N)\right]$$

Under the assumptions of Model I, the process Ξ_N jumps from ξ_N to $\xi_N + \frac{1}{N}\delta_{x_N+\eta_{N,j}}$ with intensity $l_N(x_N,x_N+\eta_{N,j})m_{x_N}\lambda_N(x_N,\xi_N)$ and to $\xi_N + \frac{1}{N}\delta_{x_N-\eta_{N,j}}$ with intensity $l_N(x_N,x_N-\eta_{N,j})m_{x_N}\lambda_N(x_N,\xi_N)$ (compare (A.3.1'), (A.3.2') and (A.3.3)). Therefore the infinitesimal operator of the Markov process corresponding to Model I is given by

$$A_N f(\xi_N) = \overline{\sum_{x_N \in S_N}} m_{x_N} \Big\{ \lambda_N(x_n, \xi_N) \times$$

$$\times \Big[\overline{\sum_{j=1}^{N}} I_N(x_N, x_N + \eta_{N,j}) \, (f(\xi_N + \tfrac{1}{N} \delta_{x_N + \eta_{N,j}}) - f(\xi_N))$$

$$+ \overline{\sum_{j=1}^{N}} I_N(x_N, x_N - \eta_{N,j}) \, (f(\xi_N + \tfrac{1}{N} \delta_{x_N - \eta_{N,j}}) - f(\xi_N)) \Big]$$

$$+ \mu_N(x_N, \xi_N)(f(\xi_N - \tfrac{1}{N} \delta_{x_N}) - f(\xi_N)) \Big\}$$

with $D(A_N) = \mathbb{C}_b(M(S))$.

In Model II, a change from ξ_N to $\xi_N - \frac{1}{N} \delta_{x_N} + \frac{2}{N} \delta_{y_N}$ occurs with intensity $J_N(x_N, y_N) m_{x_N} \lambda_N(x_N, \xi_N)$ where

$$J_N(x_N, y_N) := \delta_{y_N}(S_N) \; \sigma_N(x_N, y_N),$$

if $x_N \neq y_N$ and

$$J_N(x_N, x_N) := 1 - \overline{\sum_{y_N \neq x_N}} J_N(x_N, y_N)$$

(compare (A.3.1"), (A.3.2"), (A.3.3)). Hence the dynamics of the model is described by

$$B_N f(\xi_N) = \overline{\sum_{x_N \in S_N}} m_{x_N} \Big\{ \lambda_N(x_N, \xi_N) \times$$

$$\times \overline{\sum_{y_N \in S_N}} J_N(x_N, y_N)(f(\xi_N - \tfrac{1}{N} \delta_{x_N} + \tfrac{2}{N} \delta_{y_N}) - f(\xi_N))$$

$$+ \mu_N(x_N, \xi_N)(f(\xi_N - \tfrac{1}{N} \delta_{x_N}) - f(\xi_N)) \Big\}$$

with $D(B_N) = \mathbb{C}_b(M(S))$. The main result can now be stated as follows:

__Theorem 1__.

Let Ξ_N be a $M(S_N)$-valued Markov process with infinitesimal operator A_N. Assume

(B.1) There exist functions $\gamma_1 \in \mathbb{C}(S)_+$, $\gamma_2 \in \mathbb{C}(M(S))_+$, $\alpha \in \mathbb{C}(S \times M(S))$, such that with $\gamma(x, \xi) := \gamma_1(x) \gamma_2(\xi)$

$$\lambda_N(x_N, \xi_N) = N\gamma(x_N, \xi_N) + \tfrac{1}{2} \alpha(x_N, \xi_N).$$

$$\mu_N(x_N,\xi_N) = N\gamma(x_N,\xi_N) - \frac{1}{2}\,\alpha(x_N,\xi_N).$$

(B.2) There exists a constant $C > 0$ (permanent notation) such that

$$\langle\gamma(.,\xi),\xi\rangle \le C(1 + \|\xi\|^2),$$
$$\langle\alpha(.,\xi),\xi\rangle \le C(1 + \|\xi\|).$$

(B.3) There exists a constant $\gamma_0 > 0$ (permanent notation) such that $\gamma \ge \gamma_0$.

(B.4) Let A be a linear operator defined by

$$D(A) = \{f \in \mathbb{C}_b(M(S)):\ f(\xi) = g(\langle\varphi_1,\xi\rangle,\dots,\langle\varphi_n,\xi\rangle),$$
$$g \in \mathbb{C}_b^3(\mathbb{R}^n),\ \varphi_i \in D(\Delta),\ i=1,\dots,n,\ n \in \mathbb{N}\},$$

where

$$D(\Delta) = \{\varphi\in\mathbb{C}^2(S): \frac{\partial\varphi}{\partial x_j}(x) = 0 \text{ if } x \in \partial S\},$$

$$\Delta\varphi(x) = \sum_{j=1}^{d} \frac{\partial^2\varphi}{\partial x_j^2}(x)$$

and

$$Af(\xi) = \sum_{k=1}^{n} \frac{\partial g}{\partial x_k}(\langle\varphi_1,\xi\rangle,\dots,\langle\varphi_n,\xi\rangle)\ \langle\alpha\varphi_k + \tfrac{1}{d}\gamma\Delta\varphi_k,\xi\rangle$$
$$+ \sum_{k,l}^{n} \frac{\partial^2 g}{\partial x_k\partial x_l}(\langle\varphi_1,\xi\rangle,\dots,\langle\varphi_n,\xi\rangle)\ \langle\gamma\varphi_k\varphi_l,\xi\rangle.$$

Let P_{ξ_N} be the distribution of Ξ_N such that $P_{\xi_N}[\Xi_N(0) = \xi_N] = 1$. If $\xi_N \to \xi$, then $P_{\xi_N} \to P_\xi$ weakly. Furthermore there exists a homogeneous Feller process Ξ on Ω with distribution P_ξ such that $P_\xi[\Xi(0) = \xi] = 1$. The generator of Ξ is just the closure of A.

Theorem 2.

Let Ξ_N be a $M(S_N)$-valued Markov process with infinitesimal operator B_N. Assume (B.1)-(B.3) and

(B.5) For $j=1,\dots,d$ there exist continuous functions $0 \le \beta_j \le 1$, $\|\vartheta_j\| \le 1$ such that $\beta_j \ge \beta_0$ for some constant $\beta_0 > 0$, and

$$\sigma_N(x_N,x_N + \eta_{N,j}) = \frac{1}{2d}\,\beta_j(x_{N,j}) + \frac{N^{-1/2}}{4d}\,\vartheta_j(x_{N,j});$$

$$\sigma_N(x_N,x_N - \eta_{N,j}) = \frac{1}{2d}\,\beta_j(x_{N,j}) - \frac{N^{-1/2}}{4d}\,\vartheta_j(x_{N,j}).$$

(B.6) Let L be a linear operator defined by

$$D(L) = \{\varphi\in \mathbb{C}^2(S);\ \frac{\partial\varphi}{\partial x_j}(x) = 0,\ \text{if } x\in\partial S\}$$

$$L\varphi(x) = \frac{1}{d}\sum_{j=1}^{d}\left\{\vartheta_j(x_j)\frac{\partial\varphi}{\partial x_j}(x) + \beta_j(x_j)\frac{\partial^2\varphi}{\partial x_j^2}(x)\right\}.$$

(B.9) Let B be a linear operator defined by

$$D(B) = \{f\in\mathbb{C}_b(M(S));\ f(\xi) = g(\langle\varphi_1,\xi\rangle,\dots,\langle\varphi_n,\xi\rangle),$$

$$g\in\mathbb{C}_b^3(\mathbb{R}^n),\ \varphi_i\in D(L),\ i=1,\dots,n,\ n\in\mathbb{N}\};$$

$$Bf(\xi) = \sum_{k=1}^{n}\frac{\partial g}{\partial x_k}(\langle\varphi_1,\xi\rangle,\dots,\langle\varphi_n,\xi\rangle)\ \langle\alpha\varphi_k + \gamma L\varphi_k,\xi\rangle$$

$$+ \sum_{k,l}^{n}\frac{\partial^2 g}{\partial x_k\partial x_l}(\langle\varphi_1,\xi\rangle,\dots,\langle\varphi_n,\xi\rangle)\ \langle\gamma\varphi_k\varphi_l,\xi\rangle.$$

Let P_{ξ_N} be the distribution of Ξ_N such that $P_{\xi_N}[\Xi_N(0) = \xi_N] = 1$. If $\xi_N \to \xi$, then $P_{\xi_N} \to P_\xi$ weakly. Furthermore there exists a homogeneous Feller process Ξ on Ω with distribution P_ξ such that $P_\xi[\Xi(0) = \xi] = 1$. The generator of Ξ is just the closure of B.

3. Sketch of the proof.

The proof of the main theorems is rather long, therefore we only give an outline of the main ideas behind it. For details a preprint (Rosenkranz, 1985) which appeared in the SFB123 preprint series may be consulted. Since the proof depends essentially on martingale arguments and especially on the martingale problem formulation of Markov processes due to Stroock and Varadhan (1979), we shortly review the main definitions and results of that theory applied to measure diffusions.

Let G be some linear operator with domain $D(G) \subset \mathbb{C}_b(M(S))$, and let Ξ be the canonical measure-valued process on Ω. We say that a probability measure P_ξ on Ω is a solution of the martingale problem associated with $(G,D(G))$, if $P_\xi[\Xi(0) = \xi] = 1$ and $\left\{f(\Xi(t)) - \int_0^t Gf(\Xi(s))ds,\ F_t,\ P_\xi\right\}$ is a martingale where $F_t := F(\Xi(s),\ 0 \le s \le t)$ is the usual filtration. The key observation is the following

Theorem (Dawson, 1978).

Assume that the martingale problem associated with (G,D(G)) has a unique solution P_ξ for every $\xi \in M(S)$ and that $\{P_\xi;\ \xi \in K\}$ is tight for every compact $K \subset M(S)$. Then $\{P_\xi\}$ represents a homogeneous Feller process on Ω and the closure of G is the infinitesimal operator of the process.

With this theorem in mind we can prove the main result in several steps:

1.) Let P_{ξ_N} be the distribution of the Markov process with infintesimal operator B_N. If $\xi_N \to \xi$, $\xi_N \in M(S_N)$, then $\left\{P_{\xi_N}\right\}$ is tight. Hence there exists a weakly convergent subsequence of $\left\{P_{\xi_N}\right\}$ with limit P^*_ξ (say).

   Tightness will be proved with the aid of a theorem of Kurtz (1975). Thereby we do not demonstrate tightness of the sequence $\left\{\Xi_N\right\}$ directly but we show that the sequences $\left\{f(\Xi_N(.))\right\}$ are tight for a sufficiently large class of testfunctions f.

2.) P^*_ξ is a solution of the martingale problem associated with (B,D(B)). Since we can show that for every $\xi \in M(S)$ there exists a sequence $\{\xi_N\}$ with $\xi_N \in M(S_N)$ and $\xi_N \to \xi$, a solution P^*_ξ exists for every $\xi \in M(S)$.

3.) The martingale problem has a unique solution P_ξ (say) and $\{P_\xi;\ \xi \in K\}$ is tight for every compact $K \subset M(S)$.

   The uniqueness proof itself is done in several steps. First we show uniqueness in the case $\alpha = 0$, $\gamma_2 = 1$, extending a technique applied in Fleming and Viot (1979) to prove uniqueness of a probability measure valued martingale problem. The proof for general γ_2 is achieved by a random time change argument. An application of a theorem of Cameron-Martin-Girsanov type of Dawson (1978) completes the proof.

References.

Bailey, N.T.J. (1968). Stochastic birth, death, and migration processes for spatially distributed populations. Biometrika 50, 189-198.

Billingsley, P. (1968). Convergence of Probability Measures. New York: John Wiley.

Bramson, M. and Griffeath, D. (1980). The asymptotic behaviour of a probabilistic model of tumor growth. In: Biological Growth and Spread. Mathematical Theories and Applications. W. Jäger, H. Rost, P. Tautu (eds.). Berlin: Springer. 165-172.

Crow, J.F. and Kimura, M. (1970). An Introduction to Population Genetics Theory. New York: Harper and Row.

Dawson, D.A. (1975). Stochastic evolution equations and related measure processes. J. Multivariate Anal. 5, 1-52.

Dawson, D.A. (1977). The critical measure diffusion process. Z. Wahrscheinlichkeitstheorie Verw. Gebiete 40, 125-145.

Dawson, D.A. (1978). Geostochastic calculus. Canad. J. Statist. 6, 143-168.

Dawson, D.A. (1980) Qualitative behavior of geostochastic systems. Stoch. Proc. Appl. 10, 1-31.

Dawson, D.A. and Hochberg, K. (1979). The carrying dimension of a stochastic measure diffusion. Ann. Prob. 7, 693-703.

Dawson, D.A. and Hochberg, K. (1982). Wandering random measures in the Fleming-Viot model. Ann. Prob. 10, 554-580.

Dawson, D.A. and Hochberg, K. (1983). Qualitative behavior of a selectively neutral allelic model. Theor. Popn. Biol. 23, 1-18.

Ethier, S.N. (1976). A class of degenerate diffusion processes occuring in population genetics theory. Comm. Pure Appl. Math., 29, 483-493.

Feller, W. (1951). Diffusion processes in genetics. Proc. 2nd Berkeley Symp. Math. Statist. Prob., 227-246.

Fleming, W.H. and Viot, M. (1979). Some measure valued Markov processes in population genetics theory. Indiana Univ. Math. J. 28, 817-843.

Jagers, P. (1971). Diffusion approximations of branching processes. Ann. Math. Statist. 42, 2074-2078.

Karlin, S. and McGregor, J. (1964). On some stochastic models in genetics. In: Stochastic Models in Medicine and Biology. J. Gurland (ed.). Madison: University of Wisconsin Press, 249-279.

Keiding, N. (1975). Extinction and exponential growth in random environments. Theor. Popn. Biol. 8, 49-63.

Keiding, N. (1976). Population growth and branching processes in random environments. Proc. 9th Int. Biometric Conf. Boston, 149-165.

Kurtz, T.G. (1975). Semigroups of conditioned shifts and approximations of Markov processes. Ann. Prob. 3, 618-648.

Kurtz, T.G. (1979). Diffusion approximations of branching processes. In: Branching Processes, Adv. Prob. 5, A. Joffe and P. Ney (eds.). New York: Marcel Dekker, 269-292.

Potten, C.S. (1980). Proliferative cell populations in surface epithelia: Biological models for cell replacement. In: Biological Growth and Spread. Mathematical Theories and Applications. W. Jäger, H. Rost, P. Tautu (eds.). Berlin: Springer, 23-35.

Rosenkranz, G. (1985). Stochastic measure diffusions as models of growth and spread. Preprint. Sonderforschungsbereich 123, Universität Heidelberg.

Sato, K.J. (1976). Diffusion processes and a class of Markov chains related to population genetics. Osaka J. Math. 13, 631-659.

Schürger, K. and Tautu, P. (1976). A Markovian configuration model for carcinogenesis. In: Mathematical Models in Medicine. J. Berger, W. Bühler, R. Repges, P. Tautu (eds.). Berlin: Springer, 92-108.

Stroock, D.W. and Varadhan, S.R.S. (1979). Multidimensional Diffusion Processes. Berlin: Springer.

Tautu, P. (1980). Biological interpretation of a random configuration model for carcinogenesis. In: Biological Growth and Spread. Mathematical Theories and Applications. W. Jäger, H. Rost, P. Tautu (eds.). Berlin: Springer, 196-220.

Williams, T. and Bjerkness, R. (1972). Stochastic model for abnormal clone spread through epithelial basal layer. Nature 236, 19-21.

Wofsy, C. (1980). Behavior of limiting diffusions for density dependent branching processes. In: Biological Growth and Spread. Mathematical Theories and Applications. W. Jäger, H. Rost, P. Tautu (eds.). Berlin: Springer, 130-137.

Acknowledgement.

The author wants to thank Werner Rittgen (German Cancer Research Centre, Heidelberg) for his helpful comments and many stimulating discussions.

L^2 Convergence of Certain Random Walks on Z^d and related Diffusions

Wayne G. Sullivan
Department of Mathematics
University College
Dublin, Ireland
and
School of Theoretical Physics
Dublin Institute for Advanced Studies

One technique for studying the approach to equilibrium of a continuous time Markov process is to consider the restriction to the L^2 space of an invariant distribution. When the process is reversible with respect to this distribution, the generator is a selfadjoint operator. We study the L^2 spectrum of the generator for certain random walks on Z^d, where the reversible invariant distribution is concentrated near the origin and decays rapidly with distance to the origin. For the related diffusions on R^d we find that the generators are unitarily equivalent to Schrödinger operators.

§1 Introduction. One standard approach to continuous time Markov processes is to start with a generator D, construct a transition function $P_t(x, dy)$ which in an appropriate sense satisfies

$$P_t = \exp(tD) \tag{1}$$

and then study the properties of P_t, *e.g.* invariant measures. For certain physical systems it is more natural to start from a probability measure μ and develop processes which leave μ invariant. Of particular interest are generators which satisfy the *detailed balance condition* with respect to μ. In terms of μ and P_t this yields the relation

$$\mu(dx)P_t(x, dy) = \mu(dy)P_t(y, dx) \tag{2}$$

which is the condition of time reversibility for the Markov process with initial distribution μ and transition function P_t.

The assumption of time reversibility introduces additional techniques for the study of the process. The aspect we consider here is convergence in the L^2 sense. When μ and P_t satisfy (2), P_t gives rise to a selfadjoint contraction semigroup in $L^2(\mu)$. The generator D acts as a negative semi-definite selfadjoint operator in $L^2(\mu)$ with eigenvalue zero corresponding to the invariant distribution. If the spectrum of D is otherwise bounded away from zero, then convergence to equilibrium is exponentially fast in $L^2(\mu)$. Such convergence has been shown for a special dynamic spin system having distinct phases and is conjectured for dynamic two-dimensional Ising models in the two phase region (see [4], [5]).

Here we use the spectral approach to study the convergence of some simple random walk models. The original motivation is in the work of Buffet-Pulé-de Smedt on a boson system coupled to a heat bath [1]. Arising in their studies was a Markov process on the positive reals for which an estimate of the rate of convergence was desired. In [6] we gave estimates of convergence for processes of this type and related Markov chains on the positive integers. In the present work we extend the analysis to certain random walks on Z^d and look briefly at the diffusion limit which yields Schrödinger operators.

§2 Formulation. The L^2 approach is applicable to rather general continuous time Markov processes. We shall formulate here in terms of a countable discrete state space Ω and later specialize to Z^d. To each $x \in \Omega$ we associate a *finite* subset of Ω denoted ∂x called the neighbours of x. We require that

$$y \in \partial x \iff x \in \partial y.$$

For Z^d we define

$$\partial x = \{x \pm \mathbf{e}_i : i = 1 \cdots d\} \tag{3}$$

where $\{\mathbf{e}_i\}$ are the standard unit vectors.

We assume that we are given a strictly positive probability distribution π on Ω. We shall construct the generator D from a matrix Q which satisfies the reversibility condition

$$\pi(x)Q(x,y) = \pi(y)Q(y,x)$$

whenever $y \in \partial x$. We define

$$q(x) = \sum_{y \in \partial x} Q(x,y) \tag{4}$$

and for the real valued function f *equal to zero except on a finite subset of* Ω

$$(Qf)(x) = \sum_{y \in \partial x} Q(x,y) \tag{5}$$

$$(Df)(x) = (Qf)(x) - q(x)f(x). \tag{6}$$

D is densely defined and dissipative

$$(Df, f)_\pi \leq (f, f)_\pi$$

in $\ell^2(\pi)$ where

$$(f,g)_\pi = \sum f(x)g(x)\pi(x).$$

We now introduce some particular generators in terms of $Q_i, i = 0,1,2,3$. We need only define $Q_i(x,y)$ for $y \in \partial x$ and take q_i, Q_i and D_i as above:

$$\begin{aligned}
Q_0(x,y) &= \pi(y)/[\pi(x) + \pi(y)] \\
Q_1(x,y) &= \min\{1, \pi(y)/\pi(x)\} \\
Q_2(x,y) &= \sqrt{\pi(y)/\pi(x)} \\
Q_3(x,y) &= \max\{1, \pi(y)/\pi(x)\}
\end{aligned}$$

All of these are reversible for π and we have

$$(D_{i+1}f, f)_\pi \leq (D_i f, f)_\pi \tag{7}$$

for $i = 0,1,2$.

Proposition 1. *The closure of D_2 is the generator of a positive contraction semigroup in $\ell^1(\pi)$ provided that the cardinality of ∂x is uniformly bounded for $x \in \Omega$.*

Conclusions about the spectral properties of D presume that the closure of D is the generator of a positive contraction semigroup on $\ell^1(\pi)$. When ∂x is uniformly bounded, this obtains for D_0 and D_1 as bounded operators and for D_2 by Proposition 1. Related methods can prove that this property holds for D_3 in the case of π of the form in Theorem 2.

When the closure of D is the generator of a positive contraction semigroup on $\ell^1(\pi)$, the associated P_t is a Feller semigroup which extends to ℓ^p, $1 \leq p \leq \infty$. The action of the generator in $\ell^2(\pi)$, which we also denote by D, is a negative semidefinite selfadjoint operator. We use the following notation:

$$\lambda_1(D) = \inf_{\lambda>0} \{\lambda \in \ell^2(\pi) \text{ spectrum of } -D\}$$

$$\lambda_\infty(D) = \inf_{\lambda>0} \{\text{the } [0,\lambda] \text{ spectral projection of } -D \text{ is infinite dimensional}\}$$

The value of λ_1 is significant because it determines the rate of exponential convergence to equilibrium in $\ell^2(\pi)$. In [6] we give direct estimates for λ_1, but for $Z^d, d > 1$, the methods seem more suited for estimating λ_∞. We note that if $\lambda_\infty > 0$, then $\lambda_1 > 0$. From (7) we hav

$$\lambda_j(D_i) \leq \lambda_j(D_{i+1}) \tag{8}$$

for $j = 1, \infty$, $i = 0,1,2$.

For Z^1 we have the following

Theorem 1. *Let π be a strictly positive probability distribution on Z. If*

$$\limsup_{n\to\infty} \sum_{n}^{\infty} \pi(x)/\pi(n) < \infty$$

$$\limsup_{n\to\infty} \sum_{n}^{\infty} \pi(-x)/\pi(-n) < \infty$$

then we have $\lambda_\infty(D_i) > 0$, $i = 0, 1, 2, 3$. Conversely, if either limit superior is infinite, then $\lambda_\infty(D_i) = 0$, $i = 0, 1, 2$.

In general the condition is not sufficient for $\lambda_\infty(D_3) = 0$, but this is the case for π of the type in Theorem 2.

For $D > 1$, the results are for a specific family of examples. We consider probability distributions on Z^d of the form

$$\pi(x) = K \ \exp -c|x|^\alpha$$

$$c, \alpha > 0 \quad |x| = \sqrt{x_1^2 + \cdots + x_d^2}$$

with K the normalizing constant.

Theorem 2. *For π as given above we have*

$$0 < \alpha < 1 \Longrightarrow \lambda_\infty(D_i) = 0, \ i = 0, 1, 2, 3 \tag{9}$$

$$\alpha = 1 \Longrightarrow 0 < \lambda_\infty(D_i) < \infty, \ i = 0, 1, 2, 3 \tag{10}$$

$$\alpha > 1 \Longrightarrow 0 < \lambda_\infty(D_i) < \infty, \ i = 0, 1 \tag{11}$$

$$\alpha > 1 \Longrightarrow \lambda_\infty(D_i) = \infty, \ i = 2, 3 \tag{12}$$

§3 The Diffusion Limit. Before proving the above results we shall take an elementary look at the diffusion limit of operators like those above. The limiting diffusion operators on $L^2(R^d, \pi dx)$ when symmetrized to $L^2(R^d)$ take the form

$$\Delta - V$$

with Δ denoting the Laplace operator and V a multiplicative operator. Thus, apart from irrelevant constants, the limiting operators are Schrödinger operators (see [3]).

We shall consider limits of operators based on D_2. The reader may verify that D_1 and D_3 lead to these same limiting operators, while D_0 yields half these limits. Let $\pi : R^d \longrightarrow R$ be a strictly positive twice continuously differentiable function such that

$$\sum \pi(x/n) < \infty$$

for each positive integer n, with the sum over $x \in Z^d$. Define

$$e(x) \equiv -\frac{1}{2}\log \pi(x)$$

The generator ${}_nD$ is defined by the matrix ${}_nQ$. where

$$\partial x = \{x \pm e_j/n : j = 1 \cdots d\}$$

$${}_nQ(x,y) = n^2 \sqrt{\pi(y)/\pi(x)}$$

Proposition 2. *Let $f : R^d \longrightarrow R$ be twice continuously differentiable. Then*

$$\lim_{n \to \infty} {}_nD\, f = \Delta f - 2\nabla e \cdot \nabla f \tag{13}$$

pointwise. This operator acting on $f \in L^2(R^d, \pi dx)$ is formally unitarily equivalent to

$$\Delta g - (|\nabla e|^2 - \Delta e)\, g \tag{14}$$

acting on $g \in L^2(R^d)$.

Proof. To get (13), expand f to second order in a Taylor series and take limits. The mapping taking $f \in L^2(R^d, \pi dx)$ to $g \in L^2(R^d)$ is given by

$$g(x) = f(x)\sqrt{\pi(x)}. \tag{15}$$

Expression (14) follows from a straightforward calculation.

Thus we obtain Schrödinger operators which can be expressed with V of the form

$$V = |\nabla e|^2 - \Delta e.$$

For example, $e(x) = |x|^2$ corresponds to the harmonic oscillator potential. The following shows that quite a broad class of Schrödinger operators can be expressed in this way.

Proposition 3. *Let the Schrödinger operator $-\Delta + V$ possess a strictly positive twice continuously differentiable eigenfunction ψ:*

$$-\Delta\psi + V\psi = \lambda\psi$$

for the real constant λ. Then V can be expressed in the form

$$V = \lambda + |\nabla e|^2 - \Delta e$$

with $e = -\log\psi$.

Proof. Calculation

§4 Remaining proofs. For Proposition 1: If $\sharp\partial x$ is uniformly bounded in $x \in \Omega$, then Q_2 acts as a bounded linear operator in $\ell^2(\pi)$. Multiplication by $-q$ is the generator of a contraction semigroup in $\ell^2(\pi)$. By Theorem 13.2.1 of Hille-Phillips [2], the closure of D_2 generates a contraction semigroup in $\ell^2(\pi)$. Then for $\lambda > 0$ and any $f \in \ell^2(\pi)$ there exists $g \in \ell^2(\pi)$ with

$$(\lambda - D_2)g = f$$

Now $\ell^2(\pi) \subset \ell^1(\pi)$ and is dense. We verify that the above holds in $\ell^1(\pi)$ and the result follows from the Hille-Yosida Theorem.

Proof of Theorem 1. If we have for all $n \geq 0$ and some $\alpha > 0$

$$\sum_n^\infty \frac{\pi(x)}{\pi(n)} \leq \frac{1}{\alpha}, \; \sum_n^\infty \frac{\pi(-x)}{\pi(-n)} \leq \frac{1}{\alpha},$$

an adaptation of the proof of Theorem 1 of [6] shows that

$$\begin{aligned} \lambda_1(D_3) &\geq \alpha^2/4 \\ \lambda_1(D_2) &\geq \sqrt{\alpha/(1-\alpha)}\alpha^2/4 \\ \lambda_1(D_1) &\geq [\alpha/(1-\alpha)]\alpha^2/4 \\ \lambda_1(D_0) &\geq \alpha \cdot \alpha^2/4. \end{aligned}$$

As the null space of D is one dimensional

$$\lambda_\infty > \lambda_1 > 0$$

in each case

For the case where one of the inferior limits is infinite— for definiteness we assume

$$\liminf_{n\to\infty} \pi(n) / \sum_n^\infty \pi(x) = 0, \tag{16}$$

consider the functions

$$f_n(x) = 0 \text{ for } x_1 < n, \quad = 1 \text{ for } x_1 \geq n \tag{17}$$

As $d = 1$, $x_1 = x$, but the above definition will be used again below. We have

$$(-D_2 f_n, f_n)_\pi / (f_n, f_n)_\pi = \sqrt{\pi(n-1)\pi(n)} / \sum_n^\infty \pi(x)$$

It is not difficult to deduce from (16) and the above that

$$\liminf_{n\to\infty} (-D_2 f_n, f_n)_\pi / (f_n, f_n)_\pi = 0,$$

so $\lambda_\infty(D_2) = 0$ by Lemma 1 below. The result for D_0 and D_1 follows from (8).

Lemma 1. *If there exists a constant a and a sequence of functions $\{f_n\} \in \ell^2(\pi)$ such that*

$$f_n(x) = 0 \text{ for } |x| < n$$

$$\liminf_{n\to\infty}(-Df_n, f_n)_\pi/(f_n, f_n)_\pi \leq a \tag{18}$$

then $\lambda_\infty \leq a$.

Proof. If $b < \lambda_\infty$, then the $[0, b]$ spectral projection of $-D$ is finite dimensional, so the contribution of this part of the spectrum to (18) goes to zero as $n \to \infty$, hence $b \leq a$.

Lemma 2. *If there exists a constant a and a positive function ρ such that with*

$$\mathbf{Q}(x, y) = Q(x, y)\sqrt{\pi(x)\rho(y)/(\rho(x)\pi(y))}$$

$$s(x) = \sum_{y\in\partial x} \mathbf{Q}(x, y)$$

we have that

$$\liminf_{|x|\to\infty} q(x) - s(x) \geq a,$$

then $\lambda_\infty \geq a$.

Proof. Assume that for all $|x| \geq n$

$$q(x) - s(x) \geq b.$$

For f such that $f(x) = 0$ for $|v| < n$ and $\{x : f(x) \neq 0\}$ is finite, with $g(x) = \sqrt{\pi(x)/\rho(x)}f(x)$ we have

$$(f, f)_\pi = (g, g)_\rho$$

$$(-Df, f)_\pi = ([q - s]f, f)_\pi + (sg, g)_\rho + (-\mathbf{Q}g, g)_\rho.$$

By an adaptation of Lemma 1 of [6], the last two terms together are nonnegative so

$$(-Df, f)_\pi \geq b(f, f)_\pi.$$

This inequality extends to all f in the $\ell^2(\pi)$ domain of D which satisfy $f(x) = 0$ for $|x| < n$, which implies $\lambda_\infty \geq b$.

Proof of Theorem 2. For $0 < \alpha < 1$, if we apply D_3 to f_n given by (17) for $n > 0$

$$(-D_3 f_n, f_n)_\pi/(f_n, f_n)_\pi \leq \pi\{x_1 = n\}/\pi\{x_1 \geq n\}$$

For fixed integer k we hav

$$\lim_{n\to\infty} \pi\{x_1 = n\}/\pi\{x_1 = n + k\} = 1.$$

Then

$$\lim_{n\to\infty}(-D_3 f_n, f_n)_\pi/(f_n, f_n)_\pi = 0$$

and Lemma 1 together with (8) gives conclusion (9).

For D_2 and $\alpha \geq 1$ we employ Lemma 2 with $\rho(x) \equiv 1$. for this case we have

$$-s(x) + q(x) = -2d + \sum_{i=1}^{d}[\pi(x+e_i)^{\frac{1}{2}} + \pi(x-e_i)^{\frac{1}{2}}]/\pi(x)^{\frac{1}{2}} \tag{19}$$

For $\alpha > 1$ the right hand side of (19) approaches $+\infty$ as $|x| \to \infty$. Hence we have (12). For $\alpha = 1$ the right hand side of (19) can for large $|x|$ be approximated arbitrarily closely by a sum of d terms of the form $r + 1/r - 2$ with at least one value of r satisfying $r \leq \exp[-c/(2d^{\frac{1}{2}})]$. Also for $\alpha = 1$,

$$Q_2 \;\leq\; \cosh(c/2)\, Q_0 \;,$$

so we have conclusion (10).

Now we consider D_1 for $\alpha > 1$ and take

$$\rho(x) = \exp[-c|x|^\alpha + 2\beta|x|$$

with $\beta > 0$ constant. Then $q(x) - s(x)$ can be represented as the sum of $2d$ terms of the following two types. With $y = x \pm e_i$ for $|y| < |x|$ the term is of the form $1 - \exp\beta[|y| - |x|]$, while for $|y| > |x|$ the term is of the form

$$(\exp c[|x|^\alpha - |y|^\alpha])(1 - \exp\beta[|x| - |y|]) \tag{20}$$

It is not difficult to deduce that terms of the form (20) go to zero as $|x| \to \infty$ so that

$$\liminf_{|x|\to\infty} q(x) - s(x) \geq 1 - \exp(-\beta/d^{\frac{1}{2}}).$$

Since $\beta > 0$ is arbitrary, $\lambda_\infty(D_1) \geq 1$. Also $2Q_0 \geq Q_1$ and both D_1 and D_0 are bounded so we have (11).

Concluding remarks. The random walks on Z^d we considered showed the following behavior. If the reversible invariant distribution π decays as $|x| \to \infty$ at a rate less than exponential, then 0 is an accumulation point of the spectrum of the generator. If the rate of decay is exponential or greater, then 0 is an isolated point of the spectrum.

We have given a very elementary approach to the diffusion limit. With a bit more effort it is possible to relate the spectral properties of the limit to those of the discrete approximations. For this to be useful we need better estimates on the spectra of the discrete models. Since Schrödinger operators have received such considerable study, it might be more useful to be able to make assertions about the spectra of the discrete models based on the properties of the diffusion limit.

References

1. Buffet,E., Pulé,J., de Smedt,P.: On the dynamics of Bose-Einstein condensation. Ann. Inst. Henri Poincaré Analyse non-linéaire 1 413-451(1984)

2. Hille,E. and Phillips,R.S.: Functional Analysis and Semi-groups, 2nd ed. Providence: AMS (1957)

3. Reed,M. and Simon,B.: Methods of Modern Mathematical Physics, Vol. 4. New York: Academic Press (1978)

4. Sullivan,W.G.: Mean square relaxation times for evolution of random fields. Commun. Math. Phys. 40 249-258 (1975)

5. Sullivan,W.G.: Exponential convergence in dynamic Ising models with distinct phases. Phys. Let. 53A 441-2 (1975)

6. Sullivan,W.G.: The L^2 spectral gap of certain positive recurrent Markov chains and jump processes. Z. Wahrscheinlichkeitstheorie verw. Gebiete (in press)

RANDOM FIELDS : APPLICATIONS IN CELL BIOLOGY

Petre Tautu

German Cancer Research Center
Heidelberg

0.Introduction

0.1.This paper is divided into two parts. In the first part (Sections 1 to 4),the general framework of the theory of random fields (RFs) is drafted,with particular attention to the correspondences with some main concepts in the theory of stochastic processes. This especially touches notions such as conditional independence,regular conditional probability,global Markov property,symmetry,etc. Concepts and methods already introduced in the random fields theory should,on the other hand, suggest correspondences in the theory of stochastic processes,e.g.,local behaviour or the weakening of the Markov property by the Osterwalder-Schrader positivity condition,etc.

There are two approaches to defining RFs : one is basically a probabilistic one,defining an RF as a collection of random variables $\{\xi(x)\}$ with a multidimensional argument,the second is a measure-theoretic approach and ultimately leads to a probabilistic description. Random fields might be viewed as a particular class of doubly indexed stochastic processes $\{\xi(x,t),x\in S,t\in T\}$ where S and T primarily represent the open sets of space and time parameters (discrete or continuous). Such processes may be called "spatial-temporal processes" (e.g.,the spatial birth-and-death process:Preston,1975),so that random fields actually are multidimensional S-(or T-)indexed processes. The difficulties of ordering the (countable) set of spatial parameters make the mathematical approach difficult and different from the usual multivariate processes. Even in the case of ordered or directed sets (e.g.,trees or digraphs), there exist unsolved problems. One of these difficulties concerns the space dimensions : the d-dimensional case ($d \geq 2$) is not a straightforward generalization of the one-dimensional case in as much as "the one--dimensional case is best thought of as a misleading anomaly foisted upon us by an understandable lack of mathematical expertise"(Hammersley, 1972). If we make distinction between local and global behaviour, in a one-dimensional space the local probabilities do,in general,uniquely determine the global probabilities,while in a d-dimensional one they may or may not,corresponding to the absence(presence) of long-range order effects in the system.

If the spatial character of the RF requires emphasis as in the case $\{\xi(x), x \in R^d\}, d \geq 2$,it is called a "random surface" (Wschebor,1985). For instance,the rough surface of a metal can be modeled by a 2-dimensional RF, while a water surface may require three dimensions (the third being the time). Yet,even in metallurgy,the temporal evolution of a microstructural state of a particle system may be represented by a 1-dimensional manifold in a (d+1)-dimensional space:this representation has been defined as the growth path for the particle (DeHoff,1972). As B.C.Goodwin(1971)asserted,the four-dimensional nature[in fact,a (d+1)-dimensional one,d=3] of the developmental process is generally accepted by many embryologists. A "positioning" process establishes basic tissue structure and the geometrical relationships of one tissue to another. Thus,a "new anatomy" in terms of cell behaviour and interactions is suggested(Curtis,1978).

These considerations suggest a comparison between some random fields and set-indexed stochastic processes or processes with multidimensional indices. The horizon can be enlarged by the remark that any RF can be written as a sum of two (or more) processes,one of them being governed by a deterministic law. Moreover,one can generalize the concept of "index set" by considering its elements as generalized random functions, measures (of bounded energy),etc. : one deals with generalized RFs (see, e.g.;Dynkin,1980,Rozanov,1982;Röckner,1985).

The main purpose of the second part (Section 5) is to give an introduction to possible biological applications,especially in cell biology. Examples in the literature (e.g.,Vanmarcke,1983) deal with the areal density of a species,the dispersal of parasites in a plantation, the neuronal field,or the molecular dynamics in cell membranes. All are phenomena which summon the image of a "distributed disordered system", displaying a complex pattern of variation in space and/or in time. Other examples of RFs -e.g.,in oceanology,metallurgy,geology,seismology (Adler, 1981)- support the idea that the theory of random fields is one of the substantial approaches to the mathematical characterization of disorder.

In the present paper, the RF-approach for particular cell systems is suggested, that is, for cell systems whose dynamics are characterized by local interactions, replication and differentiation. This approach would permit us to construct, for instance, morphogenetic as well as carcinogenetic "random fields" in order to obtain a new, coherent explanation and description of some spatial complex processes in cell biology, which are still studied and interpreted disparately. The necessity of such an integrated view clearly appears in the modern biological investigations. Papers published in the last years and dealing with the "neurocrystalline" lattice of the retina (Ready et al., 1976), the 3-dimensional growth of malignant cells embedded in collagen gel (Yang et al., 1979), the "rule" of normal neighbours (Mittenthal, 1981), or the "inside-outside" hypothesis of endodermal differentiation in mouse embryos and in aggregates of embryonal carcinoma cells (Rosenstraus et al., 1983) are convincing examples in this sense.

As far back as 1956, Clifford Grobstein conceived the "inductive tissue interaction" as a process normally taking place between cell systems in "intimate association" such that "the initial inductive effect is at cellular level, involving an alteration in cell properties which leads, in consequence, to such phenomena at the tissue level as folding, contraction thickening, etc." (p.231). For the history of biological ideas it is interesting to notice that C.Grobstein definitely stated some arguments against the hypothesis of chemical diffusion "as a complete mechanism for all inductions". This is in contrast with the dogma accepted as the main hypothesis of the deterministic reaction-diffusion models. Recently G.Odell et al. (1981) argued cogently that the introduction of "poorly understood devices such as morphogens and cellular clocks" in some models for morphogenesis is not indispensable ; in their mechanical model "coordination at the population level arises from the local behavior of each cell automatically".

The quest of modern developmental biology should be the understanding of the creation of diversity and complexity in biological systems. As G. Odell (1984) pointed out, "even very simple interaction rules iterated by many subunits of an organism (for example cells, proteins or organelles), can produce collective results of astonishing complexity". Stochastic models dealing with spatial interactions between a large number of elements as well as with their characteristic changes appear as the adequate investigation implement. Their first theoretical contribution to the understanding of developmental processes is the inclusion of new and precisely definite concepts as interaction, neighbourhood, configuration, environment, etc. For instance, if one assumes that in a large cell system many interdependent subsets of cells are distinguishable, call them "clans" (following Moussouris,1974) and enounce that in a "Markov system"(or a Markov RF) the units of interactions are clans. It is known that the essential feature of a Markov RF is the "contagious" nature of its statistical dependence : in a spatial context,this can be understood with the help of the notion of "environment" (or "boundary"). In a system S the cells which interact with a clan $A \subset S$ compose the environment E of A. If the system is Markovian,once the state of the environment E surrounding A is known,any further information about situatien outside E does not influence our expectations about the clan A. This makes unnecessary long-range interactions.

C.H.Waddington(1973) used the term "pattern formation" to designate the acquisition by cells in a cell system of (i)"behavioural tendencies" causing geometrical transformations,and (ii) determinations (or "commitments"). He suggested that "the units within a pattern are to be considered as having some epigenetic relation with one another,so that,at least at some stage of development,an alteration of the position of one element will be accompanied by some shift in the other elements"(p.501). Actually,the position and the properties (or behaviour) of one element in the system may be influenced or determined by the position and the properties of some other interacting elements. The strength of interactions may be variable,"from very strong to quite weak".

The change in cell properties (e.g.,cell "type") is still an unclear process ; the definitions of "differentiation" and "cell type" are controversial. Biologists agree that differentiation is of unstable nature and that cells which had attained "definitive" phenotypes of particular kind,nevertheless retain "options" for potential but limited conversion ("transdifferentiation") into another cell type (Okada,1980). The process is a random one (Levenson and Housman,1981) and the instabilities can be analyzed by applying the methods used in the theory of interacting particle systems (e.g.,Pilz and Tautu,1984).

Such phenomena emerge not only in large cell systems but also in different cellular substructures as skeleton networks,membrane layers or gene networks. As it is known,cell membranes are conceived as multifunctional lipid-protein assemblies which form coherent,ordered spatial systems by the interaction between common and homologous structures.The instabilities in such molecular systems may have different consequences at cellular level,e.g.,the induction of malignant cells (Chernavskii et al.,1981). Such critical phenomena were anticipated by C.Grobstein(1956) who hypothesized that the induction process might result "in change in properties of the responding cell surface by alteration in the specific character and orientation of its molecular population...The special features of this mechanism...are that the process should be able to proceed in the absence of all transfer of inductively active materials,and that it should never proceed in the absence of fairly large surface contact areas".

0.2. The simplest phenomenological description of a cell system as an RF is as follows. Let us suppose that the object of our investigation is the macroscopic spatial behaviour of a large number of identical cells in a tissue. Firstly,we assume that a regular tissue can be represented by a d-dimensional integer lattice Z^d,$d\geq 1$,whose sites are occupied by cells such that a lattice site $x=(x^1,\ldots,x^d)$ can be occupied by only one cell at a time. This "exclusion of multiple occupancy" is the most frequent hypothesis but,at least theoretically,the number of cells at each site can be unrestricted. Also,the dimension d was defined as $d\geq 1$,although a biologist would limit it to $d\leq 3$.

Secondly,we assume that the situation at each point $x\in Z^d$ is specified by a random variable $\xi(x)$. By "situation" we mean a particular state of the cell or,simply,the fact that the site x is occupied by a cell or it is vacant. Thus,the random variable $\xi(x)$ represents the configuration at x. (For instance,in a lattice model of the cell membrane,a lipid chain can be in one of two states:(0) a ground state of energy zero,and (1) an excited state : Pink and Chapman,1979.)

Thirdly,we assume that the cells in this idealized tissue will be in interaction,most frequently in local interaction,such that distant cells have little effect on each other. The current word "neighbours" does not necessarily imply closeness in terms of Euclidean distance $|x-y|$,$x,y\in Z^d$; the choice as to which pairs of sites are to be neighbours is simply a postulate.

In many biological applications,the RF-models must include two specific hypotheses which are mathematically difficult to treat,namely (α)the interacting cells replicate, (β)they may be originally or become non-identical. Hypothesis (β) was considered by M.E.Fisher and D.Ruelle (1966) and R.L.Dobrushin(1967) who studied the stability properties of particle systems with $k>1$ types. Hypothesis (α) was not taken into account,and the particles were originally of different types.

In a recent paper,D.Ruelle(1981) suggested that the apparition of new biological species might be the result of the occurrence of phase transitions in a large system of DNA particles. These new species were interpreted as the extremal Gibbs states of the considered biological system. The situation will be more complicated if (β) is formulated in the sense that the apparition of different cell types is determined by a specific sequence of "metamorphoses" (i.e.changes of cell "type").

0.3. In order to make the notations permanent and to fix the idea about random fields, I will give, from the beginning, the following probabilistic definition of an RF : If $\xi(x)(\cdot)$ for every $x \in S$ is a random variable on some probability space (Ω, F, P) with values in a state space $(W, \mathcal{W})$, we call the mapping $x \mapsto \xi(x)(\omega)$ a random field with values in W. Then, $\xi(x)(\omega)$ is, for fixed $\omega \in \Omega$, a W-valued random function of $x \in S$. Let $\omega = (\hat{\pi}_x, x \in S)$ be a point in Ω, the coordinate space, where $\hat{\pi}_x$ is any real number. Then $\xi(x)(\omega) = \hat{\pi}_x$ (the projection mapping) (See Doob, 1953, p.11)

Assume $S = Z^d$ or R^d, $d \geq 1$, such that $\{\xi(x), x \in S\}$ is a family of random variables indexed by elements of S. One must distinguish between discrete (Z^d) and continuous (R^d) models and suppose S be a countably infinite set. The product $\Xi = W^S$ will be regarded as the space of all configurations ξ of particles (cells) in S. The exclusion of multiple occupancy allows us to consider that the set Ξ of all locally finite subsets of S becomes the configuration space. (The case of several types of particles, i.e. the hypothesis (β) in 0.2, can be reduced to the case of a single type by embedding several copies of S in a higher dimensional space, say Ξ^k, $k > 1$.)

Let provide $\xi(x)$ with the discrete topology and Ξ with the product topology which is compact. Throughout the paper, Ξ will be regarded as a metric space.

As it is known, for any metric space X any two of the following properties imply the third (Dieudonné, 1, Prop. 3.16.3):

(A1) X is compact;

(A2) X is discrete (more precisely, homeomorphic to a discrete space);

(A3) X is finite.

Any discrete metric space is locally compact - but not compact unless it is finite. In order to avoid confusions and repetitions, I am going to include the Schwartz's theorem (1973, Th.6) : Let X be a locally compact (LC) space. Then the following are equivalent:

(B1) X is Polish.

(B2) X is Lusin.

(B3) X is Suslin.

(B4) X has a countable base for open sets.

(B5) X is separable and metrizable.

(B6) X is locally metrizable and denumerable at infinity.

(B7) The Alexandrov compactification has any one of the above properties.

(B8) The space $C_{oo}(X)$ of real-valued continuous functions with compact support, i.e. $C_{oo}(X) = \{f \mid f \in C(X), \mathrm{supp}(f) \text{ compact}\}$, endowed with its inductive limit topology is Lusin or Suslin or separable.

(B9) The Banach space $C_o(X)$ of real-valued continuous functions vanishing at infinity is Polish or Lusin or separable.

In (B8) and (B9), $C(X)$ is the algebra of real-valued continuous functions $f:X\mapsto R$, and $C(X)\subset C_o(X)\subset C_{oo}(X)$. If X is compact, then $C_{oo}= C_o=C$. $C=C(X)$ is a Banach space with respect to the uniform(supremum)norm, and the probability measure on X forms a convex compact subset M on the weak dual C^* of C. (C^* is the space of real measures on X ; its topology is the vague topology, i.e. the topology of pointwise convergence of linear functionals on C. Also, C^* is locally convex.)

It is assumed that the reader is familiar with the basic work in this domain (e.g., Spitzer, 1971; Preston, 1974, 1976).

1. Multiparameter stochastic processes and random fields. The conditional independence property

1.1. As it is already known, the theory of random fields can be viewed as an application of the theory of multiparameter stochastic processes originated in P.Lévy's studies on Brownian motion in higher dimensional time. Centered Gaussian processes on R^d, $d>1$, as the Lévy field (or the Lévy Brownian motion) and the Brownian sheet are natural multi parameter generalizations of the Brownian motion. For example, if $\{\xi(t), t\in T\}$ denotes a continuous Brownian motion in R^d, the mapping $f \to \int_0^{\cdot} f(\xi(u))du$ defines a RF with f a periodic function (Bolthausen, 1983). Random fields with independent increments are known as multiparameter additive processes, and it is proved (Adler et al., 1983) that every additive RF is the sum of a Lévy process, a deterministic process and a countable number of degenerate additive RFs. The analysis of multiparameter subadditive processes should be of considerable interest. R.T.Smythe(1976) gave forth the conditions that a process $\{\xi(x,y), x<y\}$, $x,y\in N^2$, is subadditive, in the line with the theory developed by J.Hammersley and D.Welsh in 1965.

It has been stated (Loève, 1973) that Markov dependence, stationarity and martingales are the only three dependence concepts so far isolated, which are sufficiently general and sufficiently amenable to investigate yet with a great number of deep properties. However, the Markov property does not arise as naturally in the case of a RF that varies along the space coordinates. In some sense, a RF lacks the directionality of the time axis. Even in the case of higher dimensional time, the Markovian property is not ever available : the classical example is the Lévy field which is Markovian in odd time dimensions but not in even dimensions where no Markov-type property at all holds (McKean, 1962).

The Markov property of a multiparameter random process $\{\xi(x), x\in X\}$ can be defined by the aid of the concept of conditional independence relation for a triple of sub-σ-algebras. Let (Ω, F, P) be a complete probability space and F_1, F_2, F_3 be sub-σ-fields of F. If

$$\text{(C1)} \quad P\{A \cap B | F_3\} = P\{A|F_3\}P\{B|F_3\} \quad \text{a.s. for all } A \in F_1, B \in F_2,$$

then the Borel fields F_1 and F_2 are weakly conditionally independent given F_3 (see also Meyer, 1966; Döhler, 1980). One says that F_3 splits F_1 and F_2. The weak conditional independence will be denoted $F_1 \perp\!\!\!\perp F_2 | F_3$ (or, following the recent suggestion by van Putten and van Schuppen (1985), as $(F_1, F_3, F_2) \in CI$; CI: the conditional independence relation).

Both F_1 and F_2 are strongly conditionally independent given F_3 iff F_1 and F_2 are splitted by G, for every Borel subfield G of F_3 (Lloyd, 1962).

Now, let $\{\xi(x), x \in X\}$ with $X \subset R^d$ an open subset, defined on (Ω, F, P). Let $E \subset X$, with $\bar{E}$, the closure, and ∂E the (topological) boundary of E. We associate to this process the following σ-algebras:

$$F_E^- = \sigma\{\xi(x), x \in \bar{E}\} \quad \text{and} \quad F_E^+ = \sigma\{\xi(x), x \in E^c\}.$$

The process $\{\xi(x)\}$ has the simple Markov property if

$$\text{(C2)} \quad F_E^- \perp\!\!\!\perp F_E^+ \, | F_{\partial E}$$

(Mandrekar, 1976, 1978). This means that "the future" (the information in the exterior of E) is independent of "the past" (the information in the interior of E), conditional by "the present", that is, by the knowledge in an arbitrary small neighbourhood of the boundary ∂E (the "environment" of E). $F_{\partial E}$ is called a germ field.

1.1.1. (Remark 1) F. Knight (1970) defined five germ fields by considering the time intervals $(t-\Delta, t)$, $(t, t+\Delta)$, $(t-\Delta, t]$, $[t, t+\Delta)$, and $(t-\Delta, t+\Delta)$, and consequently

$$G_1(t) = \bigcap_{\Delta > 0} F(t-\Delta, t) \ , \ G_2(t) = \bigcap_{\Delta > 0} F(t, t+\Delta) \ , \text{ etc.}$$

A stochastic process is called Markovian relative to the germ field G_1 iff it is Markovian relative to G_2. One says that the process is Markovian relative to the germ fields if for each t, the past and the future are CI, given the germ field $G(t) = \cap_{\tau > 0} F^0(t, t+\tau)$, where $F^0(t.t+\tau) = \sigma\{\xi(u), t < u < t+\tau\}$ (Knight, 1970, 1979). Following a lemma given by V. Mandrekar (1983), $\xi(t)$ has the simple Markov property (resp. germ-Markov) on all sets iff it has the simple Markov property (resp. germ-Markov) on all open sets.

1.1.2. (Remark 2) H.P. McKean (1962) introduced the concept of minimal (smallest) splitting field. Let $F_4 = \sigma\{E(f|F_2); f \text{ bounded } F_1\text{-measurable}\}$. If $F_4 \supseteq F_1 \cap F_2$, F_4 is the minimal splitting field. (Compare with the case of minimal conditional independence relation treated by van Putten and van Schuppen, 1985, Def. 5.1 and Problem 5.2).

1.1.3. (Note 1) The notion will not be used subsequently, but it is given here as being of an independent interest.) Y. Okabe (1973) has shown that the Markovian property can be re-phrased in terms of linear manifolds. For any open set X in R^d one defines a closed subspace B(X) of $L^2(\Omega,F,P)$ as the closed linear hull of a Gaussian process $\{\xi(x), x\in X\}$. Thus for each open $X\subset R^d$ one sets in a similar way "the past" $B^-(X)$, "the future" $B^+(X)$ and "the germ" $\partial B(X)$, with

$$\partial B(X) \subset B^+(X)\cap B^-(X) \subset \hat{\Pi}_X B^+(X),$$

where $\hat{\Pi}_X$ is the projection of $B(R^d)$ onto $B^-(X)$. For the considered Lévy Brownian motion the simple Markov property (C1) is equivalent to

$$\hat{\Pi}_X B^+(X) = \partial B(X).$$

(See also Pitt, 1971; Kotani, 1973.)

1.1.4. (Note 2) Most mathematical research is focused on two-parameter processes $\{\xi(t), t\in R_+^2\}$. The reader is referred to the book edited by H. Korezlioglu et al. in 1981. Minutiae: the general theory (P.A. Meyer, 1981), the decomposition of two-parameter processes (R. Cairoli, 1971), the Markov property (H. Korezlioglu and P. Lefort, 1980), two-parameter jump processes (A. Al-Hussaini and R.J. Elliott, 1981), martingales and stochastic integrals (R. Cairoli and J. Walsh, 1974), section and projection theorems (C. Doléans and P.A. Meyer, 1979).

1.1.5. (Note 3) The special case with $\{\xi_x, x\in Z^d\}$ as a collection of iid random variables with multidimensional indices was treated by A. Gut (1978, 1979).

1.2. The Markov property (C2) of a random field can be deduced with the same arguments. Assume $S = Z^d, d\geq 1$. Let (Ξ, F) be a standard Borel space and S_o be the class of finite subsets Λ of S, i.e. $S_o=\{\Lambda\subset S : 0<|\Lambda|<<\infty\}$. Define for each $V\subseteq S$, $\hat{\pi}_V:(\xi(x))_{x\in S} \to (\xi(x))_{x\in V}$, the natural projection function from Ξ_S to $\Xi_V = W^V$, and let $\xi(V)=\hat{\pi}_V(\xi(S))$ be the corresponding projection of a generic point (configuration) $\xi(S)\in \Xi_S$. Then $F_V = \sigma\{\hat{\pi}_x, x\in V\}$. For each $V\subseteq S$ there is a set $\partial V\subseteq S\setminus V$ of neighbours of V with $\partial\Lambda\in S_o$ for each $\Lambda\in S_o$. Following 1.1, we have to construct the three necessary sub-σ-fields of F, namely F_Λ, $F_{S\setminus\Lambda\cup\partial\Lambda}$ and $F_{\partial\Lambda}$, in order to define

$$\text{(C2')} \qquad F_\Lambda \perp\!\!\!\perp F_{S\setminus\Lambda\cup\partial\Lambda} \mid F_{\partial\Lambda}$$

which has a spatial interpretation (see Preston, 1974, p.3; 1976, p.74). The RF with the simple Markov property (C2') will be called a Markov random field (MRF). Compare with Definition 3.20 given by F.Spitzer (1971).

One can distinguish a local and a global Markov property:

(a) An RF has the local Markov property (lmp) if

$$P(A|F_{S\setminus\Lambda}) = P(A|F_{\partial\Lambda}) \ , \ A\in F_{\Lambda} \ , \ \Lambda\in S_{o} \tag{1.1}$$

where $P(\cdot|F_U)$ denotes the conditional probability with respect to F_U.

(b) If (1.1) holds for arbitrary subsets V of S, then the RF has the global Markov property (gmp).

It is clear that the global property is a strictly stronger property than the local one : gmp $\Rightarrow$ lmp (the converse is not true in general). The gmp is the most suitable property for homogeneous RFs but it is difficult to verify it. Actually, gmp is for multiparameter processes what lmp is for the one-dimensional processes (Albeverio et al., 1981; Albeverio and Høegh-Krohn, 1984).

1.2.1. The connection between RFs and Markov processes will be exemplified by the one-dimensional case, $\Xi = W^{Z}$. Let Λ be a finite interval, $\Lambda=[m,n]\in S_o = \{x\in Z^1 | m\leq x\leq n\}$. We have to define the following sub-σ-fields of :

$$F_{\Lambda} = \bigvee_{x\in\Lambda} G_x \ ,$$

$$F_m = \bigvee_{x\geq m} G_x \ , \ \text{"the past up to m"} \ ,$$

$$F_n = \bigvee_{x\geq n} G_x \ , \ \text{"the future from n on"} \ ,$$

$$\hat{F}_{[m,n]} = F_m \bigvee F_n \ ,$$

$$\partial F_{\Lambda} = \hat{F}_{\Lambda} \cap F_{\Lambda} .$$

Using the classical definition of the conditional independence (e.g., Meyer, 1966, p.30), one says that an RF is an MRF if it has the two-sided Markov property (also called "local") :

$$E[f|F_{\Lambda}] = E[f|F_{\partial\Lambda}] \ , \tag{1.2}$$

where $\{f|f\in G^b\}\subseteq F$ is the class of bounded G-measurable functions f on Ξ. A random process is Markov if it has the one-sided (or "global") Markov property

$$E[f|F_n] = E[f|G_n] \ , \ f\in F_n^b \tag{1.3}$$

(Föllmer, 1975b; see also Cox, 1977; Winkler, 1981; Papangelou, 1983).

It appears that only MRFs on Z^1 that are also Markov processes have the global property. In fact, *all stationary and irreducible one-dimen-*

sional MRFs are Markov processes (Papangelou,1983,Th.1;see also Dang-Ngoc and Yor,1978). A conjecture of F.Spitzer(1975b) stating that if S is countable,there is at most one stationary MRF satisfying (1.2) was proved by H.Kesten(1976;see Cox,1979).

1.3. This paragraph deals with the Markov property of generalized random fields. Let $S=S(R^d)$,$d\geq 1$,be the Schwartz space of real-valued infinitely differentiable functions $\phi(x)$,$x\in R^d$,decreasing at infinity, together with all their derivatives,more rapidly than any negative power $|x|^{-k}$,$k=1,2,\ldots$ Also,let $S'=S'(R^d)$,$d\geq 1$,be the space of all real linear continuous functionals on $S(R^d)$ equipped with the weak topology. S' is defined as the space of tempered distributions and is dual to S.

Definition 1.(see Dobrushin and Major,1981). A generalized RF over S is a collection of random variables $\{\xi(\phi),\phi\in S\}$ such that

$$\forall\phi\in S\ ,\ \xi(c_1\phi_1 + c_2\phi_2) = c_1\xi(\phi_1) + c_2\xi(\phi_2),\ \text{with } P=1, \quad (1.4)$$

for all real numbers c_1,c_2,and

$$\xi(\phi_n) \to \xi(\phi) \text{ in measure}$$

if $\phi_n \to \phi$in the topology of S.

Example 1.3.1. A stationary Gaussian RF with mean zero is called a generalized RF if $\xi(\phi)$ is a Gaussian random variable with

$$\forall\phi\in S\ ,\ E[\xi(\phi)] = 0,$$

$$\forall\phi,\psi\in S\ ,\ E[\xi(\phi)\xi(\psi)] = E[\xi(\phi')\xi(\psi')]$$

where $\phi'(x)=\phi(x+y)$ and $\psi'(x)=\psi(x+y)$,$y\in R^d$.

1.3.2. There are different ways to define a Markovian property for generalized RFs. The subtle definition given by E.B.Dynkin(1980,Th.1.2.1 states that the Gaussian RF associated with a symmetric Markov process has the Markov property on all sets $A,B\in R^d$ iff

(C3) A path cannot reach B from A without crossing $A\cap B$.

It was shown that the conditional independence property (C1) is equivalent to the characteristic orthogonal projections of linear functionals of a Gaussian RF on some minimal subspace of a Hilbert space (Condition 2.3.C). See also Pitt,1971;Kallianpur and Mandrekar,1974.

Two other properties,namely the ε-Markov property and the germ-Markov property,were defined by V.Mandrekar(1983,Def.4.1). For global and local properties see M.Röckner(1983,1985).

2.Set-indexed and related processes. Regularity and mixing properties

2.1. This paper is intended to accomplish a "tour d'horizon" through the mathematical problems arising in the domain of RFs and related stochastic processes;in spite of the apparently redundant information, it attempts to mark out the important steps to understand and apply these processes. The definition of RFs as a collection of random variables indexed by a multiparameter set suggests a relationship with the set-indexed processes $\{\xi(A),A\in\mathcal{A}\}$,where $\mathcal{A}$ is a family of (partially ordered or directed) sets. For instance,$\mathcal{A}$ can be the family of closed convex subsets of the unit d-dimensional cube $I^d=[0,1]^d$,$d\geq 1$,or the family of closed sets with "smooth" boundaries (determined by special differentiable functions),the collection of subsets of $R_+^d=[0,\infty)^d$,$d\geq 1$,or the set of points in I^d,etc. The domain of $\mathcal{A}$ can vary from the relatively smaller families as in the case of the set-indexed Brownian motion to the largest possible size for a compound Poisson process where $\mathcal{A}$ equals all Borel subsets of I^d,$d\geq 1$ (Bass and Pyke,1984a). For example,the continuous Brownian process $\{B(A),A\in\mathcal{A}\}$ with $\mathcal{A}$ the collection of subsets of R^d,needs $\mathcal{A}$ not too large in size as to cause the divergence of the integral $\int_0^1[H(x)/x]^{1/2}dx$, where H is the log-entropy of $\mathcal{A}$ (Dudley,1973,Th.1.1). $\{B(A)\}$ exists if $\mathcal{A}$ is defined as the class of convex subsets in I^2 but not in I^d,$d\geq 3$. Consequently,$\mathcal{A}$ must satisfy certain conditions regarding closeness,boundedness,smoothness,etc.(see Bass and Pyke,1984b,1985). Also,the suitable sample space of these set-indexed processes must have a useful topology,i.e. the space $\mathcal{D}(\mathcal{A})$ of set functions that are outer continuous with inner limits (Bass and Pyke,1985).

2.2. In this paragraph tree-indexed RFs will be considered,that is, random processes $\{\xi(x),x\in V\}$ with V the set of vertices of an infinite tree T_n with n+1 edges emanating from every vertex - e.g.,$T_1=Z^1$,T_n,$n\geq 2$, the connected infinite graph without loops. This model was treated by C.Preston(1974,p.97;1976,p.79) and F.Spitzer(1975a).(See also Higuchi, 1977;Miyamoto,1982.) The considered MRF is defined as a probability measure μ on $\Xi_V=W^V$,$W=\{0,1\}$,$V\equiv T_n$,with strictly positive values for finite cylinder sets. The conditional probabilities

$$\mu\{\xi(x)=1|\xi(\cdot) \text{ on } T_n\setminus x\} \tag{2.1}$$

depend only on the values of ξ at the neighbours N_x of $x\in T_n$. If one assumes the invariance property under graph isomorphism,the measures μ are determined by n+2 parameters,that is,

$$v_k=\mu(\xi(x)=1|\xi=1 \text{ at exactly k of the neighbours } N_x),\ 0\leq k\leq n+1.$$

Because not all possible vectors $v=(v_0,v_1,\dots,v_{n+1})$ are realizable by a MRF, the class υ of realizable vectors is given by

$$\upsilon_k=[1+ba^{2k-(n+1)}]^{-1} , \qquad (2.2)$$

where a and b are positive numbers (Spitzer,1975a,Th.1). Let G be the set of all MRFs on T_n. Then there exists a class $G_\upsilon \subset G$ of MRFs with a particular vector v satisfying (2.2). Each G_υ may consist of one or many MRFs : when n=1, $|G_\upsilon|=1$ for all v. This follows from the fact that every Markovian μ is a one-dimensional MRF with conditional probabilities

$$f_{uv}(z) = \frac{M(u,z)M(z,v)}{M^2(u,v)} ,$$

where $\{M(i,j),i,j\epsilon W\}$ is a strictly positive stochastic matrix which uniquely determines the array $\{f_{uv}(z)\}$ defined by

$$f_{uv}(z) = \mu(\xi(0)=z|\xi(-1)=u,\xi(1)=v) , u,v,z\epsilon W.$$

For each strictly positive M,μ_M is called a Markov chain (Spitzer,1975, Def.4). Yet, the array $\{f_{uv}(z)\}$ cannot be chosen arbitrarily from $W=\{0,1\}$ The class of MRFs on Z^1 will coincide with the class of Markov chains (Spitzer,1971,Th.3.22) and the same subsists if the tree T_n is finite (Zachary,1983,Corollary 1). If μ is a MRF on (Ξ_{T_n},F) having a trivial tail σ-field, then μ is a Markov chain (Zachary,1983,Th.2.1).

2.2.1(Note) Partially ordered or directed sets were used for indexing martingales, submartingales or quasimartingales : S.Bochner(1955), K. Krickeberg(1956) and more recently A.Mandelbaum and R.J.Vanderbei(1981), R.B.Washburn and A.S.Willsky(1981), and H.E.Hürzeler(1982).

2.3. Partial-sum processes indexed by sets are of importance in current statistical research. Suppose, for instance, that we have to analyze a histological slide containing a large number of normal and malignant cells. Partition this picture by a grid into relatively small and equal areas, and take measurements within each area on the number of malignant cells. The resulting observation matrix can be interpreted as an array of independent random variables indexed by the two-dimensional grid coordinates (the lattice Z^2). For any subset $A\subset Z^2$, one can define the sum $S(A)=\sum_{x\epsilon A}\xi_x, x=(x^1,x^2)\epsilon Z^2$, to represent the random measure of the area A, that is, the measure of malignant occurrence in A. If the measures of disjoint areas $(A_m)_{m>1}$ are independent, the sum $S(A_m)$ would be approximately normal by the classical central limit theorem (Pyke,1983). However, $S(\cdot)$ satisfies the strong law of large numbers.

2.3.1.(Note 1) Partial-sum processes with iid random variables indexed by an integer d-dimensional lattice were studied by R.Pyke(1973) who introduced the "Brownian sheet",a multiparameter continuous Gaussian process. Two-parameter partial-sum processes weakly converge to a Brownian motion in I^2(Bickel and Wichura,1971,Th.5).

2.3.2.(Note 2) The case A : a denumerably infinite,partially ordered set,indexing a collection of iid random variables with mean zero, was examined by R.T.Smythe(1974). A was considered a local lattice.

2.4. Conditions of weak dependence can be introduced as follows: Let $(X_n)_{n\in Z^d}$ be a sequence of Z^d-indexed ($d\geq 1$),positively correlated random variables,and $V_1,V_2\subset Z^d$ two Borel sets with Euclidean distance $d(V_1,V_2)=r$. (X_n) will satisfy a strong mixing (SM) condition if there exists a non-increasing continuous function $\alpha:[1,\infty)\to(0,\infty)$, $\alpha(r)\downarrow 0$ as $r\uparrow\infty$,so that

$$\text{(C4)} \qquad |P(A\cap B) - P(A)P(B)| \leq \alpha(r) \ , \ A\in F_{V_1} \ , \ B\in F_{V_2} \ ,$$

where $\alpha(r)$ does not depend on $P(A)$ and $P(B)$,and $P(A)P(B)>0$.

A simple mixing condition is defined (Neaderhouser,1978,1980) as

$$|P(A\cap B) - P(A)P(B)| \leq \alpha(r)|V_1| .$$

It is then obvious that the Euclidean distance $d(V_1,V_2)$ as well as the size of V_1 may have different effects on the rate of mixing. Given a a certain dependence between the random variables X,one can expect that the dependence of sets A in F_{V_1} on the sets B in F_{V_2} decreases as $d(V_1,V_2)$ increases,but it may also increase as the size of V_1 increases. In fact,the strength of dependence between the random variables X is determined by the values of an interaction function (see below).

R.L.Dobrushin(1968,p.199) noticed that a one-dimensional RF possesses (C4) but the uniqueness of a RF on Z^2 with a given conditional probability is proved,however,under strong additional limitations which may be roughly interpreted as the becoming of the RF "sufficiently near" to a collection of independent random variables (see Th.6). Also,for an Ising model on Z^2 with positive nearest neighbour interactions at the critical temperature,(C4) does not hold (Neaderhouser,1980,Ex.4.2). The important consequence is that in some RF-models the critical behaviour should be linked with failure of SM condition (Hegerfeld and Nappi,1977).

2.4.1. As it was shown by A.N.Kolmogorov and Y.A.Rozanov(1960),for Gaussian sequences SM condition coincides with the condition of (linear) complete regularity (CR) (see also Ibragimov and Rozanov,1978,p.111), but in this case regularity conditions reduce to an approximation problem related to linear spectral theory (Ibragimov and Rozanov,1978,Chapter IV;see also Yaglom,1965). Let assume the one-dimensional case and consider the following σ-fields of events generated by a stationary sequence of random variables $\{\xi_k,\ k=\ldots,m\le k\le n,\ldots\}$:

$$F_o = F_m^n = \sigma(\xi_k,\ m\le k\le n),$$
$$F_1 = F_{-\infty}^m = \sigma(\xi_k,\ -\infty\le k\le m),$$

and,similarly, $F_2 = F_n^\infty$, $F_3 = F_{m+1}^{n-1}$, and $F_4 = F_{-\infty}^{n-1}$. Then the condition of <u>almost Markov regularity</u> (AMR) asserts that the following relations hold w.p.1 :

$$\text{(C5)} \qquad \sup_A |P(A|F_1) - P(A)| \le \phi(m,n)\ ,\ A\in F_2$$

$$\text{(C6)} \qquad \sup_A |P(A|F_4) - P(A|F_3)| \le \gamma(m,n)\ ,\ A\in F_2\ ,$$

where sup $(m,m+\nu)\downarrow 0$,sup$(m,m+\nu)\downarrow 0$ as $\nu\to\infty$ (Statulyavichus,1983). Condition (C5) is equivalent to the uniform strong mixing (USM) condition. The AMR condition (C6) is equivalent to any of the following condition:

$$\sup_A |P(A|\sigma(F_3,B)) - P(A|F_3)| \le \gamma(m,n)\ ,\ \text{w.p.1}$$

$$|P(A\cap B|F_3) - P(A|F_3)P(B|F_3)| \le \gamma(m,n)P(B|F_3),\ \text{w.p.1},$$

for all $A\in F_2, B\in F_1, P(B)>0$ (Statulyavichus,1983,Th.1). In the first relation,$\sigma(A,B)$ denotes the smallest σ-algebra generated by a σ-field $A\subset F$ and a set $B\in F$. The above AMR condition generalizes the Markov property.

Following M.Rosenblatt(1979,Proposition 1),the only stationary countable state Markov chains that statisfy a Markov-type regularity condition are sequences of independent random variables.

A stronger mixing condition,the <u>absolute regularity</u> (AR) condition (or "weak Bernoulli") was suggested by V.A.Volkonskii and Y.A.Rozanov in their paper on limit theorems for random functions(1959) :

$$\text{(C7)} \qquad \sup \frac{1}{2}\sum_i\sum_j |P(A_i\cap B_j) - P(A_i)P(B_j)| \le \beta(r)\ ,$$

where the supremum is taken over all pairs of partitions $\{A_1,\ldots,A_m\}$ and $\{B_1,\ldots,B_n\}$ of Ω such that each $A_i\in F_1, B_j\in F_2$,for each $i=1,\ldots,m$ and $j=1,\ldots,n$,where $\beta(r)\downarrow 0$ as $r\to\infty$ (see Bradley,1984). The known φ-mixing condition is

$$\text{(C8)} \qquad \sup_m \sup_A |P(B|A) - P(B)| = \phi(r)\ ,\ B\in F_2, A\in F_1, P(A)>0.$$

(see,e.g.,Peligrad,1985). The inequalities $\alpha(r)\le\beta(r)\le\phi(r)$ indicate that if the sequence $\{\xi_k\}$ is φ-mixing,then it is AR. SM and USM conditions were introduced as conditions under which limit theorems (e.g.,CLT) would hold for stochastic processes satisfying auxiliary moment conditions. However,a distinction can be drawn between mixing and regularity: mixing and ergodicity are preserved under some transformation condition while regularity is not preserved. As it is known,the property of ergodicity implies the indecomposability of a system into (non-trivial) invariant subsets.

A regular stochastic process is metrically transitive.

2.4.2. The invariance principle and the central limit theorem for RFs with different mixing conditions were studied by V.V.Gorodetskii (1982),B.S.Nahapetian(1980),H.Takahata(1983),etc. Some weaker conditions were given by E.Bolthausen(1982). I.Berkes and G.J.Morrow(1981) approximated the partial-sum RF with SM condition by a Brownian sheet. Weak invariance principle for a class of non-stationary mixing sequences was suggested by M.Peligrad(1981) and might be applied in the case of non-stationary RFs.

3.Lévy and Gibbs measures

3.1. This section is devoted to the construction of those locally finite and positive random measures of interest,possessing a conditional independence structure analogous to the Markov property (C2) above. Thus the Gibbs measures which are familiar in statistical mechanics will be defined as submeasures of a Lévy random measure by using the arguments of A.Karr(1978). The already known equivalence between Markov RFs and Gibbs RFs (Averintsev,1970;Spitzer,1971;Hammersley and Clifford,1971; Preston,1973;Grimmett,1973;Sherman,1973,etc.) will be mentioned. Finally the hyperfinite representation of the Radon measure space will be suggested for the construction of hyperfinite RFs.

The starting point is the following simple definition of a Radon measure (Schwartz,1973):

Definition 2.Let X be a Hausdorff topological space and $\mathcal{X}$ its Borel σ-algebra. A Radon measure on X is a measure m on $\mathcal{X}$ satisfying the following two conditions:

(i) m is locally finite;

(ii) m is inner regular (on $\mathcal{X}$),i.e. for every $B \in \mathcal{X}$

$$m(B) = \sup\{m(K) : K \subset B, K \text{ compact}\}.$$

The space $(X,\mathcal{X},m)$ is called the Radon measure space and

$$M_+(X) = \{m : m\in M(X) \text{ and } \langle m,f\rangle \geq 0, \forall f\in C_K^+(X)\}$$

is the space of positive Radon measures.

In other words (Daletskii and Smolyanov,1984), m is a σ-additive function defined on X,assuming real values and satisfying the (more precise) condition that for any $B\in X$ and $\varepsilon>0$,

(ii') $\exists K_\varepsilon \subset B$,compact, such that $\|m\| (B\backslash K)< \varepsilon$,

where $\|m\|$ $(\cdot)$ is the total variation of m on the set $(\cdot)$.

3.1.1. Some important statements will be briefly mentioned:

(1)Any positive linear form on $C_K(X)$ is a positive Radon measure.

(2)$M_+(X)$ with the vague topology is a complete,separable,metrizable space(=Polish).

A measurable map from an abstract probability space into $M_+(X)$ is a random measure. It induces a probability measure P on $M_+(X)$ which is Polish and uniquely determined by its characteristic functional. The space of probability measures on $M_+(X)$ will be denoted $M^*[M_+(X)]$.

(3)Vague probability measures are regular and tight;any tight probability on the Borel σ-algebra of a metric space is Radon.

3.1.2.(Note) Following a theorem given by A.Bose(1978,Th.4.2.3), for every non-atomic measure $\mu\in M_+(X)$ there exists a measure-valued Markov process. A.F.Karr(1979) introduced MRFs as an application of a measure-valued Markov process.

3.2. Let $\mathcal{B}$ be a class of proper subsets of X in X possessing the following characteristics:

(a) $\mathcal{B}$ is hereditary:if $B_1\in\mathcal{B}$ and $B_2\subset B_1$,then $B_2\in\mathcal{B}$.

(b) $\mathcal{B}$ is closed under countable intersections.

(c) If $B_1,B_2\in\mathcal{B}$ and $(B_1\cap B_2')\cup(B_1'\cap B_2)=\Phi$,then $B_1\cup B_2\in\mathcal{B}$.

Definition 3.(Karr,1978). If for each $B\in\mathcal{B}$ there exists a measurable subset (splitting set) B' of B^c such that

(D1) $B'\neq B^c$,

(D2) $(F_B,F_{B'},F_{B^c})\in CI$,

(D3) If $A\subset B'$ and $(F_B,F_A,F_{B'})\in CI$,then $A=B'$,

then a random measure λ on (X,X) will be called a Lêvy random measure with respect to the pair $(\mathcal{B},\{B,B'\})$.

Clearly,(D2) signifies the simple Markov property (C2),and condition (D2 expresses the minimality of the splitting set B' . The pair $(\mathcal{B},\{B,B'\})$, $B\in\mathcal{B}$,is called the Lévy space of a Lévy random measure .

The example of interest is settled as follows : Let $\{\xi(x), x\in V\}$ be a RF on $V\in S_o$, taking values in $W=\{0,1\}$. If $\Lambda\in V\backslash x$, i.e. a subset of V not containing site x but the set N_x of its 2d nearest neighbours, then

$$P\{\xi(x)=1|\xi(u),u\in\Lambda\} = P\{\xi(x)=1|\xi(u),u\in N_x\}$$

By defining

$$\lambda(\Lambda) = \sum_{x\in\Lambda} \xi(x) \quad , \tag{3.1}$$

we obtain a <u>Lêvy random measure</u> λ on V with <u>Lêvy space</u> $[(\{x\},x\in\Lambda),(\{x\},N_x)]$. The definition (3.1) is equivalent to the definition of a <u>simple</u> random measure (i.e. σ-finite and purely atomic with all atoms of mass one).

<u>3.3</u>. Now one can define a Gibbs measure as a submeasure of λ, a Lêvy submeasure on $(X,\mathcal{X})$. Let X be a nonnegative random variable in $F=\sigma(\lambda)$ such that

$$0<k=E[X] < \infty \quad . \tag{3.2}$$

A random measure μ on $\mathcal{X}$ such that for each $M\in\widetilde{M}$ (the Borel σ-algebra of $M_+(X)$),

$$P\{\mu\in M\}=k^{-1}E[X;\{\lambda\in M\}], \tag{3.3}$$

is called the submeasure of λ generated by the random variable X.

<u>Definition 4</u>. Let λ be a random measure on X. If

$$0<k(\psi,\lambda)=E[\exp\{-\psi(\lambda)\}]< \infty \tag{3.4}$$

then the submeasure γ of λ generated by $\exp\{-\psi(\lambda)\}$ is the <u>Gibbs submeasure</u> of λ. The function ψ is called the interaction function and $k(\psi,\lambda)$ is a normalization constant (called in statistical mechanics the partition function or the major statistical sum:Minlos,1967).

<u>3.3.1</u>.(Note 1) For other definitions of a Lêvy measure see (Maruyama,1970;Adler and Feigin,1984). D.Surgailis(1981) defined it as a second order measure on the space S' of tempered distributions on R^d. P.Lêvy (1937) interpreted λ as the intensity of a Poisson random measure : it has the property that for any finitely many compact disjoint sets K_1, ..., K_m, the random variables $\lambda(K_1),\ldots,\lambda(K_m)$ are independent and Poisson distributed (see Tjur,1980,p.209). As it is known, the Poisson random measure arises -in the context of integer-valued complete random measures- as a consequence of a combinatorial condition upon the random measures.

<u>3.3.2</u>.(Note 2) In order to put everything right, it must be said that some authors define the <u>Gibbs process</u> as a point process characterized in terms of its Palm measure (Nguyen and Zessin,1979;Glötzl,1983; van der Hoeven,1983,Ex.2.2.3 and §10.5). The reader is referred to C.

Preston(1976,Sect.6) for the treatment of continuous RFs as point processes. See also G.Ivanoff(1980,Def.1.2).

3.4. This is a short note about the application of nonstandard analysis to RF-models. The existence of a RF as a hyperfinite stochastic process is based on a theorem given by R.M.Anderson(1982) which states that every σ-finite Radon measure space (X,X,m) admits a hyperfinite representation,that is,in the specific terminology of nonstandard analysis,an internal measure space (Γ,G,u) [see Def.1.7 by Stoll, 1986]. The interested reader is referred to L.L.Helms and P.A.Loeb (1979),L.L.Helms(1983),and A.E.Hurd(1981) for nonstandard lattice models.

4.Random fields : local specifications and Gibbs states

4.1. This paragraph is devoted to the definition of RFs with the aid of specified conditional probabilities. Generally speaking,if a measurable space (Ω,F) is assumed to be Polish,conditional probability distributions should exist in some definite cases. Indeed,as J.L.Doob (1953,p.624) noticed,even if F is separable,a regular conditional probability on F given a sub-σ-algebra G does not always exist. A conditional probability distribution has two essential properties,namely the regularity and the properness : their definitions may be found in some basic books (e.g.,Parthasaraty,1967,Th.8.1,p.147) but we follow D.Ramachandran(1971) :

Definition 5. Let (Ω,F,P) be a probability space and let F_1,F_2 be two sub-σ-algebras of F. A regular conditional probability (rcp) on F_1 given F_2 is a map $\pi:\Omega\times F_1\to[0,1]$ satisfying the following properties for each fixed $\omega\in\Omega$:

(a) $\pi(\omega,\cdot)$ is a probability measure (countably additive) on F_1;

(b) $\pi(\cdot,A_1)$ is F_2-measurable for each fixed A_1 in F_1 (with the remark that F_2 is the Borel field of sets of the form $(\omega:\xi(\omega)\in B)$,where B is a linear Borel set);

(c) $P(A_1\cap A_2) = \int_{A_1} \pi(\omega,A_2)dP(\omega)$, for all $A_1\in F_1, A_2\in F_2$.

If $\pi(\omega,A_1)$ satisfies (a) and (c) with

(b') $\pi(\cdot,A_1)$ is F_2-measurable $P\restriction F_2$-a.s.,for every $A_1\in F_1$,

then $\pi(\omega,A_1)$ is called a rcp in Doob's sense given F_2. In addition,it is proper at $\omega_0\in\Omega$ if

(d) $\pi(\omega_0,A_2)=1$ whenever $\omega_0\in A_2\in F_2$.

The existence of rcp's has useful consequences : the conditional expectation $E[\xi|F_2]$ may be viewed as an ordinary expectation relative to the conditional probability measure (Doob,1953,Th.9.1,p.27;Chow and Teicher,1978,Th.1,p.211). If for some pair (F_1,F_2) of sub-σ-algebras $\pi(\omega,A_1)$ is a rcp on F_1 given F_2, and $\xi(\omega)$ is an F_1-measurable function with finite mean, then

$$E[\xi|F_2](\omega) = \int_\Omega \xi(\omega)dP(\omega) \quad \text{a.c.}$$

Regular conditional probabilities cannot be defined on any F_1 : this σ-field must be countably generated (Blackwell and Ryll-Nardzewski,1963; Blackwell and Dubins,1975). The condition is fulfiled in Example 1 by Y.S.Chen and H.Teicher(1978,p.211) : F_1 is said to be countably generated if there exists a sequence $(F_n,n\geq 1)\subset F_1$ such that $F_1=\sigma(F_n)_{n\geq 1}$. If F_1 is a countably generated sub-σ-algebra of F, then for any $F_2\subset F$, an rcp $\pi(\omega,A_1$ given F_2 exists and is perfect for every $\omega\in\Omega$ iff (Ω,F,P) is a perfect probability space (Sazonov,1962,Th.7). It is proved that (Ω,F) is perfect iff F is isomorphic to the Borel σ-field of a universally measurable subset of a complete separable metric space (Darst,1971). The condition is satisfied if (Ω,F) is a standard Borel space (see its definition in Parthasarathy,1967,p.132).

The properness property -more precisely:the existence of an everywhere proper rcp- has been discussed and disproved (see Blackwell and Dubins,1975,Th.1) but a necessary and sufficient condition has, however, been found (Sokal,1981,Lemma 2.2) in a selection homomorphism ϕ for F_1 with respect to F (countably generated).

These results will be applied to construct a random field (with the aid of Theorems 3.2 and 3.3 in Sokal,1981) in the following steps:

(1)Define the measurable (Ξ,F) : S is a countable index set,and for each point $x\in S$ let (Ξ_x,F_x) be a measurable space. $(\Xi,F) = (\prod_{x\in S}\Xi_x, \prod_{x\in S}F_x)$.

(2)Hold F as countably generated iff F_x is.

(3)Hold (Ξ,F) perfect iff (Ξ_x,F_x) is.

(4)Define the selection homomorphism ϕ : Denote by S^* any countable family of subsets of S and consider $\Lambda\subset S$. Let $F^*=\{F_\Lambda,\Lambda\in S^*\}$ and $f_{F_\Lambda}(\xi) =\xi(\Lambda)\times\xi^o(\Lambda^c)$, when $\xi^o\in\Xi$ is fixed. Thus $\phi_{F_\Lambda} = f^{-1}_{F_\Lambda}$ and $\{\phi_{F_\Lambda},F_\Lambda\in F^*\}$ is a compatible family of selection homomorphisms.

If μ is any perfect measure on (Ξ,F), then there exists a family $\{\pi_{F'},F'\in F^*\}$ of everywhere proper rcp's for the probability space (Ξ,F,μ) given F' which satisfies

$$\pi_1(\xi,A) = \pi_1(\xi,d\xi')\pi_2(\xi',A) \ , \quad \xi\in\Xi \ , \ A\in F \ ,$$

whenever, by putting $\pi_1=\pi_{F_1}$, $\pi_2=\pi_{F_2}$, $F_1,F_2\in F^*$, with $F_1\subset F_2$.

4.2. Definition 6. Let $(S,\leq)$ be an infinite countable index set ordered by the relation $\leq$, $(W,\mathcal{W})$ a measurable space and $\Xi = W^S$, the set of all possible configurations $\xi : S\to W$. Assume (Ξ,F) be a standard Borel space (isomorphic to the Borel σ-field on a complete separable metric space). Denote as above by S_o the set of all non-empty finite subsets of S and by F^* the ensemble of sub-σ-algebras $\{F_\Lambda, \Lambda\in S_o\}$. Let $(\hat{F}_\Lambda)_{\Lambda\in S_o}$ be a decreasing family of sub-σ-fields of F generated by $\{\xi(x), x\in\Lambda\}$. For each $\Lambda\in S_o$ let $\pi_\Lambda = \{\pi_{F_\Lambda}\}$ be a F-measurable probability kernel on (Ξ,F), that is, a mapping $\pi_\Lambda:\Xi\times F\to[0,1]$ which satisfies specifically the conditions of a proper rcp :

(E1) $\pi_\Lambda(\xi,\cdot)$ is a probability measure on F, for each $\xi\in\Xi$;

(E2) $\pi_\Lambda(\cdot,A)$ is F_Λ-measurable, for each $A\in F$;

(E3) $\pi_\Lambda(\xi,B) = I_B(\xi)$ for all $B\in\hat{F}_\Lambda$ and all $\xi\in\Xi$;

(E4) $\pi_\Lambda(\xi,A) = \int\pi_1(\xi,d\xi')\pi_2(\xi',A)$, for all $\xi,\xi'\in\Xi$ and $A\in F$

whenever $\hat{F}_1\subset\hat{F}_2$.

Assume now that the collection $\Pi=(\pi_\Lambda)_{\Lambda\in S_o}$ satisfies the consistency condition

(E5) $\pi_\Gamma\,\pi_\Lambda = \pi_\Gamma$, $\Lambda\leq\Gamma\in S_o$,

where $(\pi_\Lambda,\pi_\Lambda)(x,A)=\int\pi_\Lambda(y,A)\pi_\Lambda(x,dy)$.

Any probability measure P on (Ξ,F) which is compatible with Π, i.e.

$$P(A|F_{S\setminus\Lambda})(\cdot) = \pi_\Lambda(\cdot,A) \quad P\text{-a.s.} , \ A\in F , \Lambda\in S \qquad (4.1)$$

will be called a random field with local characteristics Π.

4.2.1. (Note 1) The compatibility condition (4.1) is equivalent to

$$P\pi_\Lambda = P , \ \Lambda\in S , \qquad (4.2)$$

where $P\pi_\Lambda$ denotes the measure $\int P(d\xi)\pi_\Lambda(\xi,\cdot)$.

4.2.2. (Note 2) For other different problems regarding the local specifications the reader is referred to C.Preston (1975), S.Goldstein (1978), E.B.Dynkin (1978), R.G.Flood and W.G.Sullivan (1980), S.E.Kuznetsov (1984), etc.

4.3. In order to define specifications in terms of interaction, the present paragraph will deal with the formal definition of an interaction. Following the examples given in statistical mechanics, an interaction function must fulfil conditions that can induce its weakness between remote domains and prevent the collapse of an infinite number of subsystems into a bounded domain. Commonly, an interaction function ψ between two spins is defined as $\psi^{\Lambda}_{xy}=\psi(|x-y|)$, where $|\cdot|$ denotes the usual Euclidean distance between the two spins located at $x,y\in\Lambda,\Lambda\in Z^d,d\geq 1$. This function ("pair potential") is some nonnegative function on $[0,\infty)$ which rapidly decreases at infinity (e.g., exponentially decreasing interactions: Ruelle, 1978, p.86). The simplest case is $\psi^{\Lambda}_{xy} = \tilde{\psi}/|\Lambda|$, for all $x,y\in\Lambda$ where $|\cdot|$ is the cardinality of domain Λ and $\tilde{\psi}$ is a fixed positive number. Thus, each spin interacts equally with all its neighbours. Generally speaking, a nonnegative function ψ on the set S_o of all finite parts of S will be considered to be an interaction function iff for every site $x\in S$, the series $(\psi(\Lambda), x\in\Lambda)$ of real numbers converges. This will be shown in Subparagraph 4.3.1. However, as D. Ruelle (1978, p.31) remarked, the interest in considering systems of conditional probabilities rather than interactions is on one hand that they are *a priori* more general, and on the other hand that they behave better under morphisms (i.e. if F is a morphism, F^* is uniquely defined on systems of conditional probabilities, but not on interactions).

Definition 7. An interaction Ψ is a collection of maps $\{\psi(A,\cdot)\}$, $\psi : \Xi_\Lambda \to R^1_+, \Lambda\in S_o$, with the following properties :

(A1) $\psi\in C^2([0,R_+],R_+)$;

(A2) $\psi(\Lambda+y) = \psi(\Lambda)$, $\Lambda\in S_o, y\in S$;

(A3) $\|\psi\| = \sum|\psi(A,\cdot)| < \infty$ ($\|\cdot\|$ the supremum norm).

The values of ψ are interpreted as the joint interaction of the random variables $\{\xi(x)\}$ inside the domain $\Lambda\in S_o$. (If no confusion is possible, $\psi(\Lambda,\xi(\Lambda))$ will be written as ψ_Λ.) If one assumes that the considered cell system has a defined number of components, (A3) will be specifically

(A3') $\|\psi\| = \sum_{\Lambda}|\psi_\Lambda|\cdot|\Lambda|^{-1}$.

Also, if several simultaneous interactions are to be assumed, ψ will be defined as the mapping $\psi:\Xi_\Lambda\to R^m_+, m>1$.

4.3.1. Invariant interactions form a Banach space $\mathcal{B}$ with respect to the supremum norm. The dense linear space $\mathcal{B}_o\subset\mathcal{B}$ consists of finite range interactions. For example, $\psi\in\mathcal{B}_o$ is a finite range interaction if there exists a finite set $A\in S_o$ such that $\psi_\Lambda=0$ unless $\Lambda\backslash x\subset A$, whenever $x\in\Lambda$. Then ψ has compact support (Ruelle, 1969, p.31). A finite range interaction possesses (A3).

The pair interaction $\psi(\{x,y\}), x,y\in\Lambda$ [also written as ψ^2_Λ] is defined as the interaction between two neighbouring components of the system such that $\psi^2_\Lambda=0$ whenever $|\Lambda|>2$. The existence of such a function is conditioned on the convergence of the series $\psi(\{x,y\})$. The sufficient condition for convergence is that

$$\sup|\psi(\Lambda,\xi(\Lambda))| \leq k_1 n[u^{k_2 d}\cdot C^{n-2}_{u^d}]^{-1} , \qquad (4.3)$$

where $n=|\Lambda|$; u:diam.Λ; $k_1,k_2(>1)$, two constants (Sinai,1982).

Positive-type pair interactions ψ^2_Λ are of particular interest. As it is known, a function ψ is of positive type iff it is the Fourier transform of a positive measure of finite total mass on S(i.e. R^d). This is Bochner's theorem which lies behind the definition of interaction given by J.Moulin Ollagnier(1985,p.95).

4.3.2. A pair interaction ψ^2_Λ which is continuous and of positive type defines a stable interaction. If ψ can be decomposed as $\psi=\psi_1+\psi_2$, where

$$\psi_1(x)\geq 0,$$

$$\psi_2(x) = \int du \exp\{iux\} f^*_2(u),$$

where $f^*_2(u)$ is the Fourier transform which is (absolutely) integrable and positive $(f_2(u)\geq 0)$, then ψ is stable. The reader is referred to R.L. Dobrushin(1964), D.Ruelle(1969,pp.38-39) or O.Lanford III(1973) for other (more general) sufficient stability conditions. An interaction which violates stability is called "catastrophic" (Ruelle,1969,pp.35-36). Catastrophic (non-thermodynamic) interactions, e.g.,

$$\psi(x) = -\log[1+(a-1)\exp\{-bx^2\}], \; x\geq 0, a\geq 0, b>0,$$

$$\psi(x) = -\log[1+(ax-1)\exp\{-bx^2\}],$$

were assumed in some applications (Ogata and Tanemura,1981,1984). They violate the simple stability condition (Ruelle,1969,Def.3.2.1,p.33)

$$\text{(A4)} \quad \Psi(x_1,\ldots,x_n) \geq -nc , \; c\geq 0$$

for different values of the parameters a and b (e.g., a>1,b>0). In (A4), $\Psi(x_1,\ldots,x_n)$ is assumed to be

$$\Psi(x_1,\ldots,x_n) = \sum_{k\geq 2}\sum_{1\leq i_1<\ldots<i_k\leq n} \psi^k(x_{i_1},\ldots,x_{i_k}).$$

For example, if $k=2$, ψ^2 is as above the pair interaction function. Condition (A4) is satisfied if $\psi^k\geq 0$ for all k.

4.3.3. An interaction function is called regular if it is bounded below and satisfies

$$\text{(A5)} \quad C(\beta) = \int dx|\exp\{-\beta\psi(x)\}-1| < \infty ,$$

for all $\beta>0$ (Ruelle,1969,Def.4.1.2,p.72). Condition (A5) insures the weak decrease of ψ at infinity and is equivalent to requiring that ψ (bounded below) is absolutely integrable outside of a set of finite Lebesgue measures. Remark : if ψ^2 is stable, then it is automatically bounded below by $-2c$.

4.3.4. If the cell system is assumed to be composed of cells of k different types, the pair-type interaction function $\psi_{ij}(\{x,y\})$ between one cell of type i and one cell of type j, $1 \le i,j \le k$, is a symmetric one: $\psi_{ij}(\{x,y\})=\psi_{ij}(\{-x,-y\})$. The operation of decomposition necessary to define stability will be analogously extended (Fisher and Ruelle, 1966, Th.I). See in the mentioned paper the situation when the particles in the system possess special properties ("charges") which influence the interaction (Th.II).

4.3.5. The interaction strength E_Λ^ψ, given ψ, is

$$E_\Lambda^\psi(\xi) = \sum_{V \subseteq \Lambda} \psi(V,\xi(V)) \;, \quad V \cap \Lambda \neq \phi \tag{4.4}$$

whenever the series converges. The F-measurable family $\{E_\Lambda^\psi(\xi)\}$ satisfies the conditions of normalization and consistency

(A6) $E_\Lambda^\psi(0)=0$,

(A7) $E_\Lambda^\psi(\xi) = E_V(\xi) \;, \quad V \subseteq \Lambda$.

In terms of (A3'), E is bounded as follows :

$$\begin{aligned} |E_\Lambda^\psi(\xi)| &\le \sum_{V \subseteq \Lambda} |\psi(V,\xi(V))| = \\ &= \sum_{V \subseteq \Lambda} \sum_{x \in V} |V|^{-1} |\psi(V,\xi(V))| \le \\ &\le \sum_{V \subseteq \Lambda} \sum_{x \in V} |V|^{-1} |\psi(\Lambda,\xi(\Lambda))| \le |\Lambda| \cdot \|\psi\| \;, \end{aligned}$$

so that

$$E_\Lambda^\psi(\xi)|\Lambda|^{-1} \le \|\psi\| \text{ for all } \Lambda \in S_o .$$

The strength of interaction between different domains can be defined as follows : Assume there are two disjoint subsets A and B of S ($=Z^d$) with A finite but B possibly finite. Then,

$$E(A,B) = \sum_{\Lambda} \psi(\Lambda) \;, \Lambda \subset A \cup B \;, \quad \Lambda \cap A \neq \phi \neq \Lambda \cap B .$$

Due to (A3) the sum converges and defines a continuous function of B on the parts of S\A. More generally, suppose $\Lambda, \Gamma \in S_o$ and the configurations $\xi \in \Xi_\Lambda$, $\zeta \in \Xi_\Gamma$, and $\eta \in \Xi_{S \setminus \Gamma}$. The strength E at configuration $\zeta(u)$, $u \in \Gamma$, given the surrounding environment η, will be represented by the following series :

$$E_\Lambda^\psi(\zeta) = \sum_{\Lambda} \psi(\Lambda,\xi^*(\Lambda)), \quad \Lambda \cap \Gamma \neq \phi,$$

where ξ^* coincides with ζ on Γ and with η on $S \setminus \Gamma$. The sum is convergent.

4.3.6. $E^{\psi}_{\Lambda}(\cdot)$ is superstable if there exist $A>0$ and $B\in R^1_+$ such that

$$E^{\psi}_{\Lambda}(\xi)\geq A\sum_{x\in\Lambda}\xi^2(x)+B|\Lambda|,\ \text{for all}\ \Lambda\in S_0. \tag{4.5}$$

In other words, an interaction is superstable if it is stable and if it remains stable when slightly perturbed by any continuous, finite-range interaction. It can be shown that to every superstable $E(\cdot)$ there is a consistent family of conditional distributions with respect to the corresponding σ-algebra.

4.3.7. The strength of interaction with boundary conditions is called Hamiltonian. It expresses the strength of interaction between a configuration $\xi(\Gamma)$ and its environment $\xi(S\setminus\Gamma)$ as boundary condition :

$$H[\xi(\Gamma)|\xi(S\setminus\Gamma)] = \sum_{\Lambda}\psi(\Lambda,\xi(\Lambda))\ ,\ \Lambda\cap\Gamma\neq\phi\neq\Lambda\cap(S\setminus\Gamma) \tag{4.6}$$

Clearly, Hamiltonians may have different expressions depending on the type of interaction and of the boundary condition. With the help of Hamiltonians, all possible conditional probability distributions of an RF inside any finite domain can be found under the condition that its values outside this domain are fixed. In fact, Hamiltonians can be thought of as a natural generalization of transition probabilities for stochastic processes. If we denote by p^* the stationary probability distribution of a one-dimensional Markov chain and assume that domain Λ contains two points, i.e. two states i and (i+1), then the Hamiltonian of the Markov chain ξ is

$$H(\xi) = -\sum_{i=-\infty}^{\infty}\ln p^*[\xi(i)\xi(i+1)] \tag{4.7}$$

(see Sinai, 1982, p.5).

4.4. In this paragraph the construction of specifications with interaction will be briefly indicated (following Preston, 1980). The main point is to take a collection of F-measurable functions $w=\{w_\Lambda\}_{\Lambda\in S_0}$, where $w_\Lambda:\Xi\to R^1_+$, for each $\Lambda\in S_0$, which satisfy some integrability conditions (Proposition 2.1 in Preston, 1980). Put

$$Z^{(w)}_{\Lambda}(x) = \int w_\Lambda(y)\pi_\Lambda(x,dy) \tag{4.8}$$

and then define $\pi^{(w)}_{\Lambda}:\Xi\times F\to R^1_+$ by

$$\pi^{(w)}_{\Lambda}(x,F) = \begin{cases}[Z^{(w)}_{\Lambda}(x)]^{-1}\int_F w_\Lambda(y)\pi_\Lambda(x,dy) & \text{if } 0<Z^{(w)}_{\Lambda}(x)<\infty\ ,F\in F\\ 0,\text{otherwise}\end{cases} \tag{4.9}$$

These local characteristics must satisfy the usual conditions (E1)-(E4) of a F^*-specification. Now, the functions w will be defined in terms of the interaction strength E of (4.4). The domain of action of E^{ψ}_{Λ} is

$$G(x) = \{v : \sum_{\Lambda\in S_o} |\psi_\Lambda(v)| < \infty \}, \quad \bigcap_{x\in S} G(x) = G.$$

Thus, if $v\in G$ and $\Lambda\in S_o$,

$$\sum_{\substack{A\in S_o \\ A\cap\Lambda\neq\phi}} |\psi_A(v)| = \sum_{x\in\Lambda} \sum_{x\in A} |\psi_A(v)|[A\cap\Lambda]^{-1} \leq$$
$$\leq \sum_{x\in\Lambda} \sum_{x\in A} |\psi_A(v)| < \infty$$

and $E_\Lambda : G\to R^1_+$ gives $E_\Lambda(v) = \sum_{\substack{A\in S_o \\ A\cap\Lambda\neq\phi}} \psi_A(v)$. Instead of (4.9)

$$\pi^\psi_\Lambda(x,F) = \begin{cases} [Z^\psi_\Lambda(x)]^{-1} \int_{F\cap G} \exp\{-E^\psi_\Lambda(y)\} \pi_\Lambda(x,dy) \ , \text{ if } 0<Z^\psi_\Lambda(x)<\infty \\ 0, \text{otherwise} \end{cases} \tag{4.10}$$

where

$$Z^\psi_\Lambda(x) = \int_G \exp\{-E^\psi_\Lambda(y)\} \pi_\Lambda(x,dy) \tag{4.11}$$

(Preston, 1980, Proposition 4.1).

The reader should compare (4.11) with (3.4).

4.5. The result (4.10) suggests the use of Gibbs measures for RFs:

Definition 8. Given a local specification π^ψ_Λ (ψ being a finite-range interaction function), a Gibbs measure μ relative to ψ is a Radon probability measure on Ξ which verifies

$$\mu(\pi^\psi_\Lambda(f)) = \mu(f),$$

for every $\Lambda\in S_o$ and every $f\in C(\Xi)$. [See Moulin Ollagnier, 1985, p.89; Albeverio et al., 1981, Def. 2.1.]

Usually, a probability measure on (Ξ,F) is called a state. Thus, a probability measure μ on (Ξ,F) is called a Gibbs state specified by $\Pi = \{\pi^\psi_\Lambda\}$ if

$$\mu(A|F) = \pi^\psi_\Lambda A \quad P\text{-a.s.}$$

for each $A\in F$ and $\Lambda\in S$.

The relationship between the Gibbs measure and the rcp can be proved as follows (Föllmer, 1982) : Let $E(\Xi)$ be the class of functions f on Ξ which satisfy the inequality

$$|f(\xi) - f(\zeta)| \leq \sum_x r[\xi(x),\zeta(x)]\rho_x f \ , \tag{4.12}$$

where $r(\cdot,\cdot)$ is a measurable metric on W and

$$\rho_x f \equiv \sup \left(\frac{|f(\xi)-f(\zeta)|}{r[\xi(x),\zeta(x)]} : \xi=\zeta \text{ off } x \right). \tag{4.13}$$

The probability measure μ on the countable product space (Ξ,F) will be called a Gibbs measure with conditional probabilities $\mu_x(\cdot|\xi), x\in S, \xi\in\Xi$, if the following three conditions hold:

(E6) $\mu_x(\cdot|\xi)$ is a probability measure which does not depend on $\xi(x)$. The corresponding product measure $\mu_x(\cdot|\xi) \otimes \prod_{y\neq x} \varepsilon_{\xi(y)}$ on W will be denoted $\mu_x^*(\cdot|\xi)$.

(E7) For each $f\in E(\Xi)$, the function

$$\xi \to \int f d\mu_x^*(\cdot|\xi)$$

is again in $E(\Xi)$.

(E8) $\int f d\mu = \int [\int f d\mu_x^*(\cdot|\xi)]\mu(d\xi)$, $f\in E(\Xi)$.

Conditions (E6)-(E8) imply that the product measure μ^* is a rcp of μ with respect to the σ-algebra generated by $\xi(y), y\neq x$. Condition (E7) contains an additional continuity requirement.

4.6. The reader certainly noticed that the integral $Z_\Lambda^\psi(x)$ of (4.11) is equivalent to the normalization constant $k(\psi,\lambda)$ in (3.4), also called the partition function, which actually is the total mass of measure defining the particle system. Instead of (4.11) we can define it with the aid of a Hamiltonian H and the product measure $\mu_x^*(\cdot|\xi)$:

$$Z_\Lambda^\psi(x) = \int \exp\{H_\Lambda(\xi)\}\mu_x^*(\cdot|\xi) ,$$

where $H_\Lambda(\xi)=H(\xi(\Lambda)+H(\xi(\Lambda)|\xi(Z^d\setminus\Lambda))$. The above relation makes easy the definition of a limit Gibbs distribution (LGD), a basic concept for an infinite system with interaction (Minlos, 1967a).

Definition 9. The probability distribution P given on space Ξ is a limit Gibbs distribution (LGD) corresponding to the Hamiltonian H, for any finite $\Lambda\subset Z^d$, if

(i) $H(\xi(\Lambda)|\xi(Z^d\setminus\Lambda))$ and $Z_\Lambda^\psi<\infty$ w.p.1 ;

(ii) Under a fixed boundary condition $\gamma=\xi(Z^d\setminus\Lambda)$, the conditional distribution induced by P on Ξ_Λ is absolutely continuous w.p.1 with respect to the product measure μ^*. Its density with respect to μ^* is

$$\phi[\xi(\Lambda)|\gamma] = [Z_\Lambda^\psi(x)]^{-1}\exp\{-H(\xi(\Lambda)-H(\xi(\Lambda)|\gamma\}.$$

(see Sinai, 1982, Def. 1.3, p. 7).

Taking into account the example in 4.3.7, one can define one-dimensional two-state Markov chains as LDGs, when μ is considered as the Bernoulli measure for which each value of $\xi(x)$ has probability 1/2. Moreover, the usual ergodic theorem for Markov chains can be formulated in such a way that it does imply the uniqueness of LGD.

LGDs possess a number of remarkable properties (e.g., ergodicity, strong mixing, regularity : Minlos, 1967b) which will be used in the next paragraph.

4.7. It is well known that the existence of a unique Gibbs measure implies the case when phase separations in a particular system are absent Following a theorem given by R.L.Dobrushin(1968b,Th.6),the necessary and sufficient condition for the uniqueness of a Gibbs distribution is that at least one of the distributions has the property of externally uniform mixing (i.e.Condition (C5)),that is,

$$\phi(r) \leq \phi_1(r),$$

where $\phi_1 = \phi_{V_1}$, for fixed V_1 and every finite V_2. The necessary and sufficient condition for the uniqueness of a LGD is that all Gibbs measures satisfy (C7). As it was stated in 2.4.1, a unique Gibbs measure is metrically transitive.

The main statement can be formulated as follows : If ψ is any interaction of finite range such that $\|\psi\| < [2e]^{-1}$ (e:the base of natural logarithms) and μ^* any product measure, then there is only and only one Gibbs measure relative to the interaction ψ and the product measure μ^* (Albeverio et al.,1981,Th.2.1).

The interaction matrix $\underset{\sim}{A} = (a_{xy})$ has the components

$$a_{xy} \equiv \frac{1}{2} \sup \left(\| \mu_x(\cdot|\xi) - \mu_x(\cdot|\zeta) \| \right),$$

where sup is taken over all configurations $\xi, \zeta \in \Xi$ which coincide everywhere except in x ($\xi = \zeta$ off x - as in (4.13)),and $\|\cdot\|$ denotes the total variation norm. The uniqueness condition (the uniform Dobrushin-Vassershtein condition : Föllmer,1979) is

$$\text{(E9)} \qquad a \equiv \sup_y \sum_* a_{xy} < 1 ,$$

which implies

$$\lim_j \sum_* a^j_{xy} = 0 \quad (j \geq 0\text{:the j-th power of } a_{xy},\ y \in S).$$

The above relationship guarantess that μ is uniquely determined by its conditional probabilities. The coefficient a_{xy} measures the influence of site x on the conditional probabilities at site y.

The uniqueness of a Gibbs state implies that there is only one invariant equilibrium state and hence that all observables converge in probability to constants in the corresponding systems. Actually,every equilibrium state is a weak limit of convex combinations of states which are approximable by unique equilibrium states (see Lanford III, 1973).

Conditions for non-uniqueness were given by R.L.Dobrushin(1968c,§4);

they should be important for the study of homogeneous systems of cells where inhomogeneities (e.g.,change of cell phenotype) appear following change of internal parameters or external perturbations. For instance, a one-parameter family $P_\beta(\beta\to\infty)$ of LGDs describes small local distortions of a given configuration $\xi\in\Xi$ if at large values of β the random set $\{x\in Z^d:\eta(x)\neq\xi(x)\}$ decomposes w.p.1 into a countable union of finite connected subsets of Z^d and

$$\lim_{\beta\to\infty}\sup_{x} P_\beta\{\eta(V_r(x))\neq\xi(V_r(x))\} = 0 ,$$

for each $r>0$, where $V_r(x)=\{y\in Z^d : |x-y|\leq r\}, V\in Z^d$. The basic concept is that of the ground state of a Hamiltonian (Sinai,1982).

4.7.1. The uniqueness of Gibbs RFs in the non-compact case was proved by R.L.Dobrushin and E.A.Pecherski(1983).

4.7.2.(Note 1) The equivalence between Gibbs RFs and Markov RFs is a well known problem; however,there are MRFs which are not Gibbsian (Moussouris,1974). The definition of local and global Markov properties can be completely given with the aid of Gibbs measures relative to inter action ψ (Albeverio et al.,1981,Th.3.2 and 3.3). Particle systems at equilibrium are specified by Gibbs states satisfying certain conditions; equivalences were found between these Gibbs states and DLR (:Dobrushin--Lanford-Ruelle)-measures (Cassandro et al.,1978),or KMS (:Kubo-Martin--Schwinger)-states (Aizenman et al.,1977). [Notice that a state can be defined as a state on the B^*-algebra of all complex continuous functions on Ξ :Lanford and Ruelle,1967.] It is already proved that an equilibrium measure for a local specification (i.e. an invariant measure verifying certain condition) is a Gibbs measure with respect to that specification (see Moulin Ollagnier and Pinchon,1981,Th.3).

4.7.3. The set $G(\Pi)$ consisting of all Gibbs measures for a given specification is non-empty. For the vague topology,it is a convex and compact subset of the set M_+ of all Radon probability measures on Ξ (see 3.1.1). G is a simplex if each $P_\mu\in G$ is a barycentre [in Meyer,1966: $b(\mu)$] of only and only one probability measure μ such that

$$P_\mu(A) = \int_E P(A)\mu(dP) , \qquad (4.14)$$

where $E=E(\Pi)$ is the (finite) set of extreme points of G. The formula (4.14) establishes a 1-1 correspondence between G and the set of all probability measures on E (Dynkin,1978). The reader is also referred to G.Choquet(1969,II,§28,particularly Corollary 28.5),P.A.Meyer(1966, Ch.XI),K.Urbanik(1975),D.Ruelle(1978,Th.4),etc.

Any state has a unique integral representation in terms of the extremal states.

4.7.4.(Note 2) Condition (E9) is used for proving the a.s. convergence of bounded martingales over an RF. This convergence is ensured if the RF satisfies (C2)- [see Appendix,Condition (F4)]. In the two-dimensional case,there is some diagonal interaction between F_x^1 and F_x^2 which does not pass through the splitting field F_x [see Appendix].
If this interaction becomes too strong,then one would expect that a.s. convergence fails. Possibly,the breakdown of martingale convergence is one of the various critical phenomena caused by strong spatial interactions (Föllmer,1984). However,there are cases of convergence failure which are not accompanied by phase transition. It was already noticed that a large part of Gibbs RF-theory is devoted to the study of RFs with weak interaction between particles close to the fields of independent random variables (Dobrushin and Sinai,1980,p.72).

5.Cell systems as random fields

5.1. Let us start with a semantic distinction. The suggestion of an "embryonal field" goes back to H.Driesch(1892) who considered it as a system of coordinates within the fate of the parts of an embryo is realized as a function of the position of these parts. In order to illustrate his concept of "harmonious equipotential system" , H.Driesch used the analogy with a magnetic field - which,despite the naive interpretation,might be viewed as a premonition to the ferromagnetic spin model suggested by Ernst Ising thirty-three years later.

The term "field" is generally used by cell biologists to specify the co-ordinated and integrated character of the sequence of complex processes (e.g.,cell-cell interactions,multiplication,movement,differentiation,etc.) implicated in the development of a cell system. Following A.Robertson and M.Cohen(1972),a field is "a collection of functionally coupled cells,the development of which is under control of a single system", and whose size should be not more than 100 cells. For example,the cells of the first embryonic stages form a single field but at later stages they will be organized in many fields. One important characteristic is that the components within the same field interact locally,but interaction across boundaries of a field does not occur.

It would be impetuous to find equivalences between the theoretic-biological concept of "field" and the presently discussed one of "random field" . The purpose of this paper is to suggest the application of this mathematical theory for improving the biological one in order to better understand and explain some phenomena arising in cell systems. It is known that for C.H.Waddington(1972) any concept of "field" was only a "descriptive convenience" but not a unifying paradigm. Indded,

> "...if one thinks of some examples of epigenetic fields,such as those of the pentadactyl limb,the dogs' skulls or fish-shapes which D'Arcy Thompson put into his coordinate nets,the Drosophila scutellar bristles,or the regenerating Hydra,not only are the geometries all different,but one cannot help asking oneself whether perhaps,one is a diffusion field of some substance or substances, another a phase-shift fiels à la Cohen-Goodwin,another perhaps produced by interaction between growth centres,and so on"(p.143).

In his reply to R.Thom's critical comments,C.H.Waddington pointed out that saying that "a developmental performance involves a field is as important as saying that a collection of words is a sentence : but just as we need to understand the grammar which generated the sentence,so we need to enquire what generated the field".

It would be plausible to think that in our case "the grammar which generates the sentence" might be space structure,neighbourhood topology and local interaction. Obviously,a universal mathematical model for all morphogenetic phenomena does not exist. By definition,a mathematical model is a theoretical construction by means of which we try to understand and explain some quested aspects of an empirical process..

> "No model can simultaneously optimize generality,realism and precision...(they) differ in the aspect of the reality preserved, in the departure from reality,and their manipulative possibilities" (Levins,1970).

Then,our problem can be formulated as follows : what can tell the RFs-theory to theoretical biologists ? Specifically,we have to replace the trivial statement "things are different,interconnected,and changing" by a mathematical theory that tells us definitely <u>which</u> things differ in <u>what</u> ways,<u>how</u> they interact,and in <u>what</u> direction they change.

First,one may think of about the application of RFs-theory in developmental biology as a stochastic alternative to Turing model and its actual variants. As it was recently said,the majority of patterns of interest to embryologists are structural rather than chemical (Harris et al.,1984). The mechanical instability described by A.Harris and his colleagues can be positively represented in terms of stochastic mechanics,so that one might adjust the statement of Martin Garstens(1970) by saying that the central problem of modern theoretical biology is to supply the missing links between purely descriptive approach to the field and the powerful stochastic models inspired by modern physics.

The second point is a consequence of the first one:we can transcend quite naturally the explanation in physical terms by regarding RFs as stochastic processes having their own language and grammar. For example, we can prove that near bifurcations the law of large numbers is no longer valid and the uniqueness of the solution of the linear master equation is lost. (The reader is referred to I.Prigogine,1980,p.134 : "near a bifurcation point,nature always finds some clever way to avoid the consequences of the law of large numbers through an appropriate nucleation process".) Obviously,a bifurcation point is always a phase transition. Also,it is known that in the vicinity of bifurcation a particle system becomes extremely sensitive to small external factors (e.g.,structural irregularities,environmental influence,etc.). In an infinite particle system the occurrence of phase transitions can be interpreted as much as the violation of the law of large numbers or as a non-ergodic behaviour. Maybe in the first case the proof is more intuitive (see Dobrushin, 1965,1967). [The alternative is based on the remark that the celebrated Birkhoff ergodic theorem(1931) is actually a theorem of the type of a strengthened law of large numbers,with the mention that it is more informative than the strong law : Krengel,1985.] The probabilistic approach will suggest not only the ways the cellular configurations may change,the influences at the boundary,the dynamics at the interface,or the different kinds of interaction,but also the role of space structure, the local scenery,and the significance of molecular-macroscopic scales.

5.2. This paragraph contains some simple examples of RFs for cell systems with defined sets S and W.

Example 1. $S=Z^d, d\geq 1, W=\{0,1\}$. The RF$\{\xi(x)\}, x\in Z^d$,describes the local spread of a cell system growing on a solid substrate with regular geometric structure. A site $x\in Z^d$ can be either vacant ($\xi(x)=0$) or occupied ($\xi(x)=1$). If the cells in the system multiply,interaction is assumed as competition for vacant sites. The fibroblasts cultivated by A.Harris et al.(1984) move on an interlocking network of polygons created by collagen fibrils and form aggregates;the size and spacing of these aggregates vary as a function of the original population density. It is impossible to ignore the analogy with the nucleation and grain growth studied in metallurgy which lead to the random division of space (Cahn,1972).

Note:Discussing the above mentioned paper,J.Hammersley pointed out that the growth of a "perfect" crystal occurs in a three-dimensional space with different distributions in the horizontal and vertical scale. He mentioned in this context the Eden model(1961).

Example 2. $S=R^d, d\geq 1, W=(0,1)^k, k\geq 1$. We deal with a k-type cell system in culture. This is also a model for two-component bilayer biological membrane. Phase transitions in such biomolecular systems are intensely studied (see,e.g.,Scott,1974;Pink,1984). Another possible model is a speculative one:it is suggested by the hypothesis that in higher eukaryotes extragenic DNA sequences of defined length are necessary for the activity of transcriptional units (Naora and Deacon,1982). This "territorial effect" can be adequately represented in a RF-model with spatial interaction. Phase transitions might be of enormous evolutionary consequences. It has also been hypothesized that in the genome there are regulatory elements whose positions and sequences influence developmental pathways in complex and co-ordinated fashions (Rose and Doolittle,1983). This genome resetting may be one of the molecular mechanisms of speciation.

Example 3. $S=Z^d, d\geq 1, W=(0,\pm 1)$. The corresponding RF represents the spatial evolution of a cell system whose components can change properties (e.g.:cells in a resting state -1).

Example 4. $S=Z^d, d\geq 1, W=\{1,\ldots,m\}\in N$. We deal with a cell system whose components change their internal states (e.g.,cycling cells with state 1:G_1-phase,...,state m:mitosis). The cell replication imposes competition for vacant places,if one assumes that a site can be occupied by only one cell at a time (exclusion of multiple occupancy).

Example 5. $S=Z^d, d\geq 1, W=R^1_+$. As the above but the cell cycle is interpreted as a continuum.

Example 6. $S=Z^d, d\geq 1, W=M\times K$, $M=\{1,\ldots,m\}, K=\{1,\ldots,k\}$. This seem to be a more realistic model for different morphogenetic cell systems. The cells run the cycle M and choose their type ("colour") in the set K,so that $\xi_{ij}(x)$ means that at site $x\in Z^d$ there is a cell of type $i\in K$ in state $j\in M$. Cellular activities as euplasia,pro- and retroplasia,as well as neoplasia can be described in the framework of the RFs theory.

In order to summarize the possible choices of sets S and W,the following picture is suggested :

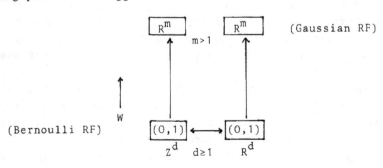

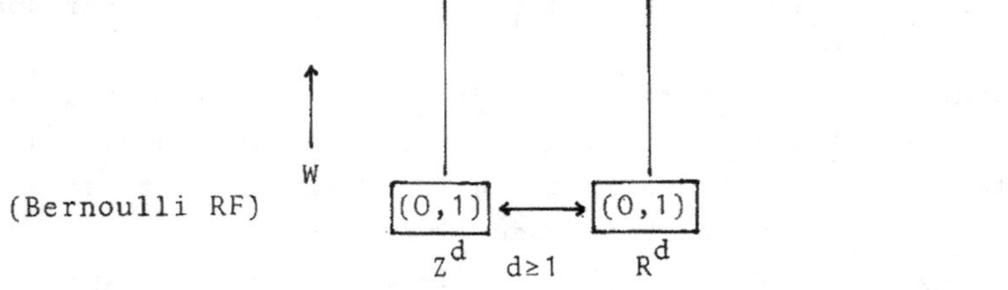

5.3. In this paragraph some examples of different cell interactions are given as the complementary experimental part to Paragraph 4.3. Generally speaking, direct cell-cell interactions are mostly realized by intercellular communication, that is, by exchange of ions or small/large molecules. Physically, this communication is realized by cell junctions, e.g., desmosomes, tight and gap junctions. Their spatial disposition plays a certain role since cells cultures in a one-dimensional space show an extensive intercellular communication such that all the cells may apparently share a common metabolite pool (see, e.g., Pitts, 1972).

In general, intercellular communication is required for many cell activities: (i) synchronization of contraction (or other mechanical behaviour) within the tissue, (ii) metabolic coordination, (iii) regulation of enzyme systems, (iv) proliferation and differentiation, (v) growth control, etc. The "competence" for communication may vary in different situations and alter the neighbourhood relationships. For example, some differentiated cells can establish communication with other highly differentiated, neighbouring cells but not with embryonic non-differentiated ones (Lo and Gilula, 1980). Also, cells being on the point to differentiate into malignant cells lose (reversibly) their communication competence (with surrounding normal cells). It appears that at least in the case of monolayer cultures small groups of neoplastic cells communicate with their normal neighbours; if the malignant cells are allowed to develop in large colonies, they form "islands" into the "sea" of surrounding normal cells. This is clearly a symmetry-breaking phenomenon appearing as a consequence of irreversibility.

This kind of "insulation" is not unusual : for example, rat pancreas cells of type B, A, D and PP establish selective contacts, having as consequence the creation of small B-type islands. There are some other examples of one-way (directed) communication as the flux of nucleotides from embryonic carcinoma cells to normally differentiated cells. A "probability model" of selective intercellular communication (Gaunt and Subak-Sharpe, 1979) is based on the (non-specific) assumption that a sequential process of gap junction formation takes place between homotypic cells - with a "greater probability per unit time" - than between heterotypic cells.

Because of different forms of interaction it is not easy to find a simple form for defining function ψ. As F. Spitzer (1969) already noticed, the problem of trying to describe all possible interactions "seems far from easy". Even in the simple cases it is not known what are "all the possible interactions". Suppose, for example, that in a cell system there are only <u>attractive</u> or <u>repulsive</u> interactions. Then the first parametrized interaction function presented in §4.3.2 as a catastrophic one, provides a variety of types, e.g., attractive interaction for $a>1$, and repulsive for $0\leq a<1$, or Poisson for $a=1$. Similarly, the second function has also both ranges of repulsive and attractive interactions, except the case $a=0$ when it represents the purely repulsive interaction.

5.4. The formation of new types of cells may be conceived as the result of apparition of critical phenomena in the system. The occurrence of phase transitions corresponds to the existence of many extremal Gibbs states. Their existence is substantiated by the following basic statement :

(*) Consider the probability space (Ξ,F,P) and let F be countably generated (see §4.1). Then the corresponding RF has a trivial tail field

$$F_{\infty} = \cap F_{\Lambda}$$

iff the field has short-range correlations. Then $\mu\in G(\Pi)$ is an extreme point of $E(\Pi)$. (See Preston, 1976, Th. 2.2)

If we intend to define them with respect to the Hamiltonian H (Sinai, 1982, p.22), let $G(H)$ be the set of LGDs for a given Hamiltonian H, and $G^{o}_{\Lambda}(H)$ be the set of finite convex combinations of conditional Gibbs distributions in a finite domain $\Lambda\in S_{o}$ with different boundary conditions. Also, let $G_{\Lambda}(H)$ be the closure of $G^{o}_{\Lambda}(H)$ in the weak topology of probability measures. Under some natural regularity conditions on ψ, $G(H)$ has the following structure :

(1) $G(H)$ is a non-empty, convex, compact set of probability distributions.

$$G(H) = \cap G_{\Lambda} \ , \ \Lambda\in S_{o} \ .$$

(2) $G(H)$ is a Choquet simplex.

(3) The extremal points of $G(H)$ are mutually singular; they are called indecomposable LGDs.

(4) $P\in G(H)$ iff there exist sequences $\Lambda_i\subset Z^d$ and $P_i\in G^{o}_{\Lambda}(H)$ such that $\Lambda_i\to\infty$ and $P=\lim P_i$.

(5) $P\in G(H)$ is indecomposable iff it is regular (in the sense of (C7)),

$$\lim_{j} |P(A\cap B_j) - P(A)P(B_j)| = 0 \ ,$$

whenever $B_i\in F_i=\sigma(\xi(x)|x\in Z^d\setminus\Lambda_i), \Lambda_i\to\infty$.

Appendix: On two-parameter martingales

Let (Ω,F,P) be a complete probability space and $S=R^2_+$ be a partially ordered set. In order to avoid subscripts, let us write $(a,b)\leq(c,d)$ iff

$a \le c$ and $b \le d$, and denote $y=(a,b)$, $z=(c,d)$ and $S_o==\{(a,b)\in R_+^2 : a=0 \text{ or } b=0\}$. Consider a filtration $\{F_x, x\in R_+^2\}$ of sub-σ-fields of F satisfying the following properties:

(F1) F_{oo} includes all the P-negligible sets of F (completeness).

(F2) F_x increases with respect to the partial order on R_+^2 (i.e. $F_x \supseteq F_y$, if $x \le y$).

(F3) F_x is right-continuous.

(F4) $\forall x=(i,j)\in R_+^2$, $F_x^1 \perp\!\!\!\perp F_x^2 | F_x$; $F_x^1 = \bigvee_{u\ge 0} F_{iu}$, $F_x^2 = \bigvee_{v\ge 0} F_{vj}$.

Clearly, (F4) is equivalent to (C2). The superscripts of F in (F4) help us to define 1- and 2-martingales; a strong martingale is both a 1- and a 2-martingale, so that (F4) will not be assumed in that case (Walsh, 1979).

REFERENCES

ADLER,R.(1981) The Geometry of Random Fields. New York:Wiley

ADLER,R.,FEIGIN,P.D.(1984) On the cadlaguity of random measures. Ann. Probab.12,615-630

ADLER,R.,MONRAD,D.,SCISSORS,R.H.,WILSON,R.J.(1983) Representations,decompositions,and sample function continuity of random fields with independent increments. Stoch.Proc.Appl.15,3-30

AIZENMAN,M.,GOLDSTEIN,S.,GRUBERG,C.,LEBOWITZ,J.L.,MARTIN,P.(1977) On the equivalence between KMS-states and equilibrium states for classical systems. Commun.Math.Phys.,53,209-220

ALBEVERIO,S.,HØEGH-KROHN,R.(1984) Local and global Markoff fields. Rep. Math.Phys.,19,225-248

ALBEVERIO,S.,HØEGH-KROHN,R.,OLSEN,G.(1981) The global Markov property J.Multivar.Anal.,11,599-607

ANDERSON,R.M.(1982) Star-finite representations of measure spaces. Trans. Amer.Math.Soc.,271,667-687

BASS,R.F.,PYKE,R.(1984a) The existence of set-indexed Lévy processes. Z.Wahrscheinlichkeitstheorie verw.Geb.,66,157-172 ; (1984b) Functional law of the iterated logarithm and uniform central limit theorem for partial-sum processes indexed by sets. Ann.Probab.,12, 13-34

BASS,R.F.,PYKE,R.(1985) The space $\mathcal{D}(A)$ and weak convergence for set-indexed processes. Ann.Probab.,13,860-884

BICKEL,P.J.,WICHURA,M.J.(1971) Convergence criteria for multiparameter stochastic processes and some applications. Ann.Math.Statist.,42, 1656-1670

BLACKWELL,D.,RYLL-NARDZEWSKI,C.(1963) Non-existence of everywhere proper conditional distributions. Ann.Math.Statist.,34,223-225

BLACKWELL,D.,DUBINS,L.E.(1975) On existence and non-existence proper regular conditional distributions. Ann.Probab.,3,741-752

BOLTHAUSEN,E.(1982) On the central limit theorem for stationary mixing random fields. Ann.Probab.,10,1047-1050

BRADLEY,R.C.(1984) Some remarks on strong mixing conditions. Proc.7th Conf.Probability Theory,Brasov 1982,pp.65-72. Bucuresti:Ed.Academiei

CAHN,J.W.(1972) The generation and characterization of shape. Adv.Appl. Probab.,4 (Suppl.),221-242

CASSANDRO,M.,OLIVIERI,E.,PELLEGRINOTTI,A.,PRESUTTI,E.(1978) Existence and uniqueness of DLR measures for unbounded spin systems. Z.Wahrscheinlichkeitstheorie verw.Geb.,41,313-334

CHERNAVSKII,D.S.,EIDUS,V.L.,POLEZHAEV,A.A.(1981) On kinetics of phase transitions in cell membranes. BioSystems 13,171-179

CHOQUET,G.(1969) Lectures on Analysis.Vol.II.Representation Theory. Reading(Mass.):W.A.Benjamin

CHOW,Y.S.,TEICHER,H.(1978) Probability Theory.Independence,Interchangeability,Martingales. New York-Heidelberg-Berlin:Springer

COX,T.(1977) An example of phase transition in countable one-dimensional Markov random fields. J.Appl.Probab.,14,205-211

COX,T.(1979) An alternate proof of a theorem of Kesten concerning Markov random fields. Ann.Probab.,7,377-378

CURTIS,A.S.G.(1978) Cell-cell recognition:Positioning and patterning systems. Symp.Soc.Exper.Biol.,32,51-82. Cambridge:Cambridge Univ. Press

DALETSKII,Y.L.,SMOLYANOV,O.G.(1984) On the weak sequential completeness of the spaces of Radon measures. Theor.Probab.Appl.,29,142-147

DANG-NGOC,N.,YOR,M.(1978) Champs markoviens et mesures de Gibbs sur R. Ann.Sci.Ecole Norm.Sup.,11,29-69

DARST,R.B.(1971) On universal measurability and perfect probability. Ann Math.Statist.,42,352-354

DeHOFF,R.T.(1972) The evolution of particulate structures. Adv.Appl. Probab.,4 (Suppl.),188-198

DIEUDONNÉ,J.(1970) Treatise on Analysis.Vol.I. New York:Academic Press

DOBRUSHIN,R.L.(1965) Existence of a phase transition in two-dimensional and three-dimensional Ising models. Theor.Probab.Appl.,10,193-213

DOBRUSHIN,R.L.(1967) Existence of phase transitions in models of a lattice gas. Proc.5th Berkeley Symp.Math.Statist.Probability,Vol.III, pp.73-87. Berkeley:Univ.California Press

DOBRUSHIN,R.L.(1968a) The description of a random field by means of conditional probabilities and conditions of its regularity. Theor. Probab.Appl.,13,197-224 ; (1984b) Gibbsian random fields for lattice systems with pairwise interactions. Funct.Anal.Appl.,2,292--301 ; (1968c) The problem of uniqueness of a Gibbsian random field and the problem of phase transitions. Funct.Anal.Appl.,2, 302-312

DOBRUSHIN,R.L.,MAJOR,P.(1981) On the asymptotic behavior of some self--similar random fields. Sel.Math.Sov.,1,265-291

DOBRUSHIN,R.L.,PECHERSKI,E.A.(1983) A criterion of the uniqueness of Gibbsian fields in the non-compact case. Proc.4th USSR-Japan Symp.(Lecture Notes in Math.,Vol.1021),pp.97-110. Berlin-Heidelberg-New York:Springer

DOBRUSHIN,R.L.,SINAI,Y,G.(1980) Mathematical problems in statistical mechanics. Math.Physical Rev.,1,55-106.

DÖHLER,R.(1980) On the conditional independence of random events. Theor. Probab.Appl.,25,628-634

DOOB,J.L.(1953) Stochastic Processes. New York:Wiley

DUDLEY,R.M.(1973) Sample functions of the Gaussian process. Ann.Probab. 1,66-103

DYNKIN,E.B.(1978) Sufficient statistics and extreme points. Ann.Probab. 6,705-730

DYNKIN,E.B.(1980) Markov processes and random fields. Bull.Amer.Math. Soc.,3,975-999

FISHER,M.E.,RUELLE,D.(1966) The stability of many-particle systems. J.Math.Phys.,7,260-270

FLOOD,R.G.,SULLIVAN,W.G.(1980) Consistency of random field specification Z.Wahrscheinlichkeitstheorie verw.Geb.,53,147-156

FÖLLMER,H.(1975) Phase transition and Markov boundary. Lecture Notes in Math.,Vol.465(Séminaire de Probabilités,IX),pp.305-317. Berlin-Heidelberg-New York:Springer

FÖLLMER,H.(1979) Macroscopic convergence of Markov chains on infinite product spaces. In:Random Fields.Rigorous Results in Statistical Mechanics and Quantum Field Theory (J.Fritz,J.L.Lebowitz,D.Szász eds.),pp.363-371. Amsterdam:North-Holland

FÖLLMER,H.(1980) On the global Markov property. In:Quantum Fields,Algebras,Processes (L.Streit ed.),pp.293-302. Wien-New York:Springer

FÖLLMER,H.(1982) A covariance estimate for Gibbs measures. J.Funct.Anal. 46,387-395

FÖLLMER,H.(1984) Almost sure convergence of multiparameter martingales for Markov random fields. Ann.Probab.,12,133-140

GARSTENS,M.A.(1970) Remarks on statistical mechanics and theoretical biology. In:Towards a Theoretical Biology (C.H.Waddington ed.), Vol.3,pp.167-173. Edinburgh:Edinburgh Univ.Press

GAUNT,S.J.,SUBAK-SHARPE,J.H.(1979) Selectivity in metabolic cooperation between cultured mammalian cells. Exper.Cell Res.,120,307-320

GLÖTZL,E.(1978) Gibbsian description of point processes. In:Point Processes and Queuing Problems (P.Bártfai,J.Tomkó eds.),pp.69-84. Amsterdam:North-Holland

GOODWIN,B.C.(1971) A model of early amphibian development. Symp.Soc. Exp.Biol.,25,417-428. Cambridge:Cambridge Univ.Press

GROBSTEIN,C.(1956) Inductive tissue interaction in development.Adv. Cancer Res.,4,187-236

GUT,A.(1978) Marcinkiewicz laws and convergence rates in the law of large numbers for random variables with multidimensional indices. Ann.Probab.,6,469-482

GUT,A.(1979) Moments of the maximum of normed partial sums of random variables with multidimensional indices. Z.Wahrscheinlichkeitstheorie verw.Geb.,46,205-220

HAMMERSLEY,J.M.(1972) Stochastic models for the distribution of particles in space. Adv.Appl.Probab.,4,(Suppl.),47-68

HAMMERSLEY,J.M.,MAZZARINO,G.(1983) Markov fields,correlated percolation, and the Ising model. In:The Mathematics and Physics of Disordered Media (B.D.Hughes,B.W.Ninham eds.),pp.201-245. Berlin-Heidelberg-New York-Tokyo:Springer

HARRIS,A.K.,STOPAK,D.,WARNER,P.(1984) Generation of spatially periodic patterns by a mechanical instability:A mechanical alternative to the Turing model. J.Embryol.Exp.Morphol.,80,1-20

HEGERFELDT,G.C.,NAPPI,C.R.(1977) Mixing properties in lattice systems. Commun.Math.Phys.,53,1-7

HELMS,L.L.(1983) Hyperfinite spin models. In:Nonstandard Analysis-Recent Developments (A.E.Hurd ed.),pp.15-26. Berlin-Heidelberg-New York-Tokyo:Springer

HELMS,L.L.,LOEB,P.A.(1979) Applications of nonstandard analysis to spin models. J.Math.Anal.Appl.,69,341-352

HIGUCHI,Y.(1977) Remarks on the limiting Gibbs states on a (d+1)-tree. Publ.RIMS Kyoto Univ.,13,335-348

HURD,A.E.(1981) Nonstandard analysis and lattice statistical mechanics: A variational principle. Trans.Amer.Math.Soc.,263,89-110

IBRAGIMOV,I.A.,ROZANOV,Y.A.(1978) Gaussian Random Processes. New York-Heidelberg-Berlin:Springer

IVANOFF,G.(1980) The branching random field. Adv.Appl.Probab.,12,825-847

KALLIANPUR,G.,MANDREKAR,V.(1974) The Markov property for generalized Gaussian random fields. Ann.Inst.Fourier 24,143-167

KARR,A.F.(1978) Lévy random measures. Ann.Probab.,6,57-71 (Correction: 1979,7,1098)

KARR,A.F.(1979) Classical limit theorems for measure-valued Markov processes. J.Multivar.Anal.,9,234-247

KESTEN,H.(1976) Existence and uniqueness of countable one-dimensional Markov random fields. Ann.Probab.,4,557-569

KNIGHT,F.B.(1970) A remark on Markovian germ fields. Z.Wahrscheinlichkeitstheorie verw.Geb.,15,291-296

KNIGHT,F.B.(1979) Prediction processes and an autonomous germ-Markov property. Ann.Probab.,7,385-405
KOREZLIOGLU,H.,MAZZIOTTO,G.,SZPIRGLAS,J.(eds.):Processus Aléatoires à Deux Indices (Lect.Notes in Math.,Vol.863). Berlin-Heidelberg-New York:Springer
KOTANI,S.(1973) On a Markov property for stationary Gaussian processes with a multidimensional parameter. Proc.2nd Japan-USSR Symp.on Probab.Theory (Lect.Notes in Math.,Vol.330),pp.239-250. Berlin-Heidelberg-New York:Springer
KRENGEL,U.(1985) Ergodic Theorems. Berlin-New York:de Gruyter
KUZNETSOV,S.E.(1984) Specifications and a stopping theorem for random fields. Theor.Probab.Appl.,29,66-78
LANFORD III,O.E.(1973) Entropy and equilibrium states in classical statistical mechanics. Lect.Notes in Physics,Vol.20,pp.1-113. Berlin-Heidelberg-New York:Springer
LANFORD III,O.E.,RUELLE,D.(1967) Integral representations of invariant states on B*algebras. J.Math.Phys.,8,1460-1463
LEVENSON,R.,HOUSMAN,D.(1981) Commitment:How do cells make the decision to differentiate ? Cell 25,5-6
LEVINS,R.(1970) Complex systems. In:Towards a Theoretical Biology (C.H. Waddington ed.),Vol.3,pp.73-88. Edinburgh:Edinburgh Univ.Press
LLOYD,S.P.(1962) On a measure of stochastic dependence. Theor.Probab. Appl.,7,301-312
LO,C.W.,GILULA,N.B.(1980) PCC4azal teratocarcinoma stem cell differentiation in culture.III.Cell-to-cell communication properties. Devel.Biol.,75,112-120
LOÈVE,M.(1973) Paul Lévy,1886-1971. Ann.Probab.,1,1-8
MANDREKAR,V.(1976) Germ-field Markov property for multiparameter processes. Lect.Notes in Math.,Vol.511(Sémin.de Probabilités X),pp. 78-85. Berlin-Heidelberg-New York:Springer
MANDREKAR,V.(1983) Markov properties for random fields. In:Probabilistic Analysis and Related Topics (A.T.Bharucha-Reid ed.),Vol.3, pp.161-193. New York:Academic Press
MARUYAMA,G.(1970) Infinitely divisible processes. Theor.Probab.Appl., 15,1-22
McKEAN(1963) Brownian motion with a several-dimensional time. Theor. Probab.Appl.,8,335-354
MEYER,P.A.(1966) Probability and Potentials. Waltham(Mass.):Blaisdell
MINLOS,R.A.(1967a) Limiting Gibbs distributions. Funct.Anal.Appl.,1, 140-150 ; (1967b) Regularity of the Gibbs limit distributions. Funct.Anal.Appl.,1,206-217
MITTENTHAL,J.E.(1981) The rule of normal neighbors:A hypothesis for morphogenetic pattern regulation. Devel.Biol.,88,15-26
MIYAMOTO,M.(1982) Spitzr's Markov chains with measurable potentials. J.Math.Kyoto Univ.,22,41-69
MOULIN OLLAGNIER,J.M.(1985) Ergodic Theory and Statistical Mechanics. (Lect.Notes in Math.,Vol.1115) Berlin-Heidelberg-New York-Tokyo: Springer
MOULIN OLLAGNIER,J.,PINCHON,D.(1981) Mesures de Gibbs invariantes et mesures d'equilibre. Z.Wahrscheinlichkeitstheorie verw.Geb.,55, 11-23
MOUSSOURIS,J.(1974) Gibbs and Markov random systems with constraints. J.Statist.Phys.,10,11-33
NAORA,H.,DEACON,N.J.(1982) Implication of the effect of extragenic territorial DNA sequences on a mechanism involving switch-onn/off of neighbouring genes by transposable elements in eukaryotes. J.Theor.Biol.,95,601-606
NEADERHOUSER,C.C.(1980) Convergence of block spins defined by a random field. J.Statist.Phys.,22,673-684
NGUYEN,X.X.,ZESSIN,H.(1979) Integral and differential characterizations of the Gibbs process. Math.Nachr.,88,105-115

OGATA,Y.,TANEMURA,M.(1981) Estimation of interaction potentials of spatial point patterns through the maximum likelihood procedure. Ann.Inst.Statist.Math.,33,315-338 ; Likelihood analysis of spatial point patterns. J.Roy.Statist.Soc.Ser.B,46,496-518

ODELL,G.M.(1984) A mathematically modelled cytogel cortex exhibits periodic Ca^{++}-modulated contraction cycles seen in Physarum shuttle streaming. J.Embryol.Exp.Morphol.,83(Suppl.),261-287

ODELL,G.M.,OSTER,G.,ALBERCH,P.,BURNSIDE,B.(1981) The mechanical basis of morphogenesis.I.Epithelial folding and invagination. Devel.Biol. 85,446-462

OKABE,Y.(1973) On a Markovian property of Gaussian processes. Proc.2nd Japan-USSR Symp.Probab.Theory (Lect.Notes in Math.,Vol.330),pp. 340-354. Berlin-Heidelberg-New York:Springer

OKADA,T.S.(1980) Cellular metaplasia or transdifferentiation as a model for retinal cell differentiation. Curr.Topics in Devel.Biol.,16, 349-380

PAPANGELOU,F.(1983) Stationary one-dimensional Markov random fields with a continuous state space. In:Probability,Statistics and Analysis (J.F.C.Kingman,G.E.H.Reuter eds.),pp.199-218. Cambridge:Cambridge Univ.Press

PARTHASARATHY,K.R.(1967) Probability Measures on Metric Spaces. New York:Academic Press

PELIGRAD,M.(1981) An invariance principle for dependent random variables. Z.Wahrscheinlichkeitstheorie verw.Geb.,57,495-507 ; (1985) An invariance principle for ϕ-mixing sequences. Ann.Probab.,13,1304-1313

PINK,D.A.(1984) Theoretical models for monolayers,bilayers,and biological membranes. In:Biomembrane Structure and Function (D.Chapman ed.),pp.319-354. Weinheim:Verlag Chemie

PINK,D.A.,CHAPMAN,D.(1979) Protein-lipid interactions in bilayer membranes:A lattice model. Proc.Natl.Acad.Sci.USA,76,1542-1546

PITT,L.D.(1971) A Markov property for Gaussian processes with a multidimensional parameter. Arch.Rational Mech.Anal.,43,367-391

PITTS,J.D.(1972) Direct interaction between animal cells. In:Cell Interactions (L.G.Silvestri ed.),pp.277-285. Amsterdam:North-Holland

PRESTON,C.J.(1974) Gibbs States on Countable Sets. London:Cambridge Univ. Press

PRESTON,C.J.(1975) Spatial birth-and-death processes. Bull.Intern.Stat. Inst.,46(book 2),371-391

PRESTON,C.J.(1976) Random Fields.(Lect.Notes in Math.,Vol.534) Berlin-Heidelberg-New York:Springer

PRESTON,C.J.(1980) Construction of specifications. In:Quantum Fields, Algebras,Processes (L.Streit ed.),pp.268-292. Wien-New York: Springer

PRIGOGINE,I.(1980) From Being to Becoming. San Francisco:W.H.Freeman

PYKE,R.(1973) Partial sums of matrix arrays and Brownian shhets. In: Stochastic Analysis (D.G.Kendall,E.F.Harding eds.),pp.331-348. London:Wiley

PYKE,R.(1983) A uniform central limit theorem for partial-sum processes indexed by sets. In:Probability,Statistics and Analysis (J.F.C. Kingman,G.E.H.Reuter eds.),pp.219-240. Cambridge:Cambridge Univ. Press

RAMACHANDRAN,D.(1981) A note on regular conditional probabilities in Doob's sense. Ann.Probab.,9,907-908

READY,D.F.,HANSON,T.E.,BENZER,S.(1976) Development of the Drosophila retina,a neurocrystalline lattice. Devel.Biol.,53,217-240

ROBERTSON,A.,COHEN,M.H.(1972) Control of developing fields. Ann.Rev. Biophys.Bioeng.,1,409-464

RÖCKNER,M.(1983) Markov property of generalized fields and axiomatic potential theory. Math.Ann.,264,153-177 ; (1985) Generalized Markov fields and Dirichlet forms. Acta Applicandae Mathematicae 3,285-311

ROSE,M.R.,DOOLITTLE,W.F.(1983) Molecular biological mechanisms of speciation. Science 220,157-162
ROSENBLATT,M.(1979) Some remarks on a mixing condition. Ann.Probab.,7, 170-172
ROSENSTRAUS,M.J.,SPADORO,J.P.,NILSSON,J.(1983) Cell position regulates endodermal differentiation in embryonal carcinoma cell aggregates. Devel.Biol.,98,110-116
ROZANOV,Y.A.(1982) Markov Random Fields. Berlin-Heidelberg-New York: Springer
RUELLE,D.(1969) Statistical Mechanics.Rigorous Results. Reading(Mass.): W.A.Benjamin
RUELLE,D.(1978) Themodynamic Formalism.The Mathematical Structures of Classical Equilibrium Statistical Mechanics. Reading(Mass.): Addison-Wesley
RUELLE,D.(1981) A mechanism for speciation based on the theory of phase transitions. Math.Biosci.,56,71-75
SAZONOV,V.(1962) On perfect measures. Amer.Math.Soc.Transl.,248,229-254
SCHWARTZ,L.(1973) Radon Measures on Arbitrary Topological Spaces and Cylindrical Measures. London:Oxford Univ.Press
SCOTT,H.L.(1974) A model of phase transitions in lipid bilayers and biologcal membranes. J.Theor.Biol.,46,241-253
SINAI,Y.G.(1982) Theory of Phase Transitions:Rigorous Results. Oxford: Pergamon Press
SMYTHE,R.T.(1976) Multiparameter subadditive processes. Ann.Probab.,4, 772-782
SOKAL,A.D.(1981) Existence and compatible families of proper regular conditional probabilities. Z.Wahrscheinlichkeitstheorie verw. Geb.,56,537-548
SPITZER,F.(1971) Random Fields and Interacting Particle Systems. Mathematical Assoc.America
SPITZER,(1975a) Markov random fields on an infinite tree. Ann.Probab., 3,387-398 ; (1975b) Phase transition in one dimensional nearest neighbor system. J.Funct.Anal.,20,240-254
STATULYAVICHUS,V.A.(1983) On a condition of almost Markov regularity. Theor.Probab.Appl.,28,379-383
STOLL,A.(1986) A nonstandard construction of Lêvy Brownian motion. Probab.Theor.Rel.Fields,71,321-334
SURGAILIS,D.(1981) On infinitely divisible self-similar random fields. Z.Wahrscheinlichkeitstheorie verw.Geb.,58,453-477
TJUR,T.(1980) Probability Based on Radon Measures. Chichester:Wiley
URBANIK,K.(1975) Extreme point method in probability theory. Lect.Notes in Math.,Vol.472,pp.169-194. Berlin-Heidelberg-New York:Springer
van den HOEVEN,P.C.T.(1983) On Point Processes (Mathematical Centre Tracts,Nr.165) Amsterdam:Mathematisch Centrum
VANMARCKE,E.(1983) Random Fields:Analysis and Synthesis. Cambridge(Mass. MIT Press
van PUTTEN,C.,van SCHUPPEN,J.H.(1985) Invariance properties of the conditional independence relation. Ann.Probab.,13,934-945
WADDINGTON,C.H.(1972) Form and information. In:Towards a Theoretical Biology (C.H.Waddington ed.),Vol.4,pp.109-145. Edinburgh:Edinburgh Univ.Press
WADDINGTON,C.H.(1973) The morphogenesis of patterns in Drosophila. In: Developmental Systems:Insects (S.J.Counce,C.H.Waddington eds.), Vol.2,pp.499-535. London:Academic Press
WALSH,J.B.(1979) Convergence and regularity of multiparameter strong martingales. Z.Wahrscheinlichkeitstheorie verw.Geb.,46,177-192
WINKLER,G.(1981) The number of phases of inhomogeneous Markov fields with finite state spaces on N and Z and their behaviour at infinit Math.Nachr.,104,101-117
WSCHEBOR,M.(1985) Surfaces Aléatoires.Mesure géométrique des Ensembles de niveau (Lect.Notes in Math.,Vol.1147).Berlin-Heidelberg-New

York-Tokyo:Springer
YANG,J.,RICHARDS,J.,BOWMAN,P.,GUZMAN,R.,ENAMI,J.,McCORMICK,K.,HAMAMOTO, S.,PITELKA,D.,NANDI,S.(1979) Sustained growth and three-dimensional organization of primary mammary tumor epithelial cells embedded in collagen gels. Proc.Natl.Acad.Sci.USA,76,3401-3405
ZACHARY,S.(1983) Countable state space Markov random fields and Markov chains on trees. Ann.Probab.,11,894-903

CORRELATED PERCOLATION AND REPULSIVE PARTICLE SYSTEMS

D.J.A. Welsh
Merton College
University of Oxford

§1. Introduction

In this paper we trace a connection from classical percolation theory via negatively correlated percolation to a repulsive particle system known as the antivoter problem and thence to a randomised algorithm for the NP-complete problem of graph colouring.

Classical percolation theory is concerned with the flow of liquid through a random medium in which the probability that a particular edge permits the flow of liquid is a given probability p, independently of the behaviour of every other edge. Here we consider percolation models in which this hypothesis of independence is replaced by a hypothesis of consistent correlation. The situation which we discuss occurs when the probability that an edge is open is either consistently increased or decreased by conditioning on the event that any other collection of edges is an open set. This concept of consistent correlation occurs naturally in physical situations, notably in the study of the Ising model.

We assume familiarity with classical percolation; for rigorous accounts of the mathematical theory see Kesten [8].

For simplicity, throughout most of this paper G will denote a finite graph with vertex set V and edge set E. Usually G is meant to represent a finite section of a lattice, or a toroidal lattice. We shall freely use 'bond' for 'edge' and 'site' for 'vertex' as is common in the literature on percolation.

Of the various guises under which percolation is usually portrayed sites being black or white, or edges being open and closed with varying probabilities we shall mainly use the more general site percolation model which includes bond percolation as a special case.

Let μ be a random field on V. We think of μ as inducing a random black/white colouring of the sites of G and $\mu(A)$ is the probability that A is the set of black sites. The complementary random field λ is define by $\lambda(A) = \mu(V\setminus A)$. We define the distribution functions α, β by

$$\alpha(A) = \sum_{X \supseteq A} \lambda(X) , \qquad \beta(A) = \sum_{X \supseteq A} \mu(X) .$$

These represent respectively the probabilities that the set A is white (black). Throughout, the underlying probability space is the obvious product space and is denoted by Ω. We first consider what properties we should expect of a positively correlated field on Ω. First we demand that each of the following, increasingly strong, statements be true for any site x and subsets A,B,C of sites:

(3) $P(x \text{ black} \mid A \text{ black}) \geq P[x \text{ black}]$

(4) $P(B \text{ black} \mid A \text{ black}) \geq P(B \text{ black})$,

(5) $P(C \text{ black} \mid A \cup B \text{ black}) \geq P(C \text{ black} \mid A \text{ black})$,

together with corresponding statements with "white" replacing "black". It is easy to see that (5) us equivalent to the statement.

(6) $\beta(A)\ \beta(B) \leq \beta(A \cap B)\ \beta(A \cup B) \qquad (A,B \subseteq V)$.

Hence a first requirement of the field μ is that (6) together with the complementary condition

(7) $\alpha(A)\ \alpha(B) \leq \alpha(A \cap B)\ \alpha(A \cup B) \qquad (A,B \subseteq V)$

holds for its distribution functions α and β.

The following example shows that conditions (6) and (7) are independent:

<u>Example</u> If G is a triangle and μ is defined by

$$\mu(X) = \begin{cases} 5/14 - \varepsilon & X = V, \\ 1/14 & |X| = 2, \\ 1/7 & |X| = 1, \\ \varepsilon & X = \emptyset, \end{cases}$$

where $\varepsilon > 0$, then it is easy to check that β satisfies (6) but α does not satisfy (7) for sufficiently small positive ε.

However in the physical situation envisaged, such as in models for the spread of infection and the like, where we think of $\alpha(A)$ as the probability that A is healthy and $\beta(A)$ as the probability that A is infected, we might also expect further inequalities of the following type to hold for all subsets A,B and C of V.

(8) $P(A \text{ black} \mid B \text{ white}) \leq P(A \text{ black})$

(9) $P(A \text{ black} \mid B \cup C \text{ white}) \leq P(A \text{ black} \mid B \text{ white})$

and so on.

Such inequalities will certainly hold when the density μ satisfies the following condition. We call μ an <u>FKG density</u> if

(10) $\mu(A\cap B)\ \mu(A\cup B) \geq \mu(A)\ \mu(B)$, $(A,B \subseteq V)$.

Using the well known theorem of Fortuin, Kasteleyn and Ginibre we know that if μ is an FKG density then each of the conditions (3) - (9) holds. Indeed more can be said. Call a function $f : V \to R^+$ <u>monotone increasing</u> if for any pair X,Y of subsets of S with $X \subseteq Y$, $f(X) \leq f(Y)$; it is monotone decreasing if -f is increasing. Then the main theorem of [3] is that if f,g are any two functions on S which are monotone in the same sense and μ is an FKG-density then

(11) $\Sigma f(A) g(A) \mu(A) \geq \Sigma f(A) \mu(A)\ \Sigma g(A) \mu(A)$,

where in all cases the summation is taken over all subsets of V.

We define a density μ to be <u>positively correlated</u> if whenever f,g are monotone in the same sense then the condition (11) is satisfied, and hence we have:

(12) An FKG-density is positively correlated.

It is well known that the converse is not true; examples of positively correlated fields which are not FKG are given in 4 and 6 below.

This concept of positive correlation is essentially the same as that of <u>associated dependence</u> introduced by Esary, Proschan and Walkup [2]. If $X = (X_1, \ldots, X_n)$ is a random vector then X is <u>associated</u> if $\mathrm{cov}(f(X), g(X)) \geq 0$ for all non-decreasing functions f,g for which the covariance exists. For a discussion of its relation with weaker forms of dependence see for example Rüschendorf [10].

§2. <u>Some positively correlated random fields</u>

<u>Example 1</u>. <u>Classical percolation</u>

For any $p \in [0,1]$ let the sites of G be independently black with probability p and white with probability 1-p. Then μ is given by

$$\mu(A) = p^{|A|}\ (1-p)^{|V \setminus A|}, \qquad (A \subseteq V).$$

<u>Example 2</u> <u>Extreme correlation</u>

Let the origin 0 (a fixed site of G) be black with probability p and then adopt the rule that each other site is black if and only if the origin is black. Then

$$\mu(\emptyset) = 1-p\ , \quad \mu(S) = p\ , \quad \mu(A) = 0 \text{ otherwise.}$$

Example 3 The Ising model

The classical Ising model, is by far the most intensively studied model in the field of correlated percolation, Its density μ is given by

$$\mu(X) = e^{-\alpha|X|} - e^{-\alpha|X|-\beta e(X)}/Z \qquad (X \subseteq V),$$

where α,β are constants, Z is a normalising constant and $e(X)$ is the number of edges having just one endpoint in X.

It is straightforward to show that for any $X,Y \subseteq V$,

$$e(X \cup Y) + e(X \cap Y \leq e(X) + e(Y)$$

and hence when $\beta \geq 0$ we know μ is an FKG density.

Example 4 The projection model on the square lattice

Consider site percolation on a toroidal section of the square lattice in which a specific site has its colour determined by the following rule. Let the integer points of the X and Y axes be independently black with probability p and white otherwise. For an arbitrary site (x,y) rule that (x,y) is black iff $(x,0)$ and $(0,y)$ are black. This is a positively correlated model and it is easy to see that if $p < 1$ then with probability one the origin is contained in a finite cluster. In this case μ is not an FKG density; consider two rectangular blocks A and B with a non-empty intersection. Although $\mu(A)$ and $\mu(B)$ have positive probability $\mu(A \cup B)$ is zero.

Example 5 Threshold percolation

Suppose numbers are assigned independently according to some random law to the edges of the graph or lattice. If the edge e_j is assigned u_j we say that site x is open if

$$\sum_{j \in \partial\{x\}} u_j > t ,$$

where t is some chosen threshold and $\partial\{x\}$ is the set of edges incident with x.

Example 6 A model for supercooled water

Consider an arbitrary graph G. Colour the edges black with probability p and white with probability 1-p. Let a site x be black only if all its incident edges are black. We sketch an argument which shows that μ is a positively correlated density. Let A', B', be the events in the probability space induced by the underlying edge percolation model which correspond to the collection of sites A and B respectively being white in the vertex model. We know that A' and B' are covariant events and since the density μ' in this edge model is the usual, non-correlated density, we know that μ' is an FKG density on the set of edges and hence using as our probability space the configuration space on the set of edges we get the desired result.

However the density μ which this problem induces will not in general be an FKG density. To see this, consider the graph consisting of a single circuit of four vertices $\{a,b,c,d\}$. Then $\mu\{a,c\} = 0$, since it is impossible that the set of black sites is exactly $\{a,b\}$. Hence $\mu\{a\}\ \mu\{\emptyset\} > \mu\{a,c\}\ \mu\{\emptyset\} = 0$.

The above example was constructed in order to find a fairly natural example of a field which was positively correlated but not FKG. For the case when G is the square lattice this model has been studied by H.E. Stanley as a model of supercooled water.

The size and shape of the percolating cluster

Consider random fields μ on a toroidal square lattice which have constant marginal densities. That is there exists a constant such that for all sites x, $\beta(x) = P[x \text{ black}] = \beta$. Then is 0 if any fixed point (the origin) of the lattice, we are interested in the size and shape of C, the (possibly vacuous) black cluster through 0. It is difficult to say anything in general about the size of C. For consider the two possible extremes where $\mu = \mu_0$ the uncorrelated measure and $\mu = \mu_1$ the extreme measure of Example 2. When β is small the expected cluster sizes satisfy

$$E_{\mu_0}(|C|) < E_{\mu_1}(|C|)$$

but for β close to 1 the inequality is reversed.

As far as shapes are concerned we know that in Example 4 the black clusters are rectangular blocks while in classical percolation the clusters are highly ramified with a true (exterior) boundary which appears to be a linear function of the size of the cluster.

Indeed heuristics based on the preceding examples and limited Monte Carlo evidence suggest that of all positively correlated μ on a given lattice the uncorrelated case gives clusters with the highest degree of remification. This idea is made more precise below.

Define the boundary ∂C of the black cluster C to be the set of white sites x which are adjacent to a member of C.

<u>Theorem</u> If μ is a positively correlated random field then if it is positively correlated with constant marginals β then $\beta E(|\partial C|) \leq (1-\beta)\ E(|C|) - \beta(1-\beta)$ with equality holding when μ is

uncorrelated.

In the uncorrelated case the above theorem is well known, and easy to prove. A slight modification of the proof using the power of the FKG condition gives the extension to the uncorrelated case, see for example [9].

Critical probabilities

We turn now to the infinite square lattice and extend the above ideas to the case where μ is an infinite product density, see [9]. Classical percolation is concerned with the existence or not of an infinite black cluster through the origin. In the correlated examples discussed earlier and in the physical problems arising in practice we usually have the situation that the density has a number of parameters; for example in Example 1 (independent percolation) there is a single parameter p. The Ising model has two parameters α, β. If the vector $\underline{a}$ is the vector of parameters of a given model μ we say that μ is monotone in its parameter set if for any $\underline{a}_1 < \underline{a}_2$ (componentwise) the associated densities μ_1 and μ_2 satisfy $E_{\mu_1}(f) < E_{\mu_2}(f)$ for any increasing function f on S. We now have the obvious analogue of the Broadbent-Hammersley result on the existence of critical probabilities: for monotone densities there exists a critical curve or critical set in the parameter space consisting of those points $\underline{a}_c$ such that if $\underline{a} > \underline{a}_c$ (componentwise) there is a positive probability of an infinite cluster but if $\underline{a} < \underline{a}_c$ then there is zero probability of an infinite cluster.

Intuitively at least one might expect positive correlation to help, and negative correlation to hinder, the growth of large clusters. Translated into a remark about critical probabilities one might therefore expect the following to be true:

Conjecture 1 Let μ be a positively correlated monotone percolation density on the square lattice with a single parameter $\beta = \beta(x)$, the probability that a particular site be black. Then the corresponding critical probability $p_c(\mu)$ is not greater than in the case where μ is the independent density with parameter β.

This was conjectured by Stoll and Domb [11] but in its full generality as stated, it is not true as Example 4 above shows.

§3 Negative correlation in a spatial process

It seems harder to define precisely the concept of a negatively correlated spatial process. We cannot just reverse inequalities above, since if for all f,g monotone in the same sense $E(fg) \leq E(f)E(g)$ then f = g gives nonsense. A minimal requirement might be that if x and y are neighbours in space then the distribution functions α,β, satisfy

(1) $\alpha(x)\ \alpha(y) \geqslant \alpha(x,y)$;

and

(2) $\beta(x)\ \beta(y) \geqslant \beta(x,y)$,

We therefore define a random field μ to be <u>negatively correlated</u> if (1) and (2) hold for all pair of neighbouring points x and y.

It is clear that classical percolation is in a very trivial sense negatively correlated. Another example is:

<u>Example 1</u> Suppose G has N sites, that k is a fixed integer and the colouring mechanism is to choose k sites to be black, the remaining N-k to be white, with all k-subsets being equally likely. It is easy to check that this simple exclusion model is negatively correlated.

<u>Example 2</u> (Randomised Exclusion)

Again with G having N sites let X be a discrete random variable with $p_k = P(X=k)$.

Then if μ is defined by

$$\mu(A) = p_k/\binom{N}{k}, \qquad |A| = k,$$

we find that μ is negatively correlated provided that $(N-1)(EX)^2 \geqslant N\,E(X^2-X)$.

<u>Example 3</u> (Perfect Disorder)

Take any bipartite graph such as the toroidal square lattice, choose a site x at random and colour it black. Then if d(x,y) the distance of y from x colour y white if d(x,y) is odd, otherwise black. This clearly gives a negative correlation.

<u>Example 4</u> The Ising Model of Anti-Ferromagnetism

The classical Ising Model with $\beta < 0$ is often regarded as the prototype of a negatively correlated model. However it is not clear when this model does actually produce a negative correlation satisfying (1) and (2),

§4 The Antivoter Model

T.E. Harris [7] has shown that the equilibrium measures of the attractive particle systems are positively correlated. In an attempt to get a better understanding of negative correlation, in [1] we study a simple model of a repulsive particle system whose equilibria measures should possess some of the characteristics of negative correlation.

Consider a random sequence of 2-colourings of a graph G which evolves as follows. At time $t = 0$ a subset A of the vertex set V is black and $V \setminus A$ is white. At random intervals of time a vertex v chooses a neighbour u at random and if u and v are similarly coloured v changes colour. If η_t^A denotes the set of black vertices at time t, then η_t^A is the process which we call the antivoter model.

Because of the "repulsive element" in its definition η_t^A should evolve towards a collection of 2-colourings of G which are negatively correlated in some sense. Certainly if G is bipartite, then η_t^A converges as $t \to \infty$ to one of the subsets of V induced by any proper 2-colouring, in the case of a square lattice the checkerboard effect.

An extensive duality theory has been developed for particle systems. The details of this, may be found in [5]. In this case η_t^A has as a dual process a system of processes $\hat{\eta}_t^B$ consisting of annihilating random walks on G. In this model, particles initially at the vertices of B perform independent random walks on V, the vertex set of G, with the added proviso that if a particle jumps to a vertex already occupied by another particle both particles are annihilated.

To define the antivoter process η_t^A formally we associate with each vertex of G a "random clock" which rings, independently for each vertex, at the instances of a Poisson process of rate 1. When the clock at a vertex v rings the vertex chooses a neighbour u at random and adopts the colour opposite to that of u. At time $t = 0$ the vertices of A are coloured black. The set of vertices which are black at time t is denoted by η_t^A. Clearly η_t^A is a Markov process with state space 2^V and we know that when G is bipartite η_t^A has two absorbing states induced by the essentially unique proper 2-colouring of G.

The annihilating random walk process $\hat{\eta}_t^B$ is defined by associating a similar set of clocks with the vertices. When the clock at vertex v rings, if there is a particle at v it jumps to a randomly chosen neighbour u. If there is a particle at u both particles are annihilated. At time $t = 0$ the particles are located exactly at the members of B. The set of vertices occupied by particles at time t is denoted by $\hat{\eta}_t^B$. The duality relationship between these two systems can be expressed as follows:

(1) If A and B are subsets of V then

(2) $P(|\eta_t^A \cap B| \text{ even}) = P(|\hat{\eta}_t^B \cap A| + \varepsilon_t^B \text{ even})$

where ε_t^B is the number of jumps made by the particles of the process $\hat{\eta}_t^B$ in the interval $(0,t)$.

The proof of (2) depends upon a method of constructing both processes using percolation substructures. The details can be found in Griffeath (1979, p. 68).

Since the state space of the Markov process η_t^A is finite it must have an equilibrium distribution ν^A as $t \to \infty$. There are 3 distinct cases:

(a) If G is bipartite with edges joining V_1 to V_2 then

$$\nu^A(X) = 0 \qquad X \neq V_1 \text{ or } V_2.$$

(b) Suppose G is a circuit with 2m+1 vertices. Then it is easy to see that η_t^A has a closed set of recurrent states. These are the 4m+2 states in which exactly one pair of adjacent vertices is the same colour and every other pair of adjacent vertices has opposite colours.

(c) When G is not bipartite and not an odd circuit we have the following:

Theorem 1 If G is not bipartite nor an odd circuit then apart from the two obvious transient states V and $\emptyset$ every other state is recurrent and the Markov chain η_t^A is irreducible for any A and in this case, ν^A is independent of A.

The only proof we know of this is surprisingly non trivial (see [1]).

Henceforth we consider graphs which are neither bipartite nor odd circuits. Because of the repellent nature of the particle model one might expect some form of negative correlation in the equilibrium distribution ν.

In general it seems difficult to obtain ν explicitly, however since we can prove that in equilibrium a vertex has equal probability of being black or white we need only concern ourselves with the similarity function $S(x,y)$ which is the probability that at equilibrium x and y are the same colour. The main result of Donnelly and Welsh [1] is:

Theorem 2 Provided the underlying graph has a group of automorphisms which is transitive on its edges, the equilibrium measure is negatively correlated.

Thus, for any of the toroidal versions of the standard lattices we know that the equilibrium measure is negatively correlated.

Indeed provided our underlying structure has an abelian group of automorphisms we can prove much more. For such graphs, let A be its adjacency matrix and ρ the number of edges incident with each vertex, then for any pair of vertices x,y we obtain an explicit form for the similarity function, namely

$$(3) \quad S(x,y) = \frac{1}{2}\left[1 + \frac{(\rho I+A)^{-1}_{xy}}{(\rho I+A)^{-1}_{yy}}\right]$$

§4 A Random Algorithm for an NP-complete Problem

The problem of deciding whether NP (nondeterministic polynomial time) is equal to P (polynomial time) is now probably the most important and well known problem in computer science and discrete mathematics. Many NP-complete problems have been specified. Each has the property that if a polynomial time algorithm could be found for any one of then then the class NP collapses to P. For an account and listing of NP-complete problems arising in many different contexts see [4].

One of the more innocent problems which is known to be NP-complete is to decide whether or not the vertices of a graph can be labelled with 3 colours so that no two adjacent vertices are the same colour. This is known as the 3-COLOURABILITY problem. Consider the following randomised algorithm for this problem which is based on the ideas of the preceding sections.

Given a graph G we initially colour the vertices randomly with 3 colours. At discrete instants of time we let L_t be the list of vertices which have a neighbour of the same colour. At each discrete instant of time t we select a member u of L_t at random and let b, r, w be the number of black, red and white neighbours respectively. At time t + 1 we change the colour of y to a random variable with density p(b,r,w). Update L_t to L_{t+1} and repeat.

This situation is a Markov chain with 3^N states, N being the number of vertices of G. Provided G is 3-colourable the chain will, with probability one, enter a state of a proper 3-colouring. This is recognised by keeping a count of L_t . By choosing p to be a repulsive mechanism of the form

$$p(c(u) = i) \sim \bar{\theta}^{-\partial_i(u)} \qquad \sim \theta^{-\partial_i(u)}$$

where ∂_i is the number of neighbours of u with colour i, the expected, first passage time to enter a state $|L_t| = 0$ appears from Monte Carlo studies to be linear in N. We should emphasize that although an algorithm based on these ideas seems to work well in practice it cannot be compared with the classical randomised algorithms such as prime testing (see [12]). There do exist recognisable (hopefully few) bad cases

We have carried out a large number of experiments for varying value of N in the range $0 < N \leq 1000$. Typically, we have been able to find a 3-colouring of a randomly generated 1000 vertex graph in well under a minute on a Perkin Elmer minicomputer. The details of these computations can be found in [13].

Conclusion: Negatively correlated spatial processes seem to be more difficult by an order of magnitude than their positively correlated counterparts. Even in the Gibbsian situation little seems to be known once one is away from the square lattice, though Hammersley and Mazzarino [6] have recently carried out extensive Monte Carlo studies. For example in the case of the Ising model with so called negative correlation on the toroidal triangular lattice is it the case that neighbouring points are more likely than not to be dissimilar? In the light of Theorem 2 which proves this result for the corresponding antivoter model the answer would seem to be yes but we know of no rigorous proof.

References

1. P.J. Donnelly and D.J.A. Welsh, The antivoter problem: random 2-colourings of graphs, Graph Theory and Combinatorics. (ed. B. Bollobas) (1984) (Academic Press), 133-144.

2. J.D. Esary, F. Proschan and D.W. Walkup, Association of random variables with applications, Ann. Math. Statist. 38, (1976), 1466-1474.

3. C.M. Fortuin, P.W. Kasteleyn and J. Ginibre, Correlation inequalities on some partially ordered sets, Comm. Math. Phys. 22 (1971) 89-103.

4. M.R. Garey and D.S. Johnson, Computers and Intractibility, (1979), W.H. Freeman & Co. San Francisco.

5. D. Griffeath, Additive and Canellative Interacting Particle Systems Springer Lecture Notes in Mathematics, 724, (1979)

6. J.M. Hammersley and G. Mazzarino, Markov fields, correlated percolation and the Ising model, The Mathematico and Physics of Disordered Media, Springer Lecture Notes in Mathematics, 1035, (1983) 201-245

7. T.E. Harris, A correlation inequality for Markov processes in partially ordered state spaces, Ann. Prob. 5 (1977), 451-454.

8. H. Kesten, Percolation Theory for Mathematicians (1982), Birkhauser Boston.

9. C.M. Newman and L.S. Schulman, Infinite clusters in percolation models, J. Stat. Physics 26, (1981), 613-628

10. L. Ruschendorf, Weak association of random variables, J. Multivariate Analysis, 11 (1981), 448-451.

11. E. Stoll and C. Domb, Shape and size of two dimensional percolation clusters with and without correlations, J. Phys. A 12 (1979), 1843-1855.

13. D.J.A. Welsh and D. Petford, A randomised algorithm for an NP-complete problem. (preprint) 1984.

Editor's note. Theorem 1 is a re-formulation of Theorem 3 in (Donnelly and Welsh, 1984 : I call it DW). The proof is obtained with the aid of three lemmas showing the possibility of change of configurations in non--trivial states. It follows from this theorem that there exists an equilibrium process η with density ν so that for any subset A of V, $P\{\eta=A\}=\nu(A)$.

Theorem 2 is a contraction of Theorems 4 and 5 in DW. By introducing a random walk ξ^x on the vertex set V, one first proves the existence of a transitive Abelian group $\mathcal{A}$ of automorphisms; further it will be shown that the equilibrium measure for the antivoter model on graphs with certain symmetry satisfies a "nearest-neighbour negative correlation". The bound $S(x,y)\leq(n-2)/(2n-3)<1/2$ holds for some graphs which are not bound-transitive (e.g., an $m\times n$ toroidal square lattice, $m\neq n$).